REFERENCE FRAMES

ASTROPHYSICS AND SPACE SCIENCE LIBRARY

A SERIES OF BOOKS ON THE RECENT DEVELOPMENTS
OF SPACE SCIENCE AND OF GENERAL GEOPHYSICS AND ASTROPHYSICS
PUBLISHED IN CONNECTION WITH THE JOURNAL
SPACE SCIENCE REVIEWS

VOLUME 154
CURRENT RESEARCH

REFERENCE FRAMES

IN ASTRONOMY AND GEOPHYSICS

edited by

JEAN KOVALEVSKY
CERGA, Grasse, France

IVAN I. MUELLER
Ohio State University, Columbus, U.S.A.

and

BARBARA KOLACZEK
Polish Academy of Sciences, Warsaw, Poland

KLUWER ACADEMIC PUBLISHERS
DORDRECHT / BOSTON / LONDON

Library of Congress Cataloging in Publication Data

Reference frames / edited by J. Kovalevsky, I.I. Mueller, and B. Kolaczek.
p. cm. -- (Astrophysics and space science library ; 154)
Bibliography: p.
Includes index.
ISBN 0-7923-0182-X
1. Celestial reference systems. 2. Geodesy. 3. Geodynamics.
4. Earth--Rotation. I. Kovalevsky, Jean. II. Mueller, Ivan Istvan, 1930- . III. Kołaczek, Barbara. IV. Series: Astrophysics and space science library ; v. 154.
QB633.R33 1989
525'.35--dc20 89-2582

ISBN 0-7923-0182-X

Published by Kluwer Academic Publishers,
P.O. Box 17, 3300 AA Dordrecht, The Netherlands.

Kluwer Academic Publishers incorporates
the publishing programmes of
D. Reidel, Martinus Nijhoff, Dr W. Junk and MTP Press.

Sold and distributed in the U.S.A. and Canada
by Kluwer Academic Publishers,
101 Philip Drive, Norwell, MA 02061, U.S.A.

In all other countries, sold and distributed
by Kluwer Academic Publishers Group,
P.O. Box 322, 3300 AH Dordrecht, The Netherlands.

printed on acid free paper

Printed in The Netherlands

PREFACE

This book on reference systems is the first comprehensive review of the problem of celestial and terrestrial reference systems and frames. Over 20 years, the importance of this problem emerged slowly as the accuracy of new observational techniques improved. The topic has already been approached in several symposia such as Stresa (1967), Morioka (1971), Perth (1973), Columbus (1975, 1978 and 1985), Kiev (1977) and San Fernando (1978). Two IAU colloquia held in Turin (1974) and in Warsaw (1980) were exclusively devoted to discuss reference systems.During this time, the problem of terrestrial and celestial reference systems has been discussed also in many astronomical and geodetic symposia, but always among other topics. Thus, a review devoted solely to the definition and practical realization of such systems was needed.

It is hoped that this book, containing modern comprehensive reviews of important facets of this problem will contribute not only to a better and wider understanding of the mathematics and the physics that are behind the concepts and the realizations, but also to future development in a field that can only expand with the rapidly increasing accuracy of geodetic and astronomical observations.

We are pleased to thank all the authors of the book who have enthusiastically agreed to contribute to the book in their field of competence and have gracefully accepted guidance from the editors in the definition of the subject and of the interfaces with other chapters. We thank Prof. Y. Kozai for his involvement in the original definition of the book. We thank Mrs Jeanne Falin who has painstakingly retyped several chapters and made a large number of editing corrections.

July 1988

The Editors.

AUTHORS

Dr Cl. BOUCHER - Institut Géographique National - Saint Mandé - France

Dr V.A. BRUMBERG - Institute of Applied Astronomy - Leningrad - USSR

Dr C. MA - Goddard Space Flight Center - Greenbelt - USA

Dr J.O. DICKEY - Jet Propulsion Laboratory - Pasadena - USA

Dr T. FUKUSHIMA - Maritime Safety Agency - Tokyo - Japan

Dr B. GUINOT - Bureau International des Poids et Mesures - Sèvres - France

Dr H. KINOSHITA - Tokyo Astronomical Observatory - Tokyo - Japan

Dr B. KOLACZEK - Space Research Centre - Warsaw - Poland

Dr S.M. KOPEJKIN - Sternberg State Astronomical Institute - Moscow - USSR

Dr J. KOVALEVSKY - Observatoire de la Côte d'Azur - Grasse - France

Prof. K. LAMBECK - Australian National University - Canberra - Australia

Prof. I.I. MUELLER - Ohio State University - Columbus - USA

Prof. R.H. RAPP - Ohio State University - Columbus - USA

Dr Ch. REIGBER - Deutsches Geodätisches Forschungsinstitut - München - FRG

Dr T. SASAO - International Latitude Observatory - Mizusawa -Japan

Dr E.M. STANDISH - Jet Propulsion Laboratory - Pasadena - USA

Dr G.A. WILKINS - Royal Greenwich Observatory - Hailsham - Great Britain

Dr J.G. WILLIAMS - Jet Propulsion Laboratory - Pasadena - USA

Dr YE. SHU-HUA - Shanghai Observatory - Shanghai - China

TABLE OF CONTENTS

INTRODUCTION

J. KOVALEVSKY
CERGA, Grasse, France

Ivan I. MUELLER
Ohio State University, Columbus, USA

1 . WHY REFERENCES FRAMES ?

The Earth, its environment and the celestial bodies in the Universe are not static: they move, rotate and undergo deformations. The study of their kinematics and the dynamics that are behind is one of the major tasks of geodesy, geophysics and astronomy. Without trying to be exhaustive, let us mention some of the dynamical phenomena that currently concern these sciences :

- Lithospheric plate motions,
- Earth and oceanic tides,
- Loading effects on the Earth's crust,
- Polar motion,
- Earth's rotation,
- Dynamical behaviour of the Earth-Moon system,
- Motion of planets and satellites,
- Rotation of planets and satellites,
- Motions of stars in the Galaxy,
- Dynamics of star clusters,
- Evolution of star association and open clusters,
- Differential rotation of the Galaxy,
- Motion in clusters of galaxies,
- etc...

Motion and position are not absolute concepts and can be described only with respect to some reference. Mathematically this is done using a system of coordinates that can be constructed at will. Physically, for the earth sciences and astronomy, it is not so simple, because the coordinates of a point should be readily accessible. This means that there should exist an observational relationship between the point and the physical objects that are to be used to obtain its coordinates. We shall call 'reference frame' the physical realization of such a reference system.

J. Kovalevsky et al. (eds.), Reference Frames, 1–12.

In order to clarify some of the conceptual aspects of various reference systems and frames, we propose to use specific terms suggested by Kovalevsky and Mueller (1981) that have been used somewhat inconsistently in the past.

The purpose of a *reference frame* is to provide the means to materialize a *reference system* so that it can be used for the quantitative description of positions and motions on the Earth (terrestrial frames), or of celestial bodies, including the Earth, in space (celestial frames). In both cases the definition is based on a general statement giving the rationale for an ideal case, i.e., for an *ideal reference system.* For example, one would have the concept of an ideal terrestrial system, through the statement that with respect to such a system the crust should have only deformations (i.e. no rotations or translations, cf. the Tisserand axes). The ideal concept for a celestial system is that of an inertial system so defined that in it the differential equations of motion may be written without including any rotational or acceleration terms. In both cases the term 'ideal' indicates the conceptual definition only and no means are proposed to actually construct the system (see Section 3).

The actual construction implies the choice of a physical structure whose motions in the ideal reference system can be described by physical theories. This implies that the environment that acts upon the structure is modelled by a chosen set of parameters. Such a choice is not unique: there are many ways to model the motions or the deformations of the Earth; there are also many celestial bodies that may be the basis of a dynamical definition of an inertial system (Moon, planets or artificial satellites). Even if the choice is based on sound scientific principles, there remains some degree of imperfection or arbitrariness. This is one of the reasons why it is suggested to use the term 'conventional' to characterize this choice. The other reason is related to the means, usually conventional, by which the reference frames are defined in practice.

At this stage, there are still two steps that are necessary to achieve the final materialization of the reference system so that one can refer coordinates of objects to it. First, one has to define in detail the model that is used in the relationship between the configuration of the basic structure and its coordinates. At this point, the coordinates are fully defined, but not necessarily accessible. Such a model is called a *conventional reference system.* The term 'system' thus includes the description of the physical environment as well as the theories used in the definition of the coordinates. For example, the FK4 (conventional) reference system is defined by the ecliptic as given by Newcomb's theory of the Sun, the values of precession and obliquity, also given by Newcomb, and the Woolard theory of nutation. Once a reference system is chosen, it is still necessary to make it available to the users. The system usually is materialized for this purpose by a number of points, objects or coordinates. In referencing any other point, object or coordinate one must tie it to them. Thus, in addition to the conventional choice of a system, it is necessary to construct a set of conventionally chosen (or arrived at) parameters (e.g., star positions or pole coordinates). The set of such parameters, materializing the system, define a *conventional reference frame*. For example, the FK4 catalogue of over 1535 star coordinates defines the FK4 frame, materializing the FK4 system.The position of a celestial body is determined using an instrument that interpolates the position of the body between the coordinates of several catalogue stars.

However, the accessibility of the reference frame is not the only requirement. It is also expected that it is defined in such a way that one can write unambiguously the equations of motion of a body whose coordinates are given in this frame. This implies, for instance in Newtonian mechanics, that there exists no additional force produced by a non-uniform motion of the reference frame. The search for a non-rotating and non-accelerating reference frame is not trivial and will be discussed in several chapters of this book.

If one is interested in all the aspects of the motion of a body of finite dimensions, it is also necessary to establish a frame of reference attached to this body so that its motion is described with respect to the external reference frame. If the body is rigid, this is a simple matter, but if this is not the case, as for the Earth, one must define some kind of averaged mean position of the axes called 'terrestrial reference frame'.

Finally, a systematic error in the time reference affects the coordinates of all moving objects in a manner proportional to the error in time and, if it is ignored, may produce a distortion in the realization of a reference frame. To give an example, the irregular retardation of planets in their orbits was correctly attributed by Newcomb to irregularities in the Earth's rotation that was used as a concept called ideal reference system clock. For this reason, time systems also belong to reference systems.

2. DEFINITION AND CONSTRUCTION OF A REFERENCE FRAME

We have mentioned in the preceding section that one wishes to have a non rotating reference frame. This is a conceptual statement that has to be physically implemented before one can realize a reference frame. Actually, between the concept and the realization of a usable reference frame, several steps must be taken. At the end of each stage, a certain state of advancement is achieved, to which a specific name is attached. We shall describe them with the example of a dynamical reference frame in Newtonian mechanics.

2.1. CONCEPT

When starting the construction of a reference frame, one has to state a basic principle that must be obeyed. This theoretical concept is called ideal reference system.More details are given in Section 3.

For example, an ideal dynamical celestial reference system also called inertial reference system, is based in Newtonian mechanics on the absence in the equations of motion of celestial bodies of an acceleration due to the rotation and the non uniform translation of the frame of reference used.

2.2. CHOICE OF A PHYSICAL STRUCTURE

Once the ideal reference system is chosen, one has to identify the physical bodies that will support the definition. The choice is made, of course, in such a way that the physical structure formed by these bodies verifies the ideal statement to the best possible approximation. The ideal reference frame and the physical system (constants, equations of motion, etc...) together constitute the reference system proper.

For example, the solar system, with its planets and satellites is an example of a physical structure to which one can apply the concept of an ideal inertial reference system. The equations of motion in this reference system must obey strictly Newton's law without any additional acceleration.

2.3. MODELLING THE STRUCTURE

Once the choice is made, one has to assign numbers to the parameters describing the physical system chosen in the reference system. In other words, one constructs a model of the physical structure, using mainly what is thought to be the best observational data. Nevertheless the values that are chosen necessarily have some arbitrariness. This is why this model representing the reference system is called conventional reference system.

For example, in the case of a dynamical reference system based on the motions in the solar system, one has to adopt a system of planetary masses and other quantities such as the constant of precession, that enter into the computation of the observed positions of planets from the theory of their motion. Other parameters may exist, but having a negligible influence in the realization of the reference system, they may be left out of the model: for instance the Earth's structure which causes the nutation. This model is described in what is called "system of fundamental constants", a conventional list periodically updated by the IAU and the IUGG (see Chapter 18).

2.4. REALIZATION OF THE SYSTEM

Applying the model described by the conventional reference system, the coordinates of a certain number of points are determined from observations together with their variations with time. The ephemerides of such fiducial points define the actual coordinates in the conventional reference system and should be sufficiently numerous and accessible to observation so as to permit the determination of the coordinates of another point with respect to them, so that they are also expressed in the conventional reference system. This set of markers of the systems realize the conventional reference frame.
For example, a numerical theory of the motion of the planets based upon the model of the conventional reference system gives at every time their coordinates in this system. The observation of these planets or of the Sun with respect to stars allows us to obtain the position of these stars in the same conventional reference system. The positions and proper motions of such stars are presented in the form of a fundamental catalogue (see Chapter 1), which are the actual realization of the conventional reference system.

2.5. EXTENSION AND DENSIFICATION

If the number of fiducial points in the conventional reference frame is not sufficient to permit the determination of the position of any point, one has to add secondary reference points in a sufficient number so that a few can be accessed simultaneously to determine the position of an unknown point.
For example, fundamental catalogues such as the FK4 or the FK5 have about 1500 stars. This is unsufficient and secondary star catalogues are constructed that contain the positions and proper motions of tens of thousands of stars in the same conventional reference frame.

3. IDEAL SYSTEMS OF REFERENCE

Several possibilities exist for defining ideal reference systems both in the case of celestial and terrestrial reference systems.

3.1. CELESTIAL SYSTEMS

There are two competing definitions of an ideal celestial reference system which are based upon different concepts but which, at least in Newtonian mechanics, should lead to the same idea of a non-rotating reference frame in space.

3.1.1. Dynamical definition

Assuming a certain number of celestial bodies, their motion must be represented in the frame of classical mechanics by a solution of a system of differential equations written in a fixed triad. The dynamics expressed by these equations define certain invariant points and directions that can be used to construct a system of coordinates. This is the case of the barycenter of the bodies which undergo no acceleration and the direction of the angular momentum axis which is fixed in space. Other invariant directions may also be derived from the equations of motion.

So, the ideal celestial reference system defined dynamically, also called ideal inertial reference system, is essentially based on the form that equations of. motion take in a non-rotating, non-accelerating frame of reference which is imposed on the actual motion of celestial bodies.

3.1.2. Kinematic definition

In this definition, one assumes that the visible Universe does not rotate, so that very distant objects in the Universe, such as quasars or very distant galaxies, do not have a group motion that could be interpreted as a rotation. Since, physically, a body cannot have a velocity reaching the speed of light, one may postulate that the apparent motion of a quasar at a distance larger than 100 Megaparsecs is necessarily smaller than 0.7 millisecond of arc per year and very probably much less. Most quasars are significantly further and there is no physical reason to assume that their tangential velocities are larger than the recession velocity. This gives an upper limit to the apparent motion of these bodies of $2\text{x}10^{-5}$ second of arc per year, far beyond the present precision of observations.

So, the ideal celestial reference system defined geometrically, also called ideal-non-rotating reference system, is essentially based on the kinematic behaviour of very distant objects assumed to be random.

3.1.3. Relativistic approach

In General Relativity, an inertial system is a freely falling coordinate system which depends upon the distribution of masses in the vicinity of the point where the system is defined. It is essentially a local system valid in a finite environment. Contrary to the classical mechanics concept in which a reference system is universal, the transport of a coordinate system from one point to another is a complicated transformation that can be defined only if the mass distribution is known everywhere in the space where the transport is realized. In order to remember this restriction to a local environment, one should speak of a quasi-inertial system.

Similarly, kinematically defined systems of reference in General Relativity also have a local meaning and the directions are affected by the presence of masses and by geodetic precession.

However, although for all practical purposes, one needs only to have access only to a local system of reference for which the relativistic effects can be modelled as corrections to the Newtonian concept, it is important to describe rigourously the effect of General Relativity. This is presented in Chapter 5.

3.2. TERRESTRIAL SYSTEMS

An ideal terrestrial reference system should represent some idealized terrestrial body on which coordinates of stations are either fixed or change in a theoretically modelled manner. If the Earth was perfectly rigid, any triad fixed to the Earth were suitable and the choice would be made only on the grounds of convenience. But since the Earth is deformable and parts of the crust move with respect to each other, one has to set up conditions for the definition of the ideal reference system. If D is the domain considered as being the surface of the Earth and dD an element of this domain in the part M and if O is the origin of the coordinates, one may postulate that D has no translational and no rotational motion. This is expressed by :

$$\int_D \frac{d\overrightarrow{OM}}{dt}\, dD = 0 \tag{1}$$

and

$$\int_D \overrightarrow{OM} \wedge \frac{d\overrightarrow{OM}}{dt}\, dD = 0 \quad . \tag{2}$$

Of course, such an ideal definition is not practical and one must replace the integrals by summations over a certain number of points on the Earth (observing stations) or consider that the crust is composed of a finite number of moving rigid plates. Each of these two possibilities lead to a different definition of the ideal terrestrial reference system (see Chapter 7).

3.3. PLANETARY SYSTEMS

Definitions similar to (1) and (2) can be applied to reference systems representing the Moon (selenodetic reference system) or any of the telluric planets. These applications are not considered in this book.

3.4. INTERMEDIATE SYSTEMS

Let us assume that one wishes to represent the motion of a solid body - say the Earth - in space. Let us designate by (T)=(o,x,y,z) the triad representing the terrestrial reference frame and (C)=(O,X,Y,Z) the triad representing the celestial reference frame. The motion is completely represented when one obtains the time expressions of coordinates of o and of the direction cosines of ox, oy, oz with respect to (C).

In practice, however, there may be advantages to somewhat complicate the link between (T) and (C) by the introduction of a triad (I)=(ω,ξ,η,ζ). The transformation (T) to (C) is then made in two steps :
(i)- a first transformation giving the motion of (T) with respect to (I).
(ii)- a second transformation giving the motion of (I) with respect (C).

The reference frame realized by (I) is called "intermediate". Such a frame is used in order to simplify the expression for the rotation of the Earth and separate for instance the polar motion from the nutation and precession (see Chapter 11). The choice of the intermediate frame is arbitrary since one can indifferently account for a given component of the motion in step (i) or in step (ii). Actually several definitions exist and the best (meaning the most efficient or the most convenient) is still being discussed. For instance for the Earth, the $\omega\zeta$ axis has been proposed to coincide with the angular momentum axis, with the position of the instantaneous axis of rotation, with the axis of figure, etc...

From this, it is clear that intermediate frames are not reference frames as defined in the preceding sections. The most logical and conceptually sound approach to this problem is to stick only to the two unavoidable reference frames: the celestial and the terrestrial frames and to express their relations directly without any intermedary.

4. TIME REFERENCES

In Newtonian mechanics a single time, valid for all the Universe is postulated. If this were still true, there would be no problem. Assuming Special Relativity, a relation between space and time would have been introduced, but since the space-time is Euclidean, there is no effect on reference systems. In General Relativity, a third element is affecting the space-time relationship. This is gravity in the presence of masses. The consequence is that the definition of time scales leads to difficulties encountered in defining reference systems in space. In fact, one can follow the procedure of Section 2 and identify five stages.

4.1. CONCEPT

The concept is naturally based on the theory of General Relativity. Many different definitions can be given corresponding to different physical situations. The resulting definition will be called 'ideal time-like argument', leaving the word 'time' alone to the ideal Newtonian absolute time.

To illustrate this, we present two cases that are of particular interest (for more details, see Chapter 15).

4.1.1. Ideal proper time

It is the time-like argument associated with an isolated mass at rest in its material environment and with respect to which the laws of quantum mechanics are satisfied.

4.1.2. Ideal coordinate time

A certain domain of space-time is described by a four dimensional Einstein's metric tensor. The one among the four coordinates that corresponds to the negative component of the diagonal is the ideal coordinate time (see Chapter 17).

4.2. CHOICE OF A PHYSICAL STRUCTURE

The definitions of the ideal time-like argument are to be applied to some physical system. In the case of proper time, it could be an isolated atom of cesium 133. In the case of coordinate time, one has to define in what domain it is valid. For example on the surface of the Earth (terrestrial coordinate time-like argument), or in the solar system (barycentric coordinate time-like argument).

4.3. MODELLING THE STRUCTURE

As in the case of reference systems, the physical environment has to be modelled using a certain number of conventionaly adopted parameters. In the case of proper time of a cesium 133 atom, one can define the number N of periods of the radiation produced by the transition between the two hyperfine levels of the fundamental state in one second (N=919631770 is actually the definition of the S.I. second). In the case of barycentric coordinate time, the same model as for the dynamical definition of the conventional reference system using the solar system is suitable.

Although this terminology has not been used, we would suggest to call these time-like arguments conventional time-like arguments by an analogy with the conventional reference systems.

4.4. REALIZATION

The realization of a conventional time-like argument is a time scale, just as a reference frame is the realization of a reference system. It involves continuous measurements and intercomparisons.

4.5. TIME DISSEMINATION

To finalize the analogy with the construction to reference frames, one may add that various clocks, synchronized to a time scale so that it is accessible anywhere where needed, are analogous to the extension and densification of a reference frame. The best example is the distribution of TAI using UTC dissemination.

4.6. REFERENCE FRAMES AND COORDINATE TIME SCALES

We have seen, in Section 4.3 that similar physical models are required to construct conventional reference frames or time scales. There is actually a close association between them. So, in order to construct a dynamical reference frame based on the dynamics of the solar system one should use the barycentric time-like argument, while the study of motion of satellites in the vicinity of the Earth must make use of terrestrial coordinate time-like argument.

5. ORGANIZATION OF THE BOOK

Based on the material presented so far, the organization of the book ought to be self-explanatory. However, to convince the reader that this book is not just a collection of independent articles but follows a logical line of thought, we describe the main parts of the book.

5.1. CELESTIAL REFERENCE FRAMES

This part describes the main quasi-inertial reference frames which are defined either kinematically (Chapters 1 and 2) or dynamically (Chapters 3 and 4), concluding with Chapter 5 on the relativistic aspects. The inclusion of this latter chapter was necessary because the high accuracy of the state of the art technology does not longer permit in some cases to use Newtonian approximation any longer.

5.2. TERRESTRIAL REFERENCE FRAMES

This part deals with reference frames that are fixed in some sense to the Earth and in which the positions of objects, instruments or points on the surface of the Earth are positioned by means of their terrestrial coordinates. These coordinates may be the traditional geodetic coordinates which are referenced to a geodetic datum (Chapter 6) or, because of the use of new geodetic technology, to a Cartesian frame (Chapter 7).

5.3. ROTATION OF THE EARTH AND THE TERRESTRIAL AND CELESTIAL FRAMES

In this part, the components of the link connecting the celestial and the terrestrial reference frames are presented. These components consist of those parameters of the Earth rotation which are largely modelable, such as precession and nutation (Chapter 8), and those which can be determined largely through monitoring such as polar motion and Universal time (Chapter 9).
Since all these parameters depend on geophysical phenomena, we deemed necessary to include a chapter on these aspects as well (Chapter 10).

5.4. RELATIONSHIPS BETWEEN SYSTEMS

This part starts with the formal description of the above mentioned link between the terrestrial and celestial frames and how the parameters of this link are used in formal transformation (Chapter 11). Then, after describing the general principles of intercomparing reference frames (Chapter 12), actual current intercomparison methods and results are presented for celestial frames (Chapter 13) as well as for terrestrial frames (Chapter 14).

5.5. TIME

Motion is a relationship between space and time. In Newtonian mechanics where an absolute time valid in all points in space is postulated, the time reference is common to all objects. However in General Relativity that governs the real word, time and space are interconnected and it is no longer possible to ignore this dependency. This is why Chapters 15, 16 and 17 of the book are devoted to various concepts of time and to the identification of the definitions to be used in different dynamical environments.

5.6. STANDARDS

As we pointed out earlier, the reference systems, as opposed to reference frames include a model and a set of constants. Although it is understood that for most scientific investigations, the best available models and constants are used for operational purposes, the uniform use of these models and constants is imperative for conventional systems. They are usually adopted by some international scientific unions and are called fundamental constants and are used as standards (Chapter 18).

REFERENCE

Kovalevsky, J. and Mueller, I.I., 1981, in *'Reference coordinate systems for Earth Dynamics'*, E.M. Gaposchkin and B. Kolaczek (eds), IAU Coll. 56, D. Reidel Publ. Co., Dordrecht, p. 375.

PART 1

CELESTIAL REFERENCE FRAMES

STELLAR REFERENCE FRAMES

J. KOVALEVSKY
CERGA, Grasse, France

1. INTRODUCTORY REMARKS

Ever since man tried to trace the motion of celestial objects or to identify their positions, he referred to stars as fiducial points This meant that one had to measure the positions of these reference stars with respect to a system of spherical coordinates. The choice of the axes was imposed by the Earth's rotation: the celestial equator and the meridian half plane crossing it at the vernal equinox point. Traditionally the equatorial system is direct, the Z-axis pointing towards North. The two angles are the right ascension α and the declination δ (fig. 1).

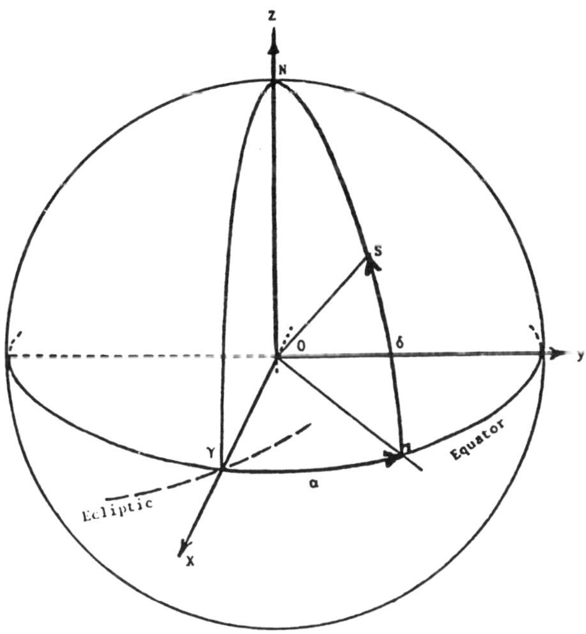

Figure 1. The equatorial system of coordinates; γ is the vernal equinox.

J. Kovalevsky et al. (eds.), Reference Frames, 15–41.

Ancient Greeks did these measurements with sufficient accuracy for Hipparchus to discover that these coordinates were changing with time. The effect of the precession was found to be of the order of one degree per century. In order that the positions of stars do not vary with time, one had to refer them to a coordinate frame as it was at a given time, called the epoch of the system. Later - in the 18-th century - when nutation and aberration were discovered, one was led to distinguish three types of positions, in addition to the position at epoch:

(i)- the apparent position for a given date corrected for both these effects as well as for precession

(ii)- the true position, as (i), but not corrected for aberration,

(iii)- the mean position, corrected only for precession.

The position at epoch referred necessarily to a mean position.

All these effects are due to various motions of the Earth and do not affect the assumed quality of stars as reference points. In order to obtain the position of a celestial body, one had only to interpolate, using some sort of observational technique, its observed position with respect to that of reference stars defining the reference frame.

The problem became much more complex when it was discovered that star positions vary because the stars themselves are moving. These proper motions - as well as the effect due to the orbital motion of the Earth (the parallax displacement) - are proper to each star and are not global effects.

From then on, one had to measure the star positions not only to improve the various parameters of the Earth motions or to assess the theory of aberration, but also to study the kinematics of the stars. It also meant that, unless some precise theory describing the motion of all reference stars could be found, stars were not fit to define a reference system.

However, stars are bound to be the fiducial points of celestial reference systems since most of astronomical instruments work in visible light. Therefore the construction of catalogues of star positions is an essential step in materializing celestial reference systems whatever definition is adopted for such systems. The present chapter is devoted to this task.

2. MEASUREMENT OF STAR POSITIONS

2.1. PRINCIPLES OF THE MEASUREMENTS

Astronomical instruments can measure the relative positions of two or more celestial objects emitting radiation in a given spectral region for which the detector of the instrument is sensitive. They may also refer their position to directions linked to the instrument or the environment (e.g. the local vertical). This has the following two consequences :

(i)- The observations are by definition always expressed in terms of differences of coordinates between celestial objects. At this level, the system of coordinates used may be closely linked to the instrument and even disconnected from the celestial reference frame.

(ii)- Independent systems of measurements (implemented for instance in optical and in radio spectral ranges) cannot be reduced to a single system unless one can observe bodies emitting in both wavelengths.

In this chapter, we shall deal only with optical observations. The next chapter is devoted to determinations of teh positions of radiosources.

One can also divide the instruments into two classes :

(i)- limited field instruments that determine relative positions of objects situated in a small region of the sky,

(ii)- global instruments that can link the positions of objects that lie in different regions of the celestial sphere.

Let us give an example of each type of astrometry on ground and in space.

2.2. SMALL FIELD ASTROMETRY

2.2.1. Ground based instruments

Astrographs are telescopes that are devised to take photographic plates of a stellar field. The dimension of the field varies from a fraction of a degree in long focus telescopes to five degrees in Schmidt telescopes. The photographic plate may now be replaced by charge coupled devices (CCD) cameras, at least in small field instruments.

The principle of the position determination is that there exists a transformation between the celestial coordinates of stars and the position measured on the photographic plate or CCD. The ideal transformation is the gnomonic projection of the sky on the plane tangential to the celestial sphere on its intersection O with the optical axis of the telescope. If α_0, δ_0 are the coordinates of this point and $\alpha_0+\Delta\alpha$, $\delta_0+\Delta\delta$ the coordinates of a star, and if the axes $O\xi$, $O\eta$ on the tangent plane are its intersections with the celestial meridian and parallel of O, one has exactly :

$$\xi = \frac{\cos(\delta_0+\Delta\delta)\sin\Delta\alpha}{\sin\delta_0\sin(\delta_0+\Delta\delta)+\cos\delta_0\cos(\delta_0+\Delta\delta)\cos\Delta\alpha} \tag{1}$$

$$\eta = \frac{\cos(\delta_0+\Delta\delta)\sin\Delta\alpha}{\sin\delta_0\sin(\delta_0+\Delta\delta)+\cos\delta_0\cos(\delta_0+\Delta\delta)\cos\Delta\alpha}$$

The metric coordinates x and y, measured on the plate are ideally proportional to ξ and η. These formulae whether expressed in ξ, η or x, y may be developed in power series of $\Delta\alpha$ and $\Delta\delta$ and, inversely, $\Delta\alpha$ and $\Delta\delta$ can similarly be developped in x, y. All perturbations to this ideal situation such as tilts, rotation and deformation of the plate, scale factors, various distortions and aberrations due to the optics and differential refraction of the atmosphere, differential astronomical aberration can be expressed in the same developed form. Generally, one expresses $\Delta\alpha$ and $\Delta\delta$ as second or third order polynomials of x and y and vice versa :

$$\begin{aligned} \Delta\alpha &= A_0 + A_1x + A_2y + A_3x^2 + A_4xy + A_5y^2 + \ldots \\ \Delta\delta &= B_0 + B_1x + B_2y + B_3x^2 + B_4xy + B_5y^2 + \ldots \end{aligned} \tag{2}$$

Small colour and magnitude effects may also exist.

For more details on plate reduction, see Eichhorn, 1974, Section 2.3 or Van de Kamp (1967, Chapter 5). It is clear that the plate constants A and B of (2) are not known *a priori* and are different for each plate. They have to be determined from the measurement of a number of stars whose positions are already known. This means that in photographic or CCD observations, positions of measured celestial bodies are referred to the same reference system as the coordinates of the stars used to determine A_i and B_i. If one changes the frame of reference of these stars, their coordinates change in some defined smooth manner, and this modifies the value of the plate constants without any possibility of identifying this particular effect from all others. This is expressed by stating that such observations are relative observations.

The present accuracy of photographic positions is usually dominated by the errors on the positions of the reference stars. The instrumental errors as such range from 0".2 or 0".3 for old astrographs of the type "Carte du Ciel" to 0".15 or 0".10 for modern astrographs or astrometric Schmidt telescopes measured with modern measuring machines (Eichhorn, 1974, p. 81) and may approach 0".02 in long focus small-field telescopes (Van de Kamp, 1981).

2.2.2. Small field space astrometry

The Hubble Space Telescope will have two small field astrometry capabilities. The main advantage of space observation is that the size of stellar images is strictly diffraction limited rather than extended to 0".5 to 2" or more by the atmospheric agitation in the case of ground-based instruments: this permits a much better definition of the centre of the image.

The Wide Field Planetary Camera will transmit f/30 images taken by four 800x800 pixel CCD camera of a 3x3 arc min field (Westphal, 1983) Repeated measurements of close objects could yield precisions of 0".01 or so.

But the main astrometric instrumentation on board of the Space Telescope are the Fine Guidance Sensors, FGS (Duncombe *et al*, 1982). Each FGS views a 90° sector of an annulus of about 17x4 arc min. In each of them, two beam deflectors whose positions are determined by the angles θ_a and θ_b and the lengths a, b of the segments can be pointed at the image of a celestial object (fig. 2).

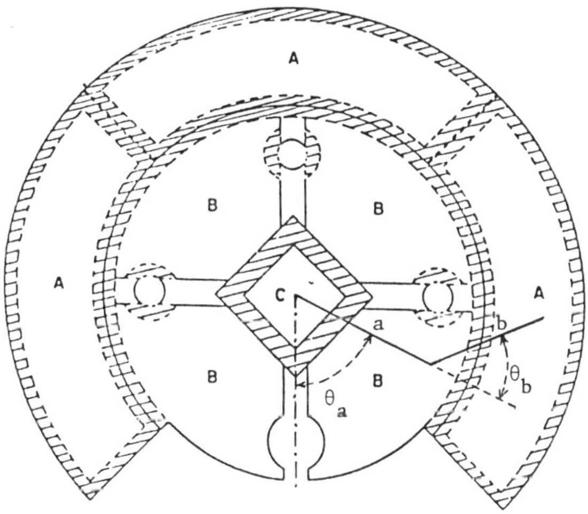

Figure 2. Distribution of fields in the focal plane of the Space Telescope :
A : Fine Guidance sensors; B : Axial instruments field; C : Wide field camera.

The light coming from each image passes through a beam splitter and is treated by two Koester's prism interferometers that control the position of the deflectors until the image is well centred. Despite the expected mechanical jitter of the satellite estimated to be of the order of 7 milliseconds of arc (0".007=7 mas), it is expected that the relative positions of the two objects will be obtained to an accuracy of 2 mas within the field of view.

The knowledge of the actual orientation of the telescope is limited by the accuracy of the *a priori* position of guide stars which will be about 3" (Russell and Egret, 1987). For stars separated by 15 arc min - this represents an error in orientation of 13 mas for the vector joining the stars. It results that the full accuracy of the FGS is achieved only for the distances between objects. The same accuracy of 2 mas is achieved for the absolute orientation of the vector only when stars are separated by 2 arc min or less.

2.3. GLOBAL ASTROMETRY

2.3.1. Ground-based global astrometry

The typical ground-based instrument that can link the positions of widely separated stars is the meridian circle. It is a classical instrument in use for almost three centuries since O. Roemer designed the first meridian circle and transit instrument. Its principle can be found for instance in Woolard and Clemence (1966) or, with more technical details, in Podobed (1968). Its goal is to determine the time of transit of a celestial body on the meridian plane of the observer and its zenith distance at the same instant through the reading of graduations on a vertical circle.

Actually, in reducing meridian observations, one has to correct for a number of instrumental imperfections such as the non coindicence of the surface scanned by the optical axis with the meridian plane, various screw and micrometer errors, flexures or bending of various mechanical parts, magnitude dependent errors, etc... Some of them may be calibrated in advance, but most are to be determined using observations of stars.

Over the last 15-20 years, there has been a great improvement in meridian instruments by a replacing the eye by a photoelectric receiver for both transit and zenith distance determination (Hoeg, 1970, 1972). In addition to a sizeable improvement of the precision of observations and to the disappearance of observer's personal systematic errors, this has led to the possibility of automation of the instrument which has considerably increased its efficiency (Helmer, 1986).

The determination of right ascensions by a transit instrument is based upon the formula :

$$\alpha = T - H \tag{3}$$

where T is the local sideral time of transit and H is the hour angle of the observation, in this case practically 0° or 180°. One can see that this determination depends upon the knowledge of the sideral time, that is of the system of reference defining the Earth rotation (Chapter 11). But it is possible to eliminate this dependency by determining differences of right ascension, assuming that the variation of T during one night is well known.

One can also assume that the right ascensions of some stars are known and determine the others with respect to them (relative observations).

Similarly, the determination of declinations from the zenith distance z is based upon the following formula :

$$z = \left|\phi - \delta\right| + R(z) \qquad (4)$$

where R is the atmospheric refraction and ϕ the instantaneous latitude (that is corrected for the polar motion, see Chapter 9). Again, one can use known declinations to obtain relative declinations of other stars.

Currently, the precision that is achieved by modern meridian circles is of the order of 0".06 to 0".10 in α and δ (Requième and Mazurier, 1986).

Other global astrometry instruments exist. The most important is the Danjon astrolabe, based upon the observation on a small circle of equal zenith distance (Débarbat and Guinot, 1972). It has also been modified for photoelectric detection (Billaud, 1986, Luo and Li, 1986).

2.3.2. HIPPARCOS programme

Only one global astrometry satellite has been scheduled up to now. This is the ESA satellite HIPPARCOS due to be launched in 1989.

The principle of the instrument was conceived more than 20 years ago by Lacroute (1968). The actual configuration is the result of many improvements to the original idea. It was frozen in 1980 when the European Space Agency decided to implement the project.

The main idca is that a double rigid plane mirror called a 'beam combiner' reflects into a single telescope two fields of view separated by a constant angle γ (see figure 3).

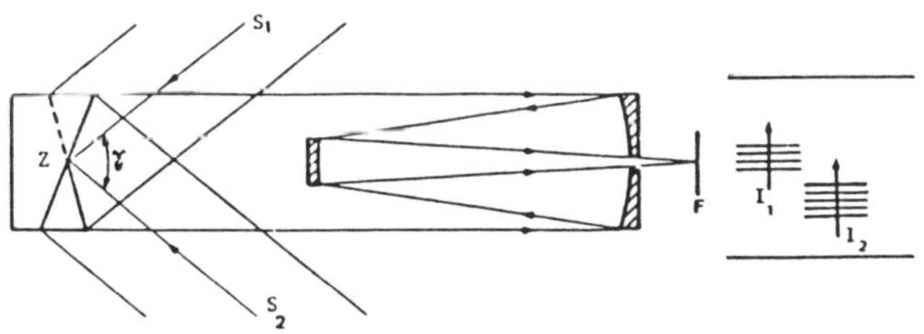

Figure 3. Principle of the HIPPARCOS telescope. While the telescope rotates around Z, the light of the images I_1 and I_2 of S_1 and S_2 in two different fields of view is modulated by the grid.

The images I_1 and I_2 of two stars S_1 and S_2 move on the focal plane, while the telescope is rotating around an axis Z parallel to the intersection of the mirrors. On the focal plane, a grid of parallel slits modulates the light of all the stars crossing one of the fields of view. An image dissector tube can select the light from only one of the stars and also shift rapidly from one image to another.

Another feature of HIPPARCOS is the possibility to determine accurately the instantaneous orientation of the satellite using gyroscope readings and a system of parallel and inclined slits, called the star-mapper, that give an aperiodic modulation of the star light crossing these slits.

From the knowledge of the attitude evolution and of γ, together with the difference of phases of the modulated light of S_1 and S_2, it is possible to determine the actual angle between the stars projected on the scanning plane.

The reduction of the data proceeds in three steps.

1st step: during 10 to 12 hours of observations, a narrow strip of the sky is scanned. All the attitude and grid modulation data are gathered and used to determine for each star the abcissa of its projection on a fixed reference great circle, with an arbitrary origin. The instrumental parameters are also determined at this stage and the attitude improved.

2nd step: during the mission, all the sky is scanned many times, so that the projection of every star is determined on 20 to 40 different reference great circles. This produces a net of connections between stars. About one third of the stars called primary stars are chosen among the brightest single stars to improve the general connection in determining the origins of the reference great circles. This step is called the sphere reconstitution and is indeed the construction of the reference frame.

3rd step: all the observations of a given star are used to determine the position at epoch, the proper motion and the parallax, at least for single stars. For double and multiple stars, some more complex data analysis is necessary. At this stage, the materialization of the reference system is achieved with the results of processing the primary stars.

Actually, the general reduction is an iterative process. After the completion of a first reduction through all the three steps, the star positions will have been improved. These will be used to update the attitude determination which, in turn, will improve the results of the first step, and so on.

A more complete description of the HIPPARCOS instrument, mission and reduction procedures can be found in Kovalevsky (1984, 1986). Other technical details on the HIPPARCOS satellite have been published in open litterature by Schrijver (1986). Detailed analyses of the reduction procedure may be found in Donati *et al.* (1986) for the attitude determination, in Van Daalen and Van der Marel (1986) for the reduction on reference great circles and in Galligani *et al.* (1986) on the sphere reconstitution.

At the end of the reduction, it is expected, at least for half of the 115.000 stars that will be observed, that the positions at epoch will be determined with an r.m.s. error smaller than 2 mas and similarly the proper motions with an r.m.s. error of the order of 2 mas per year. It is expected that these positions and proper motions will be referred to a unique consistent frame to better than 1 mas and 1 mas per year. The reasons of this remarkable consistency of the expected results are threefold :

(i)- the observations are made with a single instrument covering all the sky, unlike ground based astrometry,

(ii)- there is no disturbing systematic atmospheric effects due to unsufficiently modelled refraction and atmospheric scattering,

(iii)- the observations are independent of the parameters of the Earth orientation and are consequently not biased by errors on these parameters.

However, this "HIPPARCOS reference frame" is determined by setting arbitrarily some parameters to a given value since the sphere reconstitution problem has a rank deficiency of order six. External methods (see Section 6.4) will have to be used to force the HIPPARCOS star positions and proper motions to represent a well defined celestial reference system.

3. COMPARISON OF STAR POSITIONS

We have seen that all the observations of star positions are relative. In other terms all ground-based or space astrometric instruments refer the observed position of a star to the observed position of another star. For the ground-based astrometry, they are apparent positions that are not the direction in which the star actually is, but modified by a number of physical effects. This is also true, to a lesser extent for space observations. These effects depend both on the position of the star and on the time at which it is observed.

However, if one assigns to stars the role of fiducial points, it is on the first place necessary to be able to determine and to compare their real directions in space rather than apparent directions. It is the role of fundamental astronomy to derive the real directions of stars from the observed ones. These problems are presented in details in many recent books such as those of Green (1985), McNally (1974), Murray (1983), Podobed and Nesterov (1975) or in classical but still very useful books by Danjon (1980) or Woolard and Clemence (1966).

The corrections that have to be applied to an observed position of a star at time t in order to obtain its actual geometric direction at time t_0 are different in nature. They are briefly described in the following subsections.

3.1. PRECESSION AND NUTATION

This effect is dependent upon the definition of the equatorial system that is used in astronomy and it is very closely linked to the Earth. Let us first recollect the definition of the true and of the mean equatorial systems.

The true equatorial system at time t is defined by the direction of the Earth's rotation axis at this time as the Z axis and by the intersection of the corresponding celestial equator with the mean ecliptic of time t as the X axis. With respect to a fixed reference system, the true system has a secular motion called precession and a number of periodic terms the sum of which is the nutation.

The mean equatorial system at time t is the true system corrected for nutation. It is necessary to represent the nutation series as two rotations: $\Delta\psi$, nutation in longitude, along the ecliptic and $\Delta\varepsilon$, nutation in obliquity, around the X axis.

The rotation matrix, in the equatorial system is therefore, if ε is the obliquity of the ecliptic :

$$N = \begin{vmatrix} 1 & -\Delta\psi\cos\varepsilon & -\Delta\psi\sin\varepsilon \\ \Delta\psi\cos\varepsilon & 1 & -\Delta\varepsilon \\ \Delta\psi\sin\varepsilon & \Delta\varepsilon & 1 \end{vmatrix} \qquad (5)$$

The corrections in right ascension and declination can be deduced as:

$$\Delta\alpha = \Delta\psi(\cos\varepsilon + \sin\varepsilon\sin\alpha\tan\delta) - \Delta\varepsilon\cos\alpha\tan\delta$$

$$\Delta\delta = \delta\psi\sin\varepsilon\cos\alpha + \Delta\varepsilon\sin\alpha$$

To apply precession in order to transform mean positions from time t (pole P, vernal equinox γ) to time t_0 (pole P , vernal equinox γ_0), it is necessary to define three angles. In equatorial coordinates, the most convenient are (fig. 4):

- θ, arc PP_0 ;
- ζ, angle between P_0P and $P_0\gamma_0$
- z, angle between $P\gamma$ and P_0P_0

We have dropped the classical subscript A in the notations. The rotation matrix of this transformation is :

$$P = \begin{vmatrix} -\sin\zeta\sin z + \cos\zeta\cos z\cos\theta & -\cos\zeta\sin z - \sin\zeta\cos z\cos\theta & -\cos z\sin\theta \\ \sin\zeta\cos z + \cos\zeta\sin z\cos\theta & \cos\zeta\cos z - \sin\zeta\sin z\cos\theta & -\sin z\sin\theta \\ \cos\zeta\sin\theta & -\sin\zeta\sin\theta & \cos\theta \end{vmatrix} \quad (6)$$

Various forms of these transformations and the expressions adopted in 1976 by the IAU are given in Lieske et al. (1977). Let us note that in satellite astrometry, these effects do not exist.

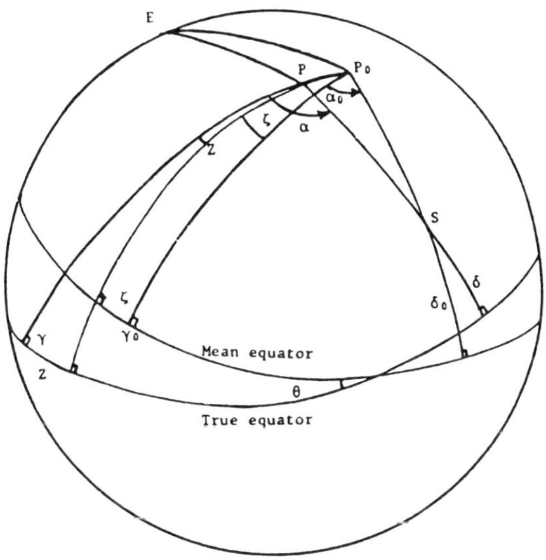

Figure 4. Precession angles, E is the pole of the ecliptic. The index zero refers to the mean positions. The transformation from mean positions α_0 and δ_0 to true positions α and δ is defined by the three angles ζ, Z and θ.

3.2. POLAR MOTION

For an Earth based observer, the motion of the pole (see Chapter 11) modifies its latitude by a quantity $\Delta\phi$ with respect to its convertional fixed latitude. This effect should be taken into account when determining the declination of stars by formula (4).

3.3. ABERRATION

The aberration is an apparent change of the direction of an object due to the relative motion of the observer with respect to the object. However, in the case of stars that heve a constant velocity with respect to the barycenter B of the solar system, and in contrast to what is done for planetary aberration, it is customary to consider only the motion of the observer with respect to B. The remaining part correspond to a fixed displacement of the position of star.

The classical, pre-relativistic conception of stellar aberration was based upon the composition of the velocity vector $\vec{V}$ of the observer with respect to B and the velocity vector of the incroming light.

If $\vec{r}'$ is the apparent position vector of the star and $\vec{r}$ the 'true' position in the barycentric frame of reference, then the classical expression is :

$$\vec{r}' = \vec{r} + r\vec{\beta} \tag{7}$$

where $\vec{\beta}=\vec{V}/c$ is the velocity vector of the observer expressed in units of speed of light.

In the case of relativity, (7) is no more exact and a second order formulation is (Kovalevsky et al., 1986) :

$$\vec{r}' = \vec{r} + \left(r + \frac{\vec{r}.\vec{\beta}}{2} \right) \vec{\beta} \tag{8}$$

If $\vec{s}$ and $\vec{s}'$ are unit vectors along $\vec{r}$ and $\vec{r}'$, one can derive from (8), the formula given by Stumpff (1979) :

$$\vec{s}' = \vec{s} + \vec{\beta} - \vec{s}(\vec{s}.\vec{\beta}) + \vec{s}\left[(\vec{s}.\vec{\beta})^2 - \beta^2/2 \right] - \vec{\beta}(\frac{\vec{s}.\vec{\beta}}{2}) \tag{9}$$

or equivalent formulae given in various textbooks.

In practice, the main part of the velocity of the observer is the motion of the Earth around the Sun and is of the order of 30 km.s^{-1}. The first term is of the order of the 'constant of aberration', k=20".496 The terms due to relativity are of the order of a millisecond of arc and cannot anymore be neglected in high accuracy astrometry.

The computation of the aberration is usually divided into steps :

(i)- the annual aberration corresponding to the motion of the Earth center of mass with respect to the barycentre of the solar system,

(ii)- for an Earth-based observer, the diurnal aberration due to the rotation of the Earth,

(iii)- for satellite-borne astrometry, the aberration produced by the orbital motion of the satellite about the Earth.

3.4. RELATIVISTIC LIGHT DEFLECTION

The deflection of the light coming from a star S due to the curvature of the space around a massive object M is a classical relativistic effect. It tends to shift the star image outwards M in the M-S-observer plane. The value of this displacement is :

$$\Delta\theta = \frac{\mu(1+\gamma)}{c^2 a} \cdot \frac{1+\cos D}{\sin D} \tag{10}$$

where μ is the gravitational constant of M, γ is one of the parametrized post Newtonian (PPN) parameters (Will, 1981), c is the speed of light, a is the distance between the observer and M, and D is the angular distance MS (see, for instance, Murray, 1981).

For the Sun, μ=0.0002959122, the units being the mass of the Sun, the day and the astronomical unit of length. Then, with a=1 and γ=1, (10) reduces to :

$$\Delta\theta = 0''.004072 \frac{1+\cos d}{\sin D}$$

3.5. PARALLAXES

The corrections for parallaxes consist in computing the apparent displacement of the direction of the star S due to the fact that the observer O is not at the barycenter B of the solar system to which all the observations have to be referred. One has (see figure 5) :

$$\vec{BS} = \vec{BO} + \vec{OS}$$

Let $\vec{s}$ and $\vec{s'}$ be the unit vectors of the star directions, and call :

$$\vec{\mathbf{BS}} = \vec{r} = \vec{s}r$$

$$\vec{\mathbf{BO}} = \vec{r'} = \vec{s'}r$$

$$\vec{\mathbf{BO}} = \vec{R} = \vec{u}R$$

The distance r of the star being very large in comparison with OB, the parallactic displacement can be represented by :

$$\Delta\vec{s} = \vec{s}' - \vec{s} = \frac{R}{r}\vec{s} \wedge (\vec{s} \wedge \vec{u}) \tag{11}$$

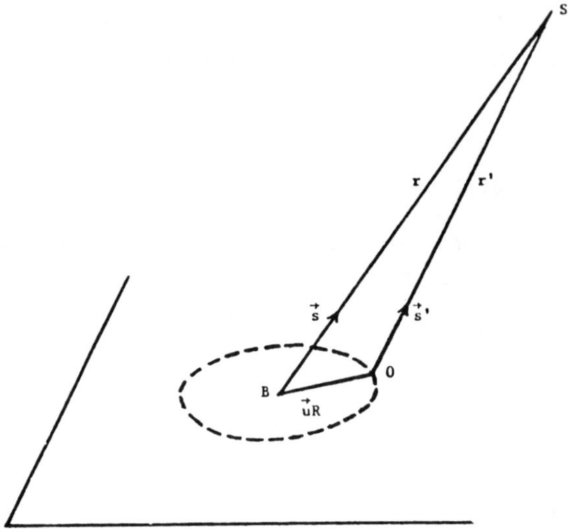

Figure 5. Geometry of the parallactic displacement $\vec{s}' - \vec{s}$.

see, for instance, Green (1985, p.60); a/R, where a is equal to one astronomical unit is the parallax of the star, usually expressed in arcseconds. No star has a parallax larger than 1".

As in the case of aberration, this effect is computed in steps :

(i)- the annual parallax for which one assumes that the observer is in the Earth's center of mass,

(ii)- for an Earth-based observer, the diurnal parallax corresponding to the displacement center of the Earth-observer. Its value is for any star smaller than 0".00004,

(iii)- for space astrometry, one has to consider the radius-vector of the orbit and the motion of the satellite.

With the present accuracies of astrometry, the second step is normally neglected.

3.6. PROPER MOTIONS OF STARS

The stars contribute, like the Sun, to the general rotation of the Galaxy.

However, this is not a solid-like rotation and every star has, in addition, a random motion. These two components of motion, as seen from the moving Sun itself, produce a linear change in the mean positions reduced to a given fixed epoch. This effect, called 'proper motion' characterizes every star and is usually expressed as a yearly variation of a mean position in differential coordinates.

The proper motion for time t starting from a position at time t_0 is :

$$\Delta\alpha = (t-t_0)\,\mu_\alpha \quad ; \quad \Delta\delta = (t-t_0)\,\mu_\delta$$

To avoid the singularity at poles, usually $\mu_\alpha \cos\delta$ is given.

In addition to this linear effect, one must, in some cases, introduce a rate of change of the proper motion, known as perspective acceleration due to a rapid change in distance measured by its radial velocity V expressed in $km.s^{-1}$. The acceleration along the proper motion direction is :

$$\frac{d\mu}{dt} = -\frac{k\mu V}{r}$$

where r is the distance to the star expressed in parsecs, μ the proper motion in arc sec per year, the constant k is 0.000117 and $d\mu/dt$ is expressed in arc-sec per year square. The correction is usually negligible, but for some stars and a long interval of time, it has to be taken into account.

3.7. GALACTIC ROTATION

The rotation of the Galaxy produces a systematic trend in the proper motions. The galactic plane being defined as a great circle with its pole at α=192°15', δ=27°24' (equinox 1950.0), the centre of the Galaxy, lies on this plane. Its approximate position, serving as the origin of the galactic longitude is the point of the celestial sphere of coordinates is: α=265°36', δ=28°55' (equinox 1950.0) and at a distance of the Sun conventionally taken as R =8500 parsccs (pc) sincc 1985 (Wielen, 1986). It rotates differentially around an axis perpendicular to the galactic plane. Around the Sun, where all astrometrically observed stars are, the rotation is retrograde and, using Oort's constants, it is given by:

$$\Omega = B - A + 2A(R/R_0 - 1) \qquad (12)$$

In this formula, R is the distance of the star to the galactic center. The present conventional values are :

$A = 0.0015\ km.s^{-1}\ .pc^{-1} = 0''.0032$ per year
$B = 0.0010\ km.s^{-1}\ .pc^{-1} = 0''.0021$ per year

(Bok, 1983). However, the newly adopted values of R =8500 pc and of the rotation speed at this distance (θ=220 km.s^{-1}) are not fully consistent with the above values of A and B and imply A+B=0.00259 (Wielen, 1986).

3.8. THE LOCAL STANDARD OF REST

We define a local system of coordinates, called local standard of rest centered at the Sun and rotating around the galactic center with $\Omega(R_0)$ defined by (12) as equal to A-B. It is an intermediate reference frame that permits to eliminate the effects of the particular stellar motion in general, and particularly that of the Sun. The motion of the Sun with respect to the local standard of rest is not very accurate mostly because of the difficulty in identifying the local standard of rest with the group motion of some kinds of stars. Classically it was assumed that the best markers are B stars, hydrogen clouds and other very young objects. This gave the following components of the solar motion in galactic rectangular coordinates (Delhaye, 1965) :

$$U = 10.4 \text{ km/s} ; V = 14.8 \text{ km/s} ; W = 7.3 \text{ km/s} \quad (13)$$

However, the assumption leading to this result are heavily dependent on the model of spiral structure that may be adopted, so that corrections of several kilometres per second may have to be introduced (Yuan, 1983).

In order to refer a proper motion of a star to the local standard of rest, one has to proceed with two corrections.

(i)- Correct for the differential rotation of the Galaxy. In galactic coordinates (l,b), the correction, expressed in seconds of arc is :

$$\Delta\mu_l = 0.211(A \cos 2l + B)$$
$$\Delta\mu_l = 0.211\, A \sin 2l \;\; \cos b \sin b \quad (14)$$

(ii)- Correct for the motion of the Sun with respect to the local standard of rest. This corrections consist in subtracting the values (13) from the space motion of the star. The latter is accessible only if, in addition to the proper motion, the distance is know.This not being the case for most stars, at least with sufficient accuracy, the correction is very uncertain.

4. STAR CATALOGUES

A star catalogue is a list of stellar positions given with respect to some reference system, generally expressed in right ascension and declination in an equatorial system of a given epoch. It used to be 1900.0, 1925.0, 1950.0. Now it is J.2000, the begining of Julian day 2000, January 1.5d. (see Chapter 15). There are several levels or types of catalogues.

4.1. INSTRUMENTAL OR OBSERVATION CATALOGUES

They are the result of a certain period of observations, generally one to three years with a given instrument (meridian circle or astrolabe). It is generally referred to the mean epoch of observations. The reduction to this epoch is made with well defined series of precession and nutations, usually the standard values adopted at that time (refer to Chapters 11 and 18) and in some cases with known values of proper motions. This means that the catalogue does not only reflect the observations, but also some reduction procedures and assumptions.

The stars in such catalogues are usually situated within some declination limits. There are different types of such catalogues.

4.1.1. Relative catalogues

Some of the observed stars are assumed to have known positions, taken from another catalogue. The positions of the other observed stars are then strictly in the system of this catalogue, provided that the same reduction standards are used. Example: astrolabe catalogues.

4.1.2. Absolute catalogues

They are obtained when celestial bodies used to define the celestial reference system are also observed by the same instrument. It is generally the Sun for dynamical reference systems based upon the dynamics of the solar system (see Chapter 3) but could also be some planets In this case, the star positions are referred to there bodies. Example: USNO meridian catalogues.

4.1.3. Quasi-absolute catalogues

We shall give this name to the catalogues that give α and/or δ without referring to objects materializing the system as in absolute catalogues, nor to stars with given coordinates as in relative catalogues. Absolute declinations can be obtained by astrolabes using at least two different zenith distances (Debarbat and Guinot, 1972) or by meridian circles observing circumpolar stars at their two culminations. Partially absolute right ascension observations are made with meridian circles at the condition that the azimuth of the instrument is known independently from circumpolar stars and that one refers to a sidereal time that has been obtained by more occurate methods (see Chapter 11). In that case, differences of right ascension of stars may be considered as absolute.

4.2. INTEGRATED CATALOGUES

Several similar instruments may join in observing a number of stars in order to cover a large portion of the sky or to obtain a dense net of stellar positions. All the instrumental relative catalogues are reduced to the same epoch with the same standards and compared in order to determine the systematic instrumental errors. After correcting these, one obtains a larger catalogue of stars which is an extension of the reference catalogue for the epoch of the observations.

This can be repeated at two different epochs. Then it is possible to determine the proper motions of stars by comparing the two integrated catalogues. Examples: AGK3R, IRS (see Section 5.4).

4.3. EXTENSIONS OF A CATALOGUE

In contrast with the integrated catalogues, extensions are constructed only in order to densify a given catalogue. The technique for observation may be the same as for the basic catalogue or different. Example: Guide Star Selection System (GSSS) catalogue extends SAO data to stars up to magnitude 15-16 from Schmidt plates, AGK3 extends the AGK3R and TYCHO will be an extension of HIPPARCOS catalogue.

4.4. COMPILATION CATALOGUES

They are obtained in compiling the maximum number of catalogues reducing them to a single system and deciding for every star what are the parameters that are to be kept from a choice provided by several sources. The best known such catalogue is the SAO. It includes 258997 stars referred to the FK4 system (see Section 4.2). It uses the positions of two fundamental catalogues (FK3 and GC), positions taken in AGK2 - AGK1, various Yale catalogues, some zones of the astrographic catalogue (Carte du Ciel), and a few meridian catalogues in the southern hemisphere.

For more details on various star catalogues, see Eichhorn (1974).

5. FUNDAMENTAL CATALOGUES

By definition, they are the catalogues that primarily represent a conventional celestial reference system.

In the process of constructing a conventional celestial reference frame, two major steps are taken (see the introduction).

(i)- The implementation of the physical model with the choice of the numerical parameters modelling it. This produces the conventional reference system.

(ii)- Producing a certain number of star positions and proper motions materializing the system.

Several chapters of this book are devoted to the first step in the case of celestial reference systems. Here, we shall assume that this system exists and see how it is transformed into a reference frame, in the form of a 'Fundamental catalogue'.

The construction of a fundamental catalogue is a tremendous work that has not been undertaken many times in the past. In addition, since one of the values of a reference system is to be a standard to which many years of observations are supposed to refer, the lifetime of such a catalogue should be large. The change to a new system or even to a new version of an unchanged system is required only if the old one becomes obsolete because the observations have a better precision than the accuracy of the fundamental catalogue.

5.1. THE MAIN FUNDAMENTAL CATALOGUES

Table 1 gives the fundamental catalogues constructed in the last 50 years. The GC has never been very much used. Actually the positions of this catalogue were very quickly degrading essentially for two reasons. The first is that thc mcan cpoch of observations was around 1900, so that proper motion uncertainty very much affected the positions in the second half of the 20-th century. Secondly, thc great number of stars retained implied that for many of them the precision was neither very good nor homogeneous.

TABLE 1

NAME	AUTHOR AND DATE	NUMBER OF STARS
Gal. Cat. GC	B. Boss, 1937	33342
FK3	A. Kopff, 1938	1535
N30	H.R. Morgan, 1952	5268
FK4	W. Fricke, A. Kopff, 1963	1535
FK5	W. Fricke, 1986	1535
FK5 Extension	in progress	about 3000

In order to cope with some of the defects, H.R. Morgan constructed a smaller catalogue, the N30, using many newer catalogues and retaining less stars. But even so, the quality of the positions of N30 were still not very good. Too many stars had not been observed in absolute catalogues and, in addition, the unhomogenities of the GC system are still present in part in N30 (Kopff, 1954).

In fact, before the computer age, the work necessary to built extended fundamental catalogues was too large, and in practice the series of catalogues produced by the Astronomisches Rechen Institut the FK3 and the FK4 with less stars, but more thoroughly analyzed, appeared to be more homogeneous and, finally, have been preferred and used as the conventional fundamental catalogues for practically all astronomical and geodynamical use.

5.2. CONSTRUCTION OF A FUNDAMENTAL CATALOGUE

One may summarize the construction of a fundamental catalogue as an integrated compilation of relative, absolute and quasi-absolute instrumental catalogues, retaining only the most accurately observed stars as isotropically distributed as possible over the celestial sphere. This is done after reducing all the catalogues with the parameters of the new conventional reference system. The link between the system and the star positions is achieved through absolute catalogues. In the general case of fundamental catalogues based on a dynamical reference system using the dynamics of the solar system (Chapter 3), the catalogues of stars observed with respect to the Sun are the basic material for this connection. However, this can be done also indirectly from the observation of relative positions of stars and of selected planets or the Moon, and this possibility is extensively used in order to fix the equinox.

A more detailed description on how a fundamental catalogue is constructed may be found in Fricke and Kopff (1964), Scott (1963), Fricke (1981) or Schwan (1986).

5.3. THE FK SERIES OF FUNDAMENTAL CATALOGUES

The FK3 is a realization of a dynamical reference system based on Newcomb's precession and dynamical theory of the motions in the solar system (see chapter 3). The FK4 is essentially an improvement of the FK3 as far as the star positions and proper motions are concerned (individual and regional errors), and thus, represents the same system of reference. No equinox or precession correction was introduced.

In order to give an example of the size of the work that is involved in compiling such catalogues, let us just state that in constructing the FK4, 86 instrumental catalogues already used for FK3 and 72 instrumental catalogues not available at the time of FK3 were analysed and used. In addition data from GC and N30 were introduced in the analysis (Fricke and Kopff, 1963). Only 1535 stars were actual fundamental stars. However the compilation work permitted them to select another 1967 stars whose positions were not as good, but still densified the catalogue. This additional list was called FK4 sup.

The compilation of the FK5 represented a major effort at the Astronomiches Rechen Institut. The comparison of 100 new catalogues with the FK4 permitted them to improve the individual proper motions of stars by a factor of 2. This part of the work is independent of the reference system. The regional errors were essentially deduced from 90 absolute or quasi-absolute catalogues (25 new ones in α and 15 in δ) including astrolabe and time catalogues (Schwan, 1986 and 1987). The system itself used the IAU.1976 values of precession (Lieske et al., 1977), a new determination of the equinox and of the equator (Fricke, 1982) and the rotation of the local standard of rest (Fricke, 1981). This last feature, which introduces the galactic rotation into the FK5 system, implies that this system is not purely dynamical but is partly kinematic. The mean precision achieved in the FK5 is 0".02 in position and 0.8 mas per year in proper motion.

5.4. EXTENSIONS OF A FUNDAMENTAL CATALOGUE

The basic stars of the FK5 are the same as in FK3 and FK4. It is not sufficient to extend the system to any small region of the sky, for instance in reducing photographic plates. So, once a fundamental catalogue is created, it is necessary to add more stars in order to represent the system in every spot of the sky. To take an example, in order to determine the position of a celestial body using classical 2° by 2° plates, it is necessary that 10 to 20 reference stars, already determined in the system of the fundamental catalogue, be present in the field. This would mean an extension to at least half a million stars, possibly more. Such an extension would be constructed by the method described in sections 3.2 and 3.3. No example of such a large extension presently exists, the main reason being that the star positions and proper motions are not know with a sufficient accuracy to transport the system without introducing sizeable errors.

It is however intended to extend the number of FK5 stars to another 3000 or so stars. It is expected that they will still well represent the system, but the random errors will be twice those of the basic stars of the FK5.

A more important extension is the International Reference Star (IRS) catalogue that is almost completed and will include about one star per square degree. It will include the AGK3R stars in the northern hemisphere, the

SRS (Southern Reference Stars) catalogue in the southern hemisphere (Zverev *et al.*, 1986) and some additional stars to ensure the homogeneity of distribution on the celestial sphere (Smith 1986). A special effort was made to obtain a homogeneous system of proper motions (Corbin, 1978).

Further extensions should be based on the IRS itself or on future larger and more homogeneous catalogues like the HIPPARCOS catalogue (see below).

6 - KINEMATIC CELESTIAL REFERENCE FRAMES

The fundamental catalogues described in the preceding section are realizations of a dynamical reference system based upon the motion of bodies in the solar system. We have seen that, for the first time, a fundamental catalogue - namely the FK5 - was also partly based upon results of stellar kinematics. One may question whether this is by accident, or whether this is the beginning of an evolution and whether, in the future, the reference system may be based on the positions of objects situated beyond the solar system.

Actually two types of objects may compete to serve as defining bodies for a kinematic system: stars and extragalactic objects such as quasars.

6.1. REFERENCE SYSTEMS DEFINED BY STARS

To define such a reference system it is necessary to postulate that the general motion of stars obey certain laws and to specify exactly these laws at least in the part of space that will include the stars used to materialize the system. Oort's model of differential rotation of the Galaxy could be adequate for stars that are within one or two kiloparsecs from the Sun, provided that this kinematic model is extended perpendicularly to the galactic plane. In addition, it should be stated how the local standard of rest, which represents the galactic rotation at Sun's position, as well as the similar systems of coordinates in the region where the differential rotation is defined can be accessed by observations.

At this point arise the first great difficulties already mentioned in sections 3.7 and 3.8. The Galaxy has a spiral structure and, in addition, stars of different classes or ages have a different kinematic behaviour (see for instance Mould, 1982 and Yuan, 1983). The choice of those that should be used to represent the actual rotation of the Galaxy is not simple as can be seen from Ovenden *et al.* (1983) and even if it was made, one will have to evaluate the corresponding rotation with respect to coordinates that will be assumed not to rotate, so that they will represent the actual system of reference. W. Fricke (1981) has evaluated to 0".06 per century the r.m.s. error of the galactic rotation. However, because of the difficulty in choising the correct markers of this rotation, the actual error could be somewhat larger.

The basic value of the rotation of the Galaxy at the level of the Sun has been determined by a number of different methods using optical or radio radial velocities of H1 regions, globular clusters, etc... The main source of error lies in the asymmetry of the outer regions of the Galaxy. The value θ=220 km.s^{-1} given in Section 2.7 has an error estimated to 20 km.s^{-1} (Kerr and Lynden-Bell, 1986). This would give an possible residual rotation of 0".05 per century to the reference system.

Let us remark that, at the moment, the precession constant has still an uncertainty of 0".15 per century (Fricke, 1981). This means that even if the numbers quoted above are optimistic and may be affected by some systematic effect, the choice of a purely kinematic reference system based only on a model of systematic motions of stars is not to be rejected *a priori.*

6.2. REALIZATION OF A STELLAR CELESTIAL REFERENCE SYSTEM

Constructing a reference system based upon definitions described in Section 6.1 would encounter major difficulties. The main reason is that every subsystem of stars has not only a systematic motion with respect to the local standard of rest, but also dispersions of velocities of the order of 20 to 30 km.s^{-1} if one excludes population II stars which may have much larger systematic motions. If r is the distance of a star in parsecs, the dispersion in proper motion would be expressed in seconds of arc per century :

$$\varepsilon_\mu = \frac{\sigma}{0.0474r}$$

One can easily see that taking r=100 as the mean distances of a star, the dispersion will be of the order of a few seconds of arc per century, so that in order to achieve a 0".05 per century accuracy in representing the system, several tens of thousands of stars will be needed.

In addition, this implies that the measurement of stellar proper motions is free from systematic errors that may be introduced in applying the procedures described in Section 3. This concerns essentially an error in precession that will inevitably be introduced by ground-based measurements. So, in practice, one would reintroduce through the observations the systematic errors of the dynamical reference frames into the kinematic frame.

This can be avoided if the observations are made from space. The

HIPPARCOS catalogue of 100.000 distances and proper motions could indeed provide the first homogeneous observational material and could permit to construct the first celestial reference frame based upon stars.

6.3. REFERENCE SYSTEMS BASED ON EXTRAGALACTIC OBJECTS

Replacing stars by extragalactic objects has an enormous theoretical advantage. Excluding nearby galaxies and more generally objects with sizeable apparent dimensions, the distance of such objects is so large, that their proper motions cannot exceed a couple of milliarc seconds per century even if they have actual velocities of the order of their recession speed. The theory defining the system reduces to the postulate that these objects are fixed on a non-rotation celestial sphere.

The optical observations of these faint objects cannot yield great accuracy, but this is not the case if they also emit in the radio spectrum. Then they can define a celestial reference frame as it is described in Chapter 2. The problem is that this reference frame will be defined in a radio spectrum and will not be accessible for optical astronomy unless by some method stars may be linked to it as it will be shown in the next section.

6.4. HIPPARCOS CATALOGUE AS AN EXTENSION CATALOGUE

We have seen that the HIPPARCOS catalogue will consist of more than 100.000 star positions and proper motions determined on a consistent celestial sphere, but with 6 arbitrary rotational parameters (Section 2.2.2). This means that it can represent any other consistent reference system on the sky by simply applying to it two rotations :

(i)- a rotation defined by the matrix $R(t_0)$ which rotates all the positions at the epoch t_0,

(ii)- a second time dependent rotation defined by the matrix R' multiplied by the difference of epochs $t-t_0$.

Since the rotation are very small one may neglect second order terms and write :

$$R = \begin{vmatrix} 1 & \alpha & \beta \\ -\alpha & 1 & \gamma \\ -\beta & -\gamma & 1 \end{vmatrix} \qquad \text{and} \qquad R' = \begin{vmatrix} 1 & \alpha' & \beta' \\ \alpha' & 1 & \gamma' \\ -\beta' & -\gamma' & 1 \end{vmatrix}$$

Let us call $\vec{X}_H$ and $\vec{X}_R$ the unit vectors of the position of a star at epoch t_0 in the reference catalogue (R) and in the catalogue to be tied to R, (H). Similarly, let $\vec{\mu}_H$ and $\vec{\mu}_R$ be the proper motions in the respective catalogues.

The position at time t in the reference catalogue (R) can be written as :

$$\vec{X}_R + (t-t_0)\vec{\mu}_R = R(t_0)\vec{X}_H + (t-t_0)R'\vec{\mu}_H$$

This equation separates into two vectorial equations, the unknowns α, β, γ and α', β', γ' being determined if (15) can be written for several stars. One can also transform (15) in the case where one observes only the distance to a reference object as it may be the case for the Fine Guidance sensors (see 2.1.2), or for any other observational conditions.

Once R and R' are determined, one will apply (15) in order to transform $\vec{X}_H$ and $\vec{\mu}_H$ into $\vec{X}_R$ and $\vec{\mu}_R$.

A simulation was made by Froeschlé and Kovalevsky (1982) assuming that both VLBI observations and HIPPARCOS observations have an r.m.s. of 0".002 and 0".002 per year. The link can be achieved to 0".001 and 0".001 per year with about 20 radio-stars. If one also uses Space Telescope observations giving the radial proper motion of a star with respect to a quasar, then the same accuracy could be achieved for the matrix R' with about 100 stars. But one can of course use simultaneously the results of both techniques and achieve better results for instance with 20 radio-stars and 50 space telescope observations.

A similar technique can be used to transfer the HIPPARCOS catalogue to the FK5 system. Since almost all FK5 stars will be observed by HIPPARCOS the link can have even a much better accuracy. In addition, the systematic regional differences, which very probably will essentially reflect the regional errors of the FK5, can also be determined if one assigns to the transformation from (H) to (R) in addition to the two rotations, a development in some kind of orthogonal functions (see Bien *et al.*, 1978).

In conclusion, one may say that once the HIPPARCOS catalogue is completed, it will be possible to use it as an extension to any existing catalogue and consequently to represent any reference system.

REFERENCES

Bien, R., Fricke, W., Lederle, T. and Schwan, H., 1978, *Verof. Astron.Rechn. Institut*, Heidelberg, n° 29.

Billaud, G., 1986, in *'Astrometric Techniques'*, H.K. Eichhorn and R.L. Leacock (eds.), IAU Symposium 109, D. Reidel Publ.Co., Dordrecht, p. 389.

Bok, B.J., 1983, in *'Kinematics, Dynamics and Structure of the Milky Way'*, W.L.J. Shuter (ed.), D. Reidel Publ.Co., Dordrecht, p. 5.

Corbin, T.E., 1978, in *'Modern Astrometry'*, F.V. Prochazka and R.H. Tucker (eds.), IAU Coll. 48, Univ. Observatory Vienna Publ., p. 505

Danjon, A., 1980, *'Astronomie Générale'*, 2nd Edition, A. Blanchard, Paris.

Débarbat, S. and Guinot, B., 1972, *'The method of equal heights'*, Gordon and Breach Science Publ., London.

Delhaye, J., 1965, in *'Galactic structure'*, Stars and Stellar System, vol. V, A. Blaauw and M. Schmidt (eds)., Univ. of Chicago Press, Chicago, p. 61.

Donati, F., Canuto, E., Fassino, B. and Belforte, P., 1986, *Manuscripta Geodetica*, **11**, p. 115.

Duncombe, R.L., Benedict, G.F., Hemenway, P.D., Jefferys, W.H. and Shelus, P.D., 1983, in *'The Space Telescope Observatory'*, D.N.B. Hall (ed.), NASA CP-2244, p. 114.

Eichhorn, H.K., 1974, *'Astronomy of star positions'*, Frederick Ungar Publish. Co., New York.

Froeschlé, M. and Kovalevsky, J., 1982, *Astron. Astrophys.*, **116**, 89.

Fricke, W. and Kopff, A., 1963, *'Fourth Fundamental Catalogue'*, Verof. des Astron. Rechen Institut, Nr 10.

Fricke, W., 1981, in *'Reference Coordinate Systems for Earth Dynamics'*, E.M. Gaposhkin and B. Kolaczek (eds.), D. Reidel Publ. Co., Dordrecht, p. 331.

Fricke, W., 1982, *Astron. Astrophys.*, **107**, 413.

Galligani, I., Betti, B. and Bernacca, P.L., 1986, *Manuscripta Geodetica*, **11**, 124.

Green, R.M., 1985, *'Spherical astronomy'*, Cambridge University Press, Cambridge.

Helmer, L., 1986, in *'Astrometric techniques'*, H.K. Eichhorn and R.L. Leacock (eds.), IAU Symposium 109, D. Reidel Publ.Co., Dordrecht, p. 429.

Hoeg, E., 1970, *Astron. Astrophys.*, **4**, 89.

Hoeg, E., 1972, *Astron. Astrophys.*, **19**, 27.

Kerr, F.J. and Lynden-Bell, D., 1986, in *'Highlights of Astronomy'*, J.P. Swings (ed.), vol. 7, p. 889.

Kopff, A., 1954, *Monthly Notices Roy. Astron. Soc.*, **114**, 478.

Kovalevsky, J., 1984, *Space Science Reviews*, **39**, 1.

Kovalevsky, J., 1986, *Manuscripta Geodetica*, **11**, 85.

Kovalevsky, J., Mignard, F. and Froeschlé, M., 1986, in *'Relativity in Celestial Mechanics and Astrometry'*, J. Kovalevsky and V.A. Brumberg (eds.), IAU Symposium 114, p. 369.

Lacroute, P., 1968, in *Transactions of the IAU*, vol. XIII B, L. Perek ed., p. 63.

Lieske, J.H., Lederle, T., Fricke, W. and Morando, B., 1977, *Astron. Astrophys.*, **56**, 1.

Luo, Ding-jong and Li Dong-ming, 1986, in *'Astrometric techniques'*, H.K. Eichhorn and R.L. Leacock (eds.), IAU Symp. 109, D. Reidel Publ. Co., Dordrecht, p. 975.

McNally, D., 1974, *'Positional astronomy'*, Frederick Mueller Ltd, London.

Mould, J.R., 1982, *Ann. Rev. Astron. and Astrophys.*, **20**, 91.

Murray, C.A., 1981, *Monthly Notices of the Roy. Astron. Soc.*, **195**, 639.

Murray, C.A., 1983, *'Vectorial astronomy'*, Adam Hilger Ltd, Bristol.

Muller, E.A. and Jappel, A., 1977, *Transactions of the IAU*, vol. XVI B, p. 58.

Ovenden, M.W., Price, M.H.L. and Shuter, W.H.L., 1983, in *'Kinematic, Dynamics and structure of the Milky Way'*, W.L.H. Shuter (ed.), D. Reidel Publ.Co, Dordrecht, p. 67.

Podobed, V.V., 1968, *'Fundamental Astrometry'*, Nauka Publ., Moscow, (in russian).

Podobed, V.V. and Nesterov, V.V., 1975, *'General Astronomy'*, Nauka Publ., Moskow, (in russian).

Requième, Y. and Mazurier, J. M., 1986, in *'Astrometric techniques'*, H.K. Eichhorn and R.L.

Leacock (eds.), D. Reidel Publ. Co., Dordrecht, p. 669.

Russel, J.L. and Egret, D., in *'Highlights of Astronomy'*, J.P. Swings (ed.),vol. 7, p.713.

Schwan, H.,1986, in *'Astrometric Techniques'*, H.K. Eichhorn and R.L. Leacock (eds.), IAU Symp. 109, D. Reidel Publ. Co. Dordrecht, p. 63.

Schwan, H., 1987, in *'Mapping the Sky'*, IAU Symp. 133, Paris, June 1987, in press.

Scott, F.P., 1963, in *'Stars and Stellar Systems'*, K. Aa. Strand (ed.), vol. 3, p. 11.

Smith, C., 1986, in *'Astrometric Techniques'*, H.K. Eichhorn and R.L. Leacock (eds.), D. Reidel Publ. Co., Dordrecht, p. 669.

Stumpff, P., 1979, *Astron. Astrophys.*, **78**, 229.

Van Daalen, D.T. and Van der Marel, H., 1986, *Manuscripta Geodetica*, **11**, 146.

Van de Kamp, P., 1967, *'Principles of Astrometry'*, W.H. Freeman and Co., San Francisco.

Van de Kamp, P., 1981, *'Stellar Paths'*, D. Reidel Publ. Co., Dordrecht, p. 13.

Walter, H.G., Mignard, F. Hering, R. Froeschlé and Falin, J.-L., 1986, *Manuscripta Geodetica*, **11**, 103.

Westphal, J.A.,1983, in *'The Space Telescope Observatory'*, D.N.B. Hall (ed.), NASA CP 2244, p.28.

Wielen,R.,1986, *Transactions of the IAU*, vol.XIX B, J.P. Swings (ed.), p.254.

Will, C.M., 1981, *'Theory and Experiment in Gravitational Physics'*, Cambridge University Press, Cambridge.

Woolard, E.W. and Clemence, G.M., 1966, *'Spherical Astronomy'*, Academic Press, New York, p. 385.

Yuan, C., 1983, in *'Kinematics, Dynamics and Structure of the Milky Way'*, W.L.H. Shuter (ed.), D. Reidel Publ. Co., Dordrecht, p.47.

Zverev, M.Z., Polozhentzev, D.D., Stepanova, E.A., Khrutskaya, E.V., Yagudin, L.I. and Polozhentzev, A.D., in *'Astrometric techniques'*, H.K. Eichhorn and R.L. Leacock (eds.), IAU Symp. 109, D. Reidel Publ. Co. Dordrecht, p. 691.

EXTRAGALACTIC REFERENCE FRAMES

C. MA
Goddard Space Flight Center, Greenbelt, USA

1. HISTORICAL DEVELOPMENT OF EXTRAGALACTIC ASTROMETRY

1.1. OPTICAL EXTRAGALACTIC OBJECTS

The idea of using extragalactic objects to define the celestial reference frame is quite old, having been discussed by Herschel and Laplace some two hundred years ago, even before there was proof that such objects existed. Any object sufficiently distant would have no detectable proper motion, thus avoiding one of the chief complications of the stellar reference frame discussed in Chapter 1. The realization of such a reference frame, however, has been dependent on scientific discovery, technological advances, and commitment of resources to a now perceived need.

The closest extragalactic objects, the Magellanic Clouds, were known to the Arabs in the eleventh century, and in the eighteenth century philosophers such as Swedenborg and Kant began to speculate about stellar systems similar to the Milky Way. In 1784 Messier published his classic catalogue of 102 "nebulae", objects unlike stars in being small areas rather than unresolved points of light, e.g., M 42 in Orion. The General Catalogue containing 5079 such objects was published by J. Herschel in 1864, and the New General Catalogue (NGC) published by Dreyer in 1888 gave positions to 0.'1. Through photography and spectroscopy it became clear that certain objects, the "spiral nebulae", were assemblages of stars even though individual stars could not be resolved. Their cosmic status was hotly debated, however, until 1925 when Hubble inferred the distance to the Andromeda galaxy (M31), now known to be 670 kilo-parsecs (kpc) or 2.2 million light-years, through the use of Cepheid variable stars.

Hubble's classification of galaxies and Oort's dynamical model of the Milky Way showed that objects like M31 and others are broadly similar to the Milky

J. Kovalevsky et al. (eds.), Reference Frames, 43–65.

Way. A comparison of distance scales is instructive. The Milky Way system is a flat disk about 30 kpc in diameter with the Sun about 8 kpc from the center in one of several spiral arms. Galaxies tend to be clustered, fewer than half being isolated, and these clusters are themselves further aggregated. The local group consisting of the Milky Way, M31, and about twenty other galaxies is about 1 Megaparsec (Mpc) in diameter, while the local supercluster occupies a space of some 40 Mpc. The Shapley-Ames catalogue of 1932 covering the entire sky shows 1249 galaxies brighter than magnitude 13, while Zwicky's catalogue of galaxies and clusters down to magnitude 15.5, completed in 1968, contains some 40,000 galaxies, roughly one per square degree. Galaxies cannot be seen, however, in the "zone of avoidance" caused by absorption in the equatorial plane of the Milky Way. While galaxies by their great distance could define a celestial reference frame, as proposed by Wright in 1950, their fuzzy, faint optical images make such an application difficult.

1.2. EXTRAGALACTIC SOURCES AT RADIO WAVELENGTHS

It is fortunate that observations in the radio frequencies revealed several other classes of extragalactic objects whose characteristics have proved amenable to astrometry of the highest precision, albeit with some limitations. In 1931 Jansky detected extraterrestrial radio signals at decametric wavelengths, and the first radio map of the Milky Way was produced by Reber at 1.87 meters in 1939. The first detection of radio emission from an extragalactic object was made by Hey using a single-dish antenna during World War II and published in 1946. Although the object was designated Cygnus A, the association with a peculiar galaxy in Cygnus was not made until 1954 by Baade. Earlier, identification of two of the strongest radio sources, Virgo A and Centaurus A, with the galaxies M87 and NGC 5128 by Bolton in 1949 showed that some extragalactic sources existed. Using various types of radio interferometers, coarse features of both extragalactic and galactic radio sources were studied, the most important result being the development of a two-component model for Cygnus A by Jennison in 1953.

Statistical studies were hampered by the absence of good source positions so considerable effort was expended in the 1950's and early 1960's in surveying larger numbers of sources with sufficient accuracy and resolution to permit optical identification. The 3C (third Cambridge) survey of 471 sources, done at 159 MHz on a 4-element interferometer in Cambridge, England in 1959, and the Parkes survey of the Southern Hemisphere, made with the 210-ft radio telescope near Parkes, Australia at several frequencies from 635 MHz (50 cm) to 11 cm in 1964, contain many of the radio sources subsequently observed in detail.

Many other survey catalogues have been published since these early efforts, and the multiple naming of radio sources is sometimes quite confusing. See

Kesteven and Bridle (1977) for a guide to radio source nomenclature. The commonly used IAU standard designation is based on the B1950.0 position of the source.

1.3. COMPACT RADIO SOURCES

By 1960 the positions of the strongest radio sources were generally known to 10 seconds of arc, and many were identified with various types of galaxies. Most radio galaxies had bright emission lines so their red shifts could be relatively easily determined even though they were optically faint. For example, the distant galaxy 3C 295 has a red shift of 0.46 and an apparent magnitude of 21. On the assumption that small diameter and high radio surface brightness would allow identification at the greatest distance, the search for distant sources in the early 1960's concentrated on a few objects, the primary candidate being 3C 48, which was less than 1 arcsec in radio size. It was identified with a 16th magnitude stellar object. Three other strong sources were soon identified with starlike objects, but their optical and radio properties were quite different from each other and no clear interpretation was made. Lunar occultation measurements of 3C 273 by Hazard in 1962 showed that it consisted of a small component coincident with a 13th magnitude stellar object and a long component coincident with a jet-like extension to the star. The optical spectrum of 3C 273 showed a series of bright emission lines identified as the Balmer hydrogen series with a red shift of 0.16. A re-examination of the spectrum of 3C 48 gave a red shift of 0.37. Radio sources of this type were designated as quasi-stellar radio sources or quasars. Following identification of the first quasars, it was discovered that they were all exceedingly blue and further searches were made optically on this basis. The large majority of optically identified quasars, i.e., stellar objects with broad emission lines and high red shifts, are not strong radio sources, however.

The refinement of the positions of compact radio sources, largely quasars, continues to the present. Positions derived from connected element interferometers (CEI) were better than 1" in 1970 (Wade, 1970) and as good as ."02 by 1976 (Wade and Johnston, 1977). The first very-long-baseline interferometer (VLBI) positions (Cohen and Shaffer, 1971) had uncertainties of 1" by 3" and soon improved to the level of CEI (Clark *et al.*, 1976). The most recent source catalogues derived from VLBI observations (Robertson et al., 1986, Ma, 1988, Sovers et al., 1988) have positions with uncertainties considerably less than 0."005 or 5 milliarcseconds (mas).

It is the class of quasars and similar compact strong radio sources that provides the best means for defining the extragalactic, quasi-inertial reference frame. Their two most important characteristics for this purpose are the small size of their cores, < 1 mas when observed by interferometers with high resolution, and their great distance; most are further away than 1000 Mpc if their red shifts are cosmological. Two factors make their direct use for general astrometry difficult: their relative scarcity and optical faintness.

2. QUASARS AND OTHER COMPACT EXTRAGALACTIC OBJECTS

2.1. OPTICAL PROPERTIES

The optical properties of quasars have been summarized by Kellerman (1974) as follows:
(i) They are points on photographic plates although a few have faint streamers, jets, or underlying optical emission.
(ii) They have large red shifts (some now measured in excess of 4).
(iii) If their red shifts are cosmological, then their optical luminosities exceed those of bright galaxies by a factor of 100 to 1000.
(iv) Many have variations in optical brightness with time scales of hours to a few years.
(v) Their optical emission rises sharply toward the infrared, where most of the energy is radiated. Compared to galaxies there is an excess of ultraviolet emission.
(vi) Their spectra show strong, broad emission lines with line widths indicating turbulent velocities up to 4000 km/sec.
(vii) Some show families of narrow absorption lines with different red shifts.
(viii) A class generally similar to quasars, BL Lacertae objects, lack strong emission lines.

2.2. RADIO PROPERTIES

The radio properties of quasars and other strong compact radio sources vary considerably. The spectra show no sharp features and the shapes can be divided only roughly into various classes. Of greatest interest for astrometric work are those with flat or rising spectra, i.e., the spectral power flux density, measured in Janskys (1 Jy = 10^{-26} W/m ^{2} Hz), remains constant or increases with increasing frequency. Such spectra indicate objects with the most compact cores. This is well understood as arising from the superposition of the spectra of many compact components, each being characterized by synchrotron self-absorption, but having maxima at different frequencies. Many of the compact sources show pronounced intensity variations on time scales from less than a week to several years, but there is no evidence of periodic behavior. The variations appear as bursts first at short wavelengths and then at longer wavelengths with reduced amplitude and longer duration. There is usually no simple correlation between optical emission and radio variability.

2.3. STRUCTURE OF QUASARS

The structures of quasars also show wide variation. For example, the quasar 3C 273 consists of a jet extending 20" from the core, which itself can be resolved into several mas-sized components. On the interferometers now used to make the most precise astrometric measurements, the jet is essentially invisible. The core components can still give rise to variations in observed position as the resolution and orientation of the interferometer change with respect to the source. Other quasars may show two components, a compact core with a halo, or an elliptical overlay. The overlay structure may extend 5 to 10 mas outside the compact core. Usually the small scale and large scale structure are aligned in orientation. Ulvestad (1987) estimates that the effect of structure on position may be several tenths of mas for some sources now listed in astrometric catalogues.

Quasar structure often varies with time, sometimes associated with bursts of radio emission, with expansion of the core or with the appearance of compact blobs being rapidly ejected from the core. These blobs often appear to be moving with superluminal velocity up to 10 c. Application of special relativity to a model of relativistic bulk motion of the radiating plasma shows, however, that the apparent superluminal motions can be derived from real subluminal motions aimed very nearly along the line of sight. For the purpose of defining a reference frame, it is desirable to choose sources with no observable structure. However, this criterion leads to a practical limitation since the simplest objects are in general the weakest radio sources because they are either more distant or linearly smaller. Hence, they may be too weak to be seen by the instruments regularly available for astrometry.

2.4. ASTROMETRIC USE OF QUASARS

While there is activity in the core of quasars, there is as yet no evidence of real proper motion. Bartel *et al.* (1986) give upper limits of 20 microarcsec/yr in right ascension and 50 microarcsec/yr in declination for the relative position change between the close pair 3C 345 and NRAO 512. If treated as a limit on parallax, these data provide a geometric proof that these quasars are extragalactic. Morabito et al. (1985), Marcaide et al. (1983) and Shapiro et al. (1979) have used VLBI to measure the relative positions of very close quasar pairs with precision as good as 3 microarcsec. More than determining proper motion, such studies are useful for testing the limits of reference frame stability and VLBI astrometry.

There are some 100-200 compact radio sources with correlated flux densities >1 Jy when observed at 8 GHz with a 4000-km long interferometer. The number of sources with flux density greater than some threshold flux S is approximately proportional to $S^{-3/2}$ so that an increase of interferometer sensitivity by a factor of two would increase the number of observable sources by a factor of three. It is clear, however, that no more than a few hundred radio sources will be available to define the extragalactic reference frame with the interferometers now available for extensive astrometric programs. Argue *et al.* (1984) have compiled a list of 234 strong compact extragalactic radio sources with optical counterparts. The problem of connecting the radio and optical frames is discussed in Chapter 12.

3. OPTICAL OBSERVATIONS OF EXTRAGALACTIC OBJECTS

There is a continuing effort to determine the positions of optical counterparts of compact radio sources. de Vegt and Prochazka (1985), Walter and West (1986) and de Vegt *et al.* (1987) illustrate the current state of optical position measurements and their limitations. The first paper measured 17 extragalactic objects using the 1.5 m L. Figl telescope of the Vienna Observatory. The plates were scanned with the 422F Mann comparator with a linear accuracy of 2.3 microns. A limiting magnitude of mv = 17.0 could be achieved with Kodak 098-04 emulsion, but this would be insufficient for most of the objects in the Argue *et al.* list. The average uncertainty was 40 mas relative to the nearby stars in the AGK3RN catalogue in the B1950.0 system.

Walter and West measured 50 sources between -45 deg and +5 deg declination using existing European Southern Observatory (ESO) Schmidt plates with the ESO S-3000 Measuring Machine. With a plate precision of 1 micron, the total optical errors were less than 250 mas relative to the Perth 70 catalogue. A comparison of 46 optical positions with the radio positions of Fanselow *et al.* (1984) gave average arclength discrepancies of 360 mas in right ascension and 330 mas in declination with perhaps a sinusoidal systematic difference in declination as a function of right ascension. Thus there are observations available to link the optical and radio frames at < 0."3.

De Vegt *et al.* (1987) showed that improvements in optical positions are still possible, albeit with some greater effort. They measured the position of the quasar 1928+738 using both plates and a CCD camera at the Kitt Peak National Observatory 4-m telescope. A system of 40 secondary sources of m_v = 12-14 and 20 tertiary sources with m_v = 16-18 was used to relate the quasar to the AGK3RN catalogue. The internal accuracy of a single plate was 40 mas. The tertiary sources were needed because of the small field of the CCD, 2' by 3' on the sky, with no more than 5 reference stars imaged with the quasar. The final, internal uncertainty of the CCD position of the quasar was 20 mas. The quasar was also measured at 6 and 20 cm using the Very Large Array (VLA) with an uncertainty of 20 mas. The radio and optical B1950.0 positions differed significantly, probably because of a systematic optical catalogue offset in this region of the sky.

4. RADIO MEASUREMENTS

4.1. CONNECTED ELEMENT INTERFEROMETRY

While the very earliest radio observations were made with single antenna systems, radio interferometers have been the primary instruments for astrometric work. The two types of interferometers are the connected element interferometer (CEI) and the very-long-baseline interferometer (VLBI). While they share certain characteristics, there are differences in practice which affect their application to the problems of radio astrometry.

An elementary radio interferometer consists of a pair of antennas. Each antenna is equipped with a receiving system which covers the same narrow band of frequency and observes a point source which is sufficiently distant so that the incoming radiation can be treated as plane waves. Now let the voltages from the two antennas be multiplied together and the resulting voltage passed through a lowpass filter whose output is recorded as a function of time. The combination of the multiplying circuit and the lowpass filter is referred to as a correlator. As the Earth rotates, the two antennas will have different velocities relative to the source. For a narrow-band receiving system, the received signal will then have different Doppler shifts at the two antennas and the output of the lowpass filter will be a sinusoid whose frequency is the beat between the two Doppler-shifted received frequencies, assuming identical electronics and local oscillator chains. This fringe frequency varies as the orientation of the baseline with respect to the source changes and is a function of various interferometer and source parameters. The amplitude and phase of the interference fringes are the CEI data from which astronomical and astrometric results can be derived. The phase of the fringes is

$$\phi = 2\pi D \bullet s$$

where D is the baseline vector between the two antennas expressed in wavelengths and s is the unit vector to the observed source. By observing at least three sources over a day (taking five or more observations) it is possible to estimate all the source positions (the right ascension of one source being arbitrary), the baseline vector and an instrumental phase.

Two CEI instruments are now regularly used for astrometric measurements. The National Radio Astronomy Observatory interferometer in Green Bank, West Virginia has a 35-km baseline and operates continuously at 2.7 GHz and 8.1 GHz as part of a US Naval Observatory program to monitor UT1. The Very Large Array near Socorro, New Mexico, while primarily a mapping instrument, is also used for differential and absolute astrometry. It consists of 27 25-m antennas laid out in a Y pattern with the longest arm 21 km. The observing frequencies range from 1.3 GHz to 24 GHz.

4.2. VERY-LONG-BASELINE INTERFEROMETRY

VLBI differs fundamentally from CEI because the separate interferometer elements operate independently. Instead of correlating the received signals in real time using a direct link, the received signals are converted to a low frequency and recorded. The conversion is dependent on a high precision frequency standard at each station (usually a hydrogen maser) with sufficient stability to allow the signals to be cross-correlated at a later time. The standard supplies both the frequency for down-conversion and the time for aligning the cross-correlation. The presence of time and frequency offsets between different frequency standards, as well as inherent geometric uncertainties, requires a search in geometric delay and fringe frequency in order to maximize the correlation. Fringe phase is generally not used in VLBI astrometry because a priori uncertainties in the geometry of the interferometer are usually larger than a wavelength, typically a few centimeters. The primary observable is the geometric group delay, which contains the same information as the fringe phase without the ambiguity associated with determining the exact fringe but which has lower inherent precision.

VLBI instrumentation has undergone considerable development since the initial efforts in the mid 1960's. Table 1 describes the primary recording systems.

TABLE 1 VLBI RECORDING SYSTEMS

System	Basic design	Sample rate Megabit/sec	Tape time (min)	Reference
Mark I 1967-78	digital recording on computer tape	0.72	3	Bare (1967) Whitney (1976)
Mark II 1971-	digital recording on various TV recorders	4	64-246	Clark (1973)
Mark III 1977-	digital recording instrumentation	112	13	Rogers (1983)
Mark IIIA 1984-	recorder	112	164	Clark (1985)

Thompson *et al.* (1986) is a comprehensive treatment of radio interferometry. Chapters by Moran, Rogers, and Shapiro in Meeks (1976) give more details about VLBI theory and data analysis.

VLBI networks, since they are composed of independent elements, vary with time and availability. Table 2 shows the stations which have contributed significantly to the current astrometric data base. Several networks currently operate on a regular basis to monitor Earth orientation, and these data can be also used for astrometry. Other networks are coordinated as necessary for specific observing purposes. Essentially all astrometric VLBI data have been taken at 2 GHz, 8 GHz, or simultaneously at both frequencies, and all the stations shown have or are planned to have compatible receiving and recording equipment.

TABLE 2 VLBI ANTENNAS USED FOR ASTROMETRY

Location	Size
Gilmore Creek, Alaska, USA	26m
Goldstone Deep Space Station, Calif., USA	64
Hartebeesthoek Radio Observatory, S. Africa	26
Hat Creek Radio Observatory, Calif., USA	26
Harvard Radio Astronomy Station, Texas, USA	26
Haystack Observatory, Mass., USA	37
Kashima Space Research Center, Japan	26
Kokee Tracking Station, Hawaii, USA	9
Madrid Deep Space Station, Spain	64
Maryland Point, Md., USA	26
Mojave Base Station, Calif., USA	12
National Radio Astronomy Observatory, W. Va., USA	43
Onsala Space Observatory, Sweden	20
Owens Valley Radio Observatory, Calif., USA	40
Richmond, Florida, USA	18
Tidbinbilla Deep Space Station, Australia	64
Westford, Mass., USA	18
Wettzell, FR Germany	20

5. ASTRONOMICAL AND GEOPHYSICAL MODELS

The models applied in the analysis of astrometric interferometer data can be divided into several general types:
(i) models of the global motion of the Earth in space, including precession to the J2000.0 system (Lieske *et al.*, 1977) and the related sidereal time equation (Aoki

et al., 1982), the IAU 1980 nutation series (Seidelmann, 1982), and the Earth's orbital motion, derived from a planetary ephemeris such as DE 200 or PEP. These are treated in detail in Chapters 11 and 13.
(ii) models of the motion of the stations relative to a rigid terrestrial system, including polar motion and UT1, solid Earth tides, pole tide, and ocean loading, to be treated in Chapters 9 and 11.
(iii) models of the delays caused by the propagation media, generally separated into the ionosphere and the dry and wet troposphere.
(iv) relativistic effects, including deflection of radio waves by massive bodies in the solar system and effects on clocks in a terrestrial system, to be treated in Chapter 5.
(v) miscellaneous models such as the geometry of the antenna mount.
(vi) a theoretical model for the interferometric observables based on a relativistic development of the underlying celestial coordinate frame with its origin at the solar system barycenter, to be discussed in Chapter 5.

While there has been an attempt to coordinate the models in common use in order to make comparison of results more meaningful, particularly in the Project MERIT Standards (Melbourne *et al.* 1983 ; see Annex), there have been independent developments by the two primary VLBI groups. The models used at the Jet Propulsion Laboratory (JPL) which are embodied in the MASTERFIT program are well documented (Sovers and Fanselow, 1987). MASTERFIT has been adopted by several European groups. The models used by Goddard Space Flight Center, Harvard-Smithsonian Center for Astrophysics, and National Geodetic Survey embodied in the CALC 6.0 program and adapted by groups in Sweden, Japan, Germany and Italy are described in Robertson (1975), Ma (1978), and Gordon (1985). Finkelstein *et al.* (1983) give an independent derivation of the theoretical delay model. Hellings (1986) has developed a more complete model for relativistic effects on astronomical time measurements, but his work has not been included in any published analysis. While both MASTERFIT and CALC 6.0 adhere generally to the MERIT standards, there are some outstanding differences, notably in the troposphere and relativistic deflection models (Murray, 1986). Sovers and Ma (1985) have compared the two sets of software and find agreement in the theoretical delays at the 15 - 50 picosec level, not including the troposphere model.

The mathematical technique used to estimate astrometric positions is quite straightforward, generally a weighted least squares adjustment of the relevant geodetic (station positions and/or Earth orientation), astrometric (source right ascension and declination), astronomical (nutation, precession), tropospheric and clock parameters. All the applicable data are included in a single large solution, although the solution may proceed sequentially because of the number of parameters involved. Application of Kalman filtering to model the stochastic parameters (troposphere and clock behavior) is quite promising but has not yet been used in astrometric work.

5.1 PRECESSION AND NUTATION

The models that have the largest effect on the absolute astrometric positions are those defining the J2000.0 reference frame (precession, nutation, and sidereal time) and the models of the propagation media. It is clear (Herring *et al.*, 1986) that the IAU 1980 nutation series is detectably incorrect in several terms, the largest error being ~2 mas in the annual term. There is not yet a sufficient time span of precise interferometric data to determine whether the longer nutation terms and the precession constant should also be corrected.

Different techniques for improving upon the precession/nutation model cause different rotations of the frame defined by the estimated source positions. One method is to estimate offsets in obliquity and longitude relative to a reference day which defines the celestial pole in a manner analogous to monitoring terrestrial polar motion (Herring *et al.*, 1986). This method requires that all data sets included in a single catalogue have sufficient overlap of reference sources to link the positions from different epochs into a coherent coordinate frame. Another method is to adjust the precession constant and the coefficients of the standard nutation series (Himwich and Harder, 1988). With the data now available, the long-period nutation terms and the precession constant cannot be separated without additional geophysical constraints although highly correlated values with small errors can be estimated. While the two methods give different source positions, the difference is essentially only a few mas rotation of the catalogues. Not adjusting the nutation model at all, however, will cause systematic errors in the positions of sources observed at only a few epochs.

5.2 PROPAGATION MODELS

The propagation models have presented the greatest difficulties in theory and practice. Elgered (1983) and Davis (1986) provide detailed discussions.

The charged particle effects in the ionosphere can be essentially completely calibrated using simultaneous observations at two frequencies (Herring, 1983).

The effect of the dry component of the troposphere at the common observing frequencies (~7 nanosec at zenith) can be largely calibrated from ground measurements of pressure and temperature, although horizontal pressure gradient effects (Van Dam and Wahr, 1987) have not yet been modeled in VLBI analysis.

The wet component, with effects on the delay observable up to 10% that of the dry, is much more variable temporally and spatially. Attempts have been made to measure it directly using microwave water vapor radiometers (Davis *et al.*, 1985), but it has not been conclusively demonstrated that astrometric or geodetic results are generally improved.

The ultimate goal is absolute calibration of the troposphere in order to avoid the loss of geometric strength that occurs when any parameters related to propagation delays are adjusted. There are two main areas that must be properly handled, the mapping of the zenith delay to the line of sight (which depends on the model of the stratification of the troposphere) and the conversion of observed radiometer brightness temperatures to wet path delay (which depends on stratification and the absorption coefficient of water vapor).

Inaccurate models for the propagation media will cause random and systematic errors in the declinations of the sources. For example, an error which systematically underestimates the propagation delay, such as neglecting the ionosphere, will cause estimated baseline lengths to be too long and source declinations to be too far from the equator. Errors in the troposphere mapping function, which likewise become more pronounced at lower elevations, cause similar effects, which may vary seasonally.

6. THE EXTRAGALACTIC RADIO REFERENCE FRAME

It is difficult to define the extragalactic reference frame with exactness since the realization must depend on both the observations and the theoretical models. Intuitively, a reference frame attached to quasars satisfies the desirable characteristics of being inertial and unchanging, at least in so far as the great distance and lack of proper motion of quasars have been demonstrated. The quasars are directly observable, and the primary data, the interferometric delays, can be acquired from the whole sky. The analysis of such a data set can lead to a truly fundamental reference frame since there is no need to step differentially from one part of the sky to another. The set of arclengths between sources defines the reference frame. In practice, given current conventions, the source positions are the realization of the reference frame.

While there are no dedicated astrometric radio interferometers corresponding to optical transit circles, the ensemble of VLBI networks can be considered a single instrument provided the raw data are consistently processed. It is necessary to make geodetic as well as astrometric adjustments, but it is not necessary to connect the networks geometrically or temporally. Data from one network from one epoch can be combined with data from a different, even disjoint, network from another epoch. The radio source positions and astronomical models can be adjusted from all the data while the geodetic parameters can be adjusted separately for each observing session, thus freeing the astrometric parameters from the effects of station motion, either relative to the spin axis, as in polar motion, or relative to each other, as in tectonic plate motion. It should be noted that a better solution would estimate the set of source positions, station positions, and station velocities simultaneously to reduce unwanted degrees of freedom. Otherwise for a mixture of networks the correlations between tropospheric errors, baseline lengths (particularly in the z-direction), and source declinations will have greater, albeit still small, effects from day to day.

The models used in the reduction of raw VLBI data are much simpler than those used in astrometric analysis and have negligible effect on the delay observables. The reduction models merely provide a starting point for the search for interferometric fringes or, alternately stated, for delay and fringe rate. The data used for analysis are the total delays, i.e., the initial model added to the residual found by correlation and fringe fitting. Thus the problem of irreversible processing, the contamination of observations by inaccuracies in the reduction models, is avoided. The entire data base of VLBI observations can be accumulated and reanalyzed as models, calibration techniques or mathematical methods improve.

6.1 CATALOGUES OF EXTRAGALACTIC RADIO SOURCES

There are at present several catalogues of extragalactic radio sources in the J2000.0 system. They vary considerably in number of sources, distribution over the sky, and precision. Table 3 gives a summary. The earliest are Wade and Johnston (1977) and Kaplan *et al.* (1982), both made with the 35-km CEI at Green Bank. The Wade and Johnston catalogue was published before the J2000.0 system was formally defined and contains some errors which can be traced to the conversion from B1950.0.

TABLE 3 J2000.0 RADIO CATALOGUES

Author	Instrument	Baseline Length km	# Sources	Uncertainties mas
Wade	CEI	35	36	20-40
Kaplan	CEI	35	16	10
Morabito	MarkII	8000-11000	836	300
Perley	VLA	<27	700	20-100
Fanselow	MarkII	8000-11000	117	1-5
Ma 1986	MarkIII	800-6000	85	0.3-13
Robertson	MarkIII	800-6000	26	0.5
Ma 1988	MarkIII	800-11000	101	0.2-9
Sovers	MarkII	8000-11000	128	0.5-7

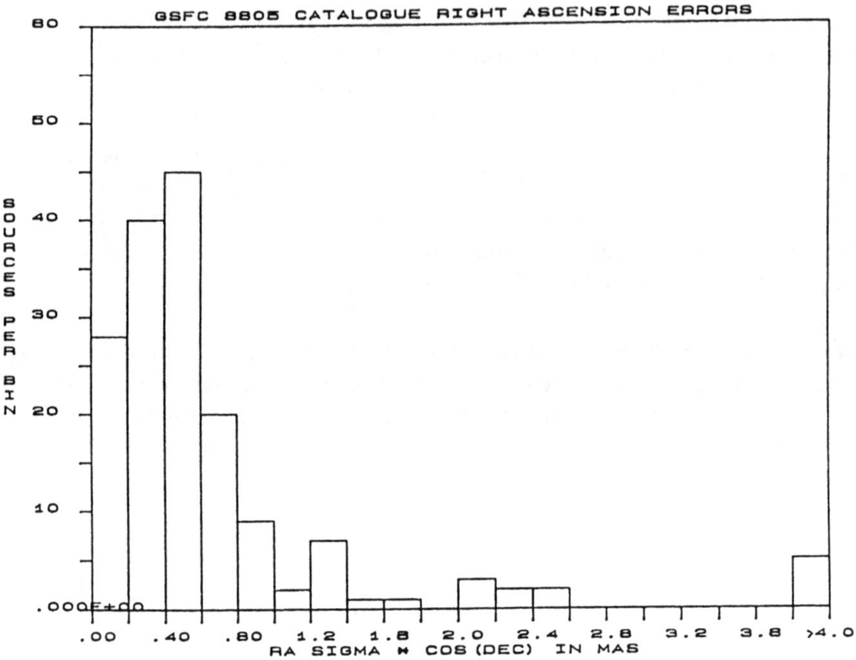

Figure 1.

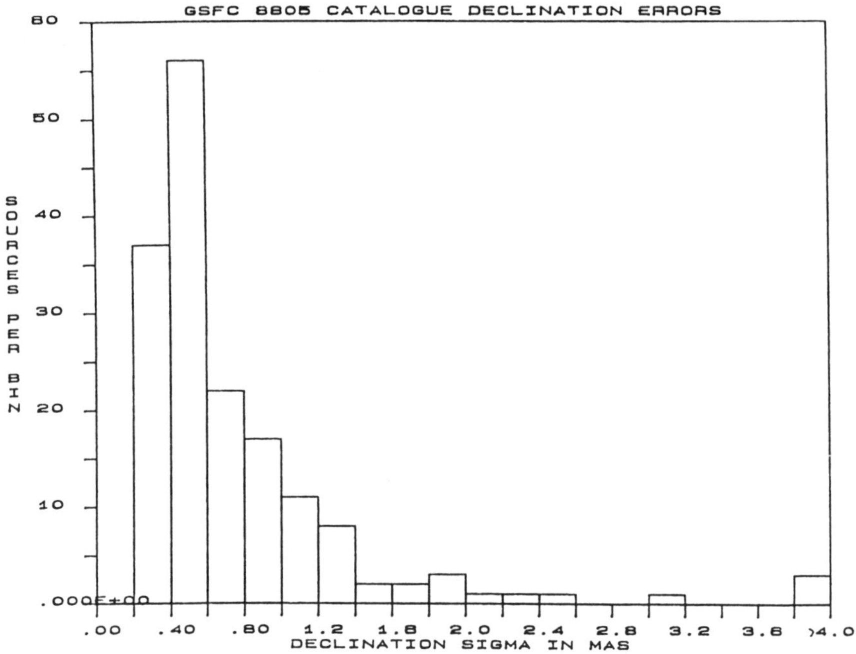

Figure 2.

The long-term VLBI survey work done at JPL (Morabito *et al.*, 1986a, 1986b) includes 836 sources with typical position uncertainties of 300 mas. The declinations span –80 deg to +80 deg. The VLA catalogue of 700 sources is fairly complete for flux densities >0.75 Jy and declinations >–40 deg (Perley, 1982). The best source positions have uncertainties of 20 mas, although there is considerable variation. Since the baselines of the VLA are less than 30 km, there are a number of sources which show significant structure when observed by interferometers with higher resolution.

The VLBI catalogues of Ma and Robertson *et al.* (1986) are all largely limited to northern hemisphere sources. Sovers *et al.* (1988) is an update of Fanselow *et al.* (1984) in which the typical errors are less than 1 mas. It should be emphasized that these catalogues are only snapshots of continuing observing programs and that improvement in coverage and precision is expected. Figures 1 and 2, for example, show the histograms of formal position errors for the 165 sources of the Goddard 8805 catalogue.

While declinations can be determined absolutely in theory (if not in practice), only relative right ascensions can be measured. The right ascension origin of the extragalactic reference frame cannot be defined solely by observations of extragalactic objects. The common method to define the origin has been to assign a right ascension value to one source, which can be chosen arbitrarily. In the past it has been defined by the assigned position of the compact core of the quasar 3C 273 (Kaplan *et al.*, 1982) based on lunar occultation measurements by Hazard *et al.* (1971). This source is not an ideal reference origin since it shows mas scale structure. Changing the reference source rotates the entire coordinate system without any other effect provided there are sufficient observations on the reference source. Another method is to define the origin by the average right ascensions of several sources, e.g., such that the average adjustment in right ascension of the sources is zero. If the a priori right ascensions are determined by optical positions, then this definition forces the optical and radio origins to be in good agreement. The difference between the optical and radio origins using the assigned right ascension of 3C 273 in J2000.0 is no worse than 200 mas (Johnston *et al.*, 1985) and may be as good as 50 mas (Lestrade *et al.*, 1988). Another method of defining the origin is to use pulsar timing to measure the ecliptic directly as described below. It should be emphasized that the extragalactic reference frame is not tied to the ecliptic except by convention.

6.2 COMPARISON BETWEEN CATALOGUES

Figures 3 and 4 show a comparison between two VLBI catalogues in progress, the Goddard GSFC870518 catalogue with 114 sources observed with Mark III and

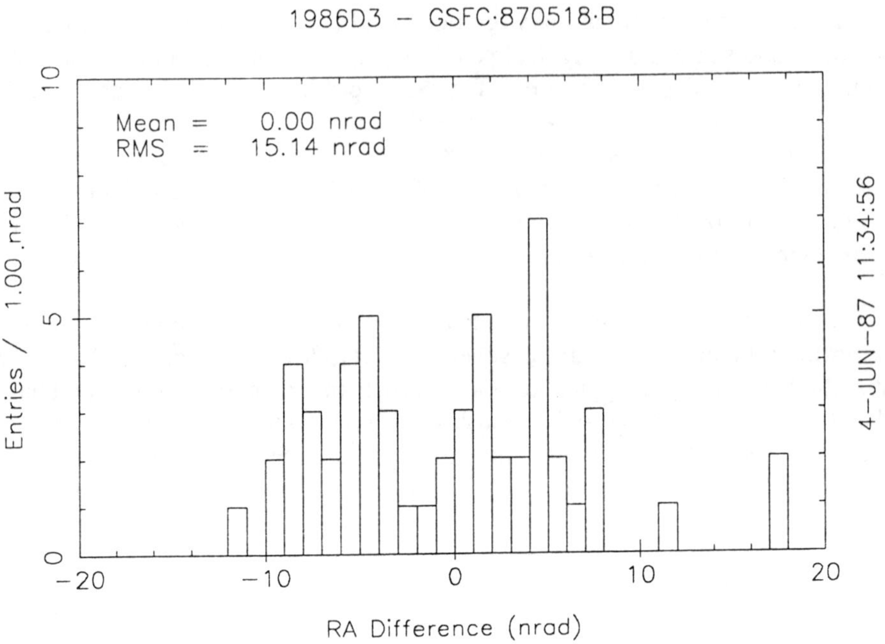

Figure 3.

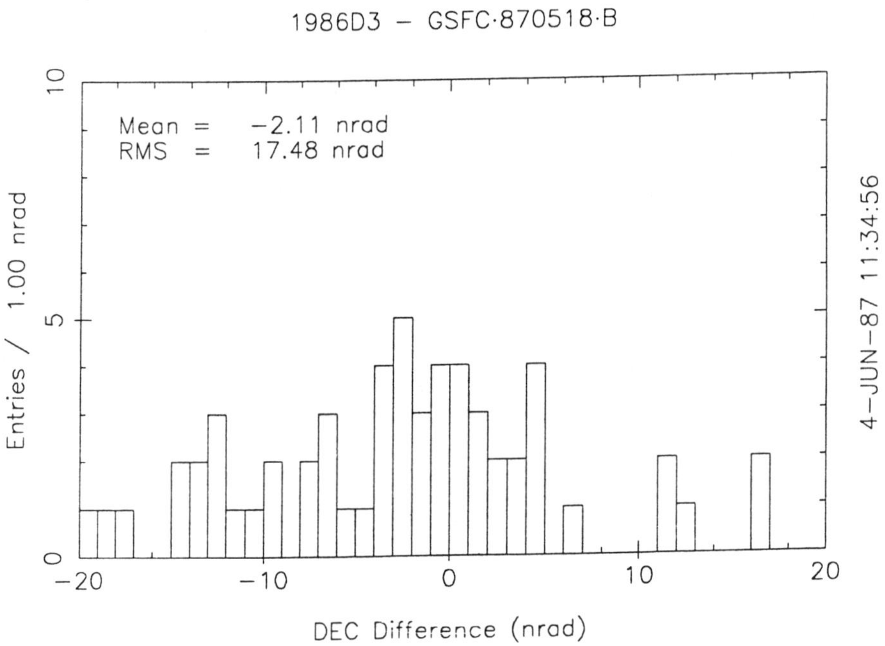

Figure 4.

the JPL JPL1986D3 catalogue of 106 sources observed with Mark II. There are 61 sources in common. The recording instrumentation, baselines, data, correlation facility, tropospheric calibration, software, and parametrization are completely disjoint so this comparison indicates the degree of agreement between two fundamental extragalactic reference frames as realized by their respective source catalogues.

The root-mean-square (rms) position uncertainties of the two catalogues are comparable, ~2 mas, with typical uncertainties ~1 mas. The figures show the differences in positions after rotations about three axes have been applied to bring the catalogues into best-fit agreement in right ascension and declination. (1 mas = 4.8 nrad). There is a rotation between the two catalogues of 2.2 mas about the x-axis directed at 0 hr RA and much smaller rotations about the other orthogonal axes. This rotation is in large part the result of a difference in nutation models. In the GSFC870518 solution daily nutation offsets were estimated relative to a reference day while no nutation adjustment was made in JPL1986D3. After rotation, the rms difference in right ascension is 3.2 mas and the rms difference in declination is 3.6 mas. (The rms values include five right ascension differences and three declination differences which are outside the limits of the plots.) The differences normalized by the root-sum-squared uncertainties are Gaussian. There is a systematic difference in declination of 15 microarcsec/deg between the two catalogues as well as the rotation. Sovers *et al.* (1988) gives a similar comparison between the JPL 1987-1 catalogue and the catalogue of Robertson et al. in which the rms difference for 18 sources is 2mas in both right ascension and declination.

6.3 SYSTEMATIC BEHAVIOR OF REFERENCE FRAMES

There are systematic differences between catalogues, particularly of declinations as a function of declination. The JPL1986D3 catalogue used data from one long east-west baseline between California and Spain and one long north-south baseline between California and Australia. The GSFC870518 catalogue used data from many baselines predominantly in the northern hemisphere with much smaller polar components than the California to Australia line. While declinations are theoretically determined relative to the instantaneous spin axis, there is a strong geometric correlation between the polar component of a baseline and the declinations of the sources. In order to achieve astrometric precision in the mas range, it is necessary to adjust geodetic parameters and so the correlation in the estimated parameters is unavoidable. Another effect that might contribute to the systematic difference is inadequacies in the troposphere models. Because the JPL1986D3 baselines are so long, the average observing elevation is lower than for GSFC870518.

The stability of the extragalactic reference frame can be tested in several ways.

Robertson *et al.* divided the data set extending from Sep. 1980 to Sep 1985 into subsets. One group of three subsets contained every third observing session. A second group had the observing sessions binned into successive two-month intervals. The third group had four subsets binned by season. The variations of right ascension and declination for each source in the various groups of subsets were less than 1 mas. Ma (1988) divided data from 1979 through 1986 into one-year subsets. The rotations between the positions derived from the different annual solutions were typically less than 1 mas. Sovers *et al.* divided data spanning 1978–1985 into three time intervals and found changes in right ascension and declination of 0.5 mas/yr and 0.6 mas/yr, respectively. For catalogues derived from radio sources observed frequently and over a period of several years, the overall system appears to be stable at the level of 1 mas or better.

The source positions in VLBI catalogues are not of uniform quality, and certain errors may not be completely taken into account in the published uncertainties. For example, in Ma *et al.* (1986), about 70% of the sources were observed in only one or two sessions. These sources may be systematically affected by errors in the nutation series since the IAU 1980 nutation model was used without modification. In Fanselow *et al.* (1984), there are 13 sources which were either observed on short baselines or with a minimum number of observations. While the stated errors are large, there may also be a systematic effect because of the distribution of observations in time or because of differences in the baseline polar components compared to the baselines used for the majority of sources. Likewise the uniformity of the networks and relatively short z-components of the baselines used by Robertson *et al.* might cause systematic errors in declination without affecting consistency. Some improvement in Ma (1988) arises from adjustment of the nutation model, while Sovers et al. discards some early data of poor quality.

The current limitations in the highest precision catalogues arise from several areas. The number of sources is small and the distribution is not uniform. The southern sky is not as well covered as the north, and the position uncertainties there tend to be larger. The number of observations on any given source differs by several orders of magnitude and the observations are not uniformly distributed in time, leading to possible systematic errors caused by errors in nutation even if nutation parameters are estimated. The set of baselines used to observe the sources is not uniform, especially in the cases of Ma *et al.* (1986) and Ma (1988), and the baselines are different between different catalogues. The calibration and modeling of the tropospheric delay also differ between catalogues, and the effects may be exaggerated by different distributions of observing elevations. A few of the sources have structure at the mas level which may have an effect on their estimated positions that is dependent on the baselines used. The structure is likely to be different at different observing frequencies with a resulting small error in registration in the application of the dual frequency ionosphere correction. If the structure changes, which component is the brightest and hence the referenced object may shift. Even with the complications described, however, the accuracy and stability of the extragalactic reference frame are probably now at the 1-2 mas level.

7. EXTENSION TO OTHER FRAMES

The extension of the extragalactic reference frame to other frames can be accomplished variously. This section will only treat those available in the radiofrequency domain.

While the first radio detections of flare stars was in 1963, the first systematic studies of radio stars did not begin until 1970 using the Green Bank CEI and the Westerbork interferometer in Holland. Walter (1977) provided a radio star catalogue which has been used as a starting point for astrometric work. Another stellar source of radio signals was discovered by Bell (Hewish *et al.* 1968) in 1967, the pulsar, a rotating neutron star with a strong magnetic field. Both classes of objects can be tracked in the radiofrequency domain and placed in the extragalactic reference frame. Froeschlé and Kovalevsky (1982) describe how such a connection can be made. In addition, pulsar timing measurements, which are dependent on the orbital position of the Earth, can be used determine the obliquity of the ecliptic and the vernal equinox in the extragalactic reference frame (Shapiro, 1970).

In a series of papers Florkowski *et al.* (1985), de Vegt *et al.* (1985) and Johnston *et al.* (1985) describe the use of 20 radio stars to link the extragalactic radio and stellar reference frames. The radio stars were observed with the VLA and their positions measured relative to nearby extragalactic sources drawn largely from Argue *et al.* (1984). The internal errors of the radio positions were on the order of 30 mas. Precise optical positions for the stars were determined from plates taken on the 23-cm astrograph and 60-cm refractor of the Hamburg Observatory. The accuracy of the positions in the northern hemisphere was 50 mas relative to the AGK3RN catalogue as the realization of the FK4 system. Since FK5 stellar positions were as yet undefined, the radio positions were also estimated in B1950.0. From the two sets of coordinates and two other radio stars, it was found that the extragalactic and stellar systems agree at the level of a few tens of mas although there may be a right ascension offset of 20 mas. There are local areas in the sky, however, where there are distortions in the optical catalogue of up to 200 mas.

Even more precise radio positions have been measured by VLBI (Lestrade *et al.*, 1986). They used Mark III VLBI and large, sensitive antennas to measure the positions of 8 radio stars. The uncertainties for the positions ranged from 2 mas to several tens of mas, the limitation being the very low flux from quiescent radio stars. For very weak sources (< 10 mJy), a technique of using a nearby strong source as a phase reference and integrating many scans, similar to CEI, made a detection possible. If 1 mJy sources could be detected by this technique on even more sensitive VLBI networks, up to 50 radio stars could be measured, albeit at the cost of 5 hours of a large network devoted to a single star. Two measurements at an interval of three or more years would be sufficient to measure proper motion except, perhaps, in the case of multiple stellar systems. In addition, some care must be given to the possibility of radio structure at the few mas size. Observations for the most precise positions may need to be restricted to the longest baselines sensitive only to emission from the surface of one star.

Backer *et al.* (1985) made VLA observations of two pulsars, PSR 1937+21 and PSR 1913+16, and compared these positions with those obtained from pulse time of arrival. While the timing positions were accurate to 1 mas and 3 mas (Davis *et al.*, 1985), respectively, the VLA positions relative to the extragalactic frame were good to only 50 mas and 150 mas and were thus the limiting factor in the tie between the extragalactic and solar system frames. Using the position of PSR 1937+21 reduced in the J2000.0 system, it was found that the discrepancy between the extragalactic and solar system frames is 10 mas.

Bartel *et al.* (1985) used Mark III VLBI to measure the positions of PSR 0329+54 and PSR 1133+16 directly in the extragalactic reference frame with uncertainties of 5 mas and 230 mas, respectively. While the positions agreed with those from the VLA and the Green Bank CEI, they disagreed with pulse timing measurements at a significant level (450 mas in declination). They suggest that the discrepancy is probably caused by unmodeled systematic errors affecting the timing data.

8. FUTURE

The premier instrument for future radio astrometry will be the Very Long Baseline Array, currently under construction. It will consist of ten 25-m antennas spaced from Hawaii to Puerto Rico, each equipped with ten receivers from .33 GHz to 43 GHz. The receiver sensitivities and recording bandwidth will be comparable to the best systems now in regular use for geodesy and astrometry with a design sensitivity of 0.1 mJy for an 8-hr integration. Given the number of sources usually observed in an astrometric schedule and the limitation the number puts on integration time, it may be possible to observe several thousand compact sources above the 50 mJy threshold. Since it is necessary to continually calibrate the array geodetically and astrometrically, a large, uniform, continuous data base of interferometric observables should develop over the lifetime of the array.

Until the VLBA becomes fully operational in the late 1990's, there are several ongoing programs which will continue to expand and refine the extragalactic catalogue. The NASA Crustal Dynamics Project has a VLBI survey program to expand its catalogue of unresolved sources to take advantage of improvements in interferometer sensitivity and to extend geodetic measurements to the southern hemisphere. The US Naval Research Laboratory and Naval Observatory have begun an astrometric program using VLBI stations in both the northern and southern hemispheres to increase the number of precisely located radio sources with optical counterparts. It is expected that the JPL survey work will be further refined to support planetary spacecraft navigation using differential VLBI. The interferometric observables from the various observing programs will be archived and made available for detailed analysis so that improvements in geophysical, astronomical, and other models can be reflected in the astrometric results.

The ultimate limitations of the extragalactic reference frame may be logistical rather than theoretical, although at this writing the limit is set by systematic errors that can be ascribed to the modeling of the troposphere, the nutation and precession models, and the distribution of the baselines in the astrometric network.

BIBLIOGRAPHY

Aoki, S., Guinot, B., Kaplan, G. H., Kinoshita, H., McCarthy, D. D. and Seidelmann, P. K., 1982, *Astron. Astrophys.*, **105**, 359.

Argue, A. N., de Vegt, C., Elsmore, B., Fanselow, J., Harrington, R., Hemenway, P., Johnston, K. J., Kuhr, H., Kumkova, I., Niell, A. E., Walter, H. and Witzel, A., 1984, *Astron. Astrophys.*, **130**, 191.

Baade, W. and Minkowski, R., 1954, *Astrophys. J.*, **119**, 206.

Backer, D. C., Fomalont, E. B., Gross, W. M., Taylor, J. H. and Weisberg, J. M., 1985, *Astron. J.*, **90**, 2275.

Bare, C., Clark, B. G., Kellerman, K. I., Cohen, M. H. and Jauncey, D. L., 1967, *Science*, **157**, 189.

Bartel, N., Ratner, M. I., Shapiro, I. I., Cappallo, R. J., Rogers, A. E. E. and Whitney, A. R., 1985, *Astron. J.*, **90**, 318.

Bartel, N., Herring, T. A., Ratner, M. I., Shapiro, I. I. and Corey, B. E., 1986, *Nature*, **319**, 733.

Bolton, J. G., Stanley, G. J. and Slee, O. B., 1949, *Nature*, **164**, 101.

Clark, B. G., 1973, *IEEE Trans. Antennas Propag.*, **61**, 1242.

Clark, T. A., Hutton, L. K., Marandino, G. E., Counselman, C. C., Robertson, D. S., Shapiro, I. I., Wittels, J. J., Hinteregger, H. F., Knight, C. A., Rogers, A. E. E., Whitney, A. R., Niell, A. E., Ronnang, B. O. and Rydbeck O. E. H., 1976, *Astron. J.*, *81*, 599.

Clark, T. A., Corey, B. E., Davis, J. L., Elgered, G., Herring, T. A., Hinteregger, H. F., Knight, C. A., Levine, J. I., Lundqvist, G., Ma, C., Nesman, E. F., Phillips, R. B., Rogers, A. E. E., Ronnang, B. O., Ryan, J. W., Schupler, B. R., Shaffer, D. B., Shapiro, I. I., Vandenberg, N. R., Webber, J. C. and Whitney, A. R., 1985, *IEEE Trans. Geosci. Remote Sensing*, GE-23, 438.

Cohen, M. H. and Shaffer, D. B., 1971, *Astron. J.*, **76**, 91.

Davis, M. M., Taylor, J. H., Weisberg, J. M. and Backer, D. C., 1985, *Nature*, **315**, 547.

Davis, J. L., Herring, T. A., Shapiro, I. I., Rogers, A. E. E. and Elgered, G., 1985, *Radio Science*, **20**, 1593.

Davis, J. L., 1986, Ph. D. dissertation, Massachusetts Institute of Technology, also AFGL-TR-86-0242, Air Force Geophysics Lab., Lexington.
de Vegt, C. and Prochazka, F., 1985, *Astron. Astrophys.*, **148**, 226.
de Vegt, C., Florkowski, D. R., Johnston, K. J. and Wade, C. M., 1985, *Astron. J.*, **90**, 2387.
de Vegt, C., Schramm, J. and Johnston, K. J., 1987, *Astron. J.*, **93**, 261.
Elgered, G., 1983, Ph. D. dissertation, Chalmers University of Technology.
Fanselow, J. L., Sovers, O. J., Thomas, J. B., Purcell, G. H., Cohen, E. J., Rogstad, D. H., Skjerve, L. J. and Spitzmesser D. J., 1984, *Astron. J.*, **89**, 987.
Finkelstein, A. M., Kreinovich, V. J. and Pandey, S. N., 1983, *Astrophys. Space. Sci.*, **94**, 233.
Florkowski, D. R., Johnston, K. J., Wade, C. M. and de Vegt, C., 1985, *Astron J.*, **90**, 2381.
Froeschlé, M. and J. Kovalevsky, 1982, *Astron. Astrophys.*, **116**, 89.
Gordon, D., 1985, NASA internal memorandum, Goddard Space Flight Center, Greenbelt.
Hazard, C., Mackey, M. B. and Shimmins, A. J., 1963, *Nature*, **197**, 1037.
Hazard, C., Sutton, J., Argue, A .N., Kenworthy, C .M., Morrison, L .V. and Murray, C .A., 1971, *Nature London Phys, Sci.*, **233**, 89.
Hellings, R. W., 1986, *Astron. J.*, **91**, 650.
Herring, T. A., 1983, Ph. D. dissertation, Massachusetts Institute of Technology, also AFGL-TR-84-0182, Air Force Geophysics Lab., Lexington.
Herring, T. A., Gwinn, C. R. and Shapiro, I. I., 1986, *J. Geophys. Res.*, **91**, 4745.
Hewish, A., Bell, S. J., Pilkington, J. D. H., Scott, P. F. and Collins, R. A., 1968, *Nature*, **217**, 709.
Hey, J. S., Parsons, S. J. and Phillips, J. W., 1946, *Nature*, **158**, 234.
Himwich, W. E. and Harder, E. J., 1988, in '*The Earth's Rotation and Reference Frames for Geodesy and Geodynamics*', A. K. Babcock and G. A. Wilkins eds., IAU Symp. 128, Reidel, Dordrecht, 301.
Hubble, E. P., 1925, *Observatory*, 48, 139.
Jansky, K. G., 1933, *Proc. IRE*, **21**, 1387.
Jennison, R. C. and Das Gupta, M. K., 1953, *Nature*, **172**, 996.
Johnston, K. J., de Vegt, C., Florkowski, D. R. and Wade, C. M., 1985, *Astron. J.*, **90**, 2390.
Kaplan, G. H., Josties, F. J., Angerhofer, P. E., Johnston, K. J., and Spencer, J. H., 1982, *Astron. J.*, **87**, 570.
Kellerman, K. I., 1974, in '*Galactic and Extra-galactic Radio Astronomy*', G. L. Verschuur and K. I. Kellerman eds., Springer-Verlag, New York, 320.
Kesteven, M. J. L. and Bridle, A. H., 1977, *J. R. Astron. Soc. Can.*, **71**, 21.
Lestrade, J.-F., Preston, R. A. and Niell, A. E., 1986, *JPL Astrophysics preprint* **136**, Jet Propulsion Laboratory, Pasadena.
Lestrade, J.-F., Requieme, Y., Rapaport, M. and Preston, R. A., 1988, in 'The Earth's Rotation and Reference Frames for Geodesy and Geodynamics', A. K. Babcock and G. A. Wilkins eds., IAU Symp. 128, Reidel, Dordrecht, 67.
Lieske, J. H., Lederle, T., Fricke, W. and Morando, B., 1977, *Astron. Astrophys.*, **58**, 1.
Ma, C., Ph. D. dissertation, 1978, University of Maryland, also NASA TM 79582, Goddard Space Flight Center, Greenbelt.
Ma, C., Clark, T. A., Ryan, J. W., Herring, T. A., Shapiro, I. I., Corey, B. E., Hinteregger, H. F., Rogers, A. E. E., Whitney, A. R., Knight, C. A., Lundqvist, G. L., Shaffer, D. B., Vandenberg, N. R., Pigg, J C, Schupler, B. R. and Ronnang, B. O., 1986, *Astron. J.*, **92**, 1020.
Ma, C., 1988, in '*The Earth's Rotation and Reference Frames for Geodesy and Geodynamics*', A. K. Babcock and G. A. Wilkins eds., IAU Symp. 128, Reidel, Dordrecht, 73.
Marcaide, J. M. and Shapiro, I. I., 1983, *Astron. J.*, **88**, 1133.
Meeks, M. L. ed., 1976, '*Methods of Experimental Physics*', vol. 12, part C, Academic Press, New York.
Melbourne, W., Anderle, R., Feissel, M., King, R., McCarthy, D., Smith, D., Tapley, B. and Vicente, R., 1983, "Project Merit Standards", USNO Circular 167, US Naval Observatory,

Washington.
Morabito, D. D., 1985, *Astron. J.*, **90**, 1004
Morabito, D. D., Niell, A. E., Preston, R. A., Linfield, R. P., Wehrle, A. E. and Faulkner, J., 1986a, *Astron. J.*, **91**, 1038.
Morabito, D. D., Preston, R. A., Linfield, R. P., Slade, M. A. and Jauncey, D. L., 1986b, *Astron. J.*, **92**, 546.
Murray, C. A., 1986, in '*Relativity in Celestial Mechanics and Astrometry*', J. Kovalevsky and V. A. Brumberg eds., IAU Symp. 114, Reidel, Dordrecht, 169.
Perley, R.A., 1982, *Astron. J.*, **87**, 859.
Reber, G., 1940, Astrophys. J., **91**, 621.
Robertson, D. S., 1975, Ph. D. dissertation, Massachusetts Institute of Technology, also X922-77-228, Goddard Space Flight Center, Greenbelt.
Robertson, D. S., Fallon, F. W. and Carter, W. E., 1986, *Astron. J.*, **91**, 1456.
Rogers, A. E. E., Cappallo, R. J., Hinteregger, H. F., Levine, J. I., Nesman, E. F., Webber, J. C., Whitney, A. R., Clark, T. A., Ma, C., Ryan, J. W., Corey, B. E., Counselman, C. C., Herring, T. A., Shapiro, I..I., Knight, C. A., Shaffer, D. B., Vandenberg, N. R., Lacasse, R., Mauzy, R., Rayhrer, B., Schupler, B. R., Pigg, J C, 1983, Science, 219, 51.
Seidelmann, P. K., 1982, *Celest. Mech.*, **27**, 79.
Shapiro, I. I., 1970, *Eos*, **51**, 266.
Shapiro, I. I., Wittels, J. J., Counselman, C. C., Robertson, D. S., Whitney, A. R., Hinteregger, H. F., Knight, C. A., Rogers, A. E. E., Clark, T. A., Hutton, L. K. and Niell, A. E., 1979, *Astron. J.*, **84**, 1459.
Sovers, O. J. and Ma, C., 1985, *Eos*, **66**, 858.
Sovers, O. J. and Fanselow, J. L., 1987, '*Observation Model and Parameter Partials for the JPL VLBI Parameter Estimation Software MASTERFIT - 1987*', JPL Pub. 83-39, rev. 3, Jet Propulsion Laboratory, Pasadena.
Sovers, O.J., Edwards, C. D., Jacobs, C. S., Lanyi, G. E., Liewer, K. M. and Treuhaft, R. N., 1988, *Astron. J.*, **95**, 1647.
Thompson, A. R., Moran, J. and Swenson, G.W., 1986, '*Interferometry and Synthesis in Radio Astronomy*' Wiley, New York.
Ulvestad, J. S., 1987, in '*The Impact of VLBI on Astrophysics and Geophysics*', IAU Symp. 129, Cambridge, May 1987, M. Reid and J. Moran eds., in press.
Van Dam, T. M. and Wahr, J. M., 1987, *Geophys. Res.*, **92**, 1281.
Wade, C. M., 1970, *Astrophys. J.*, **162**, 381.
Wade, C. M. and Johnston, K. J., 1977, *Astron. J.*, **82**, 791.
Walter, H. G., 1977, *Astron. Astrophys. Sup.*, **30**, 381.
Walter, H. G. and West, R. M., 1986, *Astron. Astrophys.*, **156**, 1.
Whitney, A. R., Rogers, A. E. E., Hinteregger, H. F., Knight, C. A., Lippincott, S., Clark, T. A., Shapiro I. I. and Robertson, D. S., 1976, *Radio Sci.*, **11**, 421.
Wright, W. H., 1950, *Proc. Am. Philos. Soc.*, **94**, 1.

DYNAMICAL REFERENCE FRAMES IN THE PLANETARY AND EARTH-MOON SYSTEMS

J. G. WILLIAMS AND E. M. STANDISH
CALTECH /Jet Propulsion Laboratory
Pasadena, CA 91109, USA

1. INTRODUCTION

The practical determination of any reference frame is a question of accuracy. If there were no errors present in the coordinates of the objects computed from a given reference system, then the corresponding reference frame would be perfectly determined. In reality, however, a reference frame cannot be uniquely specified; it is determined in a somewhat arbitrary manner - implied by some sort of weighted average of coordinates derived from the system. Furthermore, if the errors in the coordinates are time-dependent, so will the determination be time-dependent.

This chapter presents an assessment of the accuracy of modern-day ephemerides, which, in turn, relates directly to the accuracy and/or consistency of the present dynamical reference frame.

We first discuss the accuracy of the optically-based ephemerides, the forerunners of the more modern ephemerides. We then describe why the modern ephemerides, based primarily upon ranging observations, are independent of any other celestial reference frame. Next, we briefly list the different types of observations to which the modern ephemerides are now being fit, and we discuss the main sources of errors in each of these data types. Section 5 presents an analysis of the form of the ranging data; it shows which particular elements of the dynamical system are well-determined and gives an estimate of the sensitivity of the ranging data to uncertainties in each of these elements. This analysis is then compared to numerical covariance studies in Section 6. The results are summarized in Section 7.

J. Kovalevsky et al. (eds.), Reference Frames, 67–90.

2. OPTICALLY-BASED EPHEMERIDES

Until the advent of the more modern observational data types (radar, laser and spacecraft ranging), planetary ephemerides were based strictly upon optical observations, primarily meridian transit timings. The observational campaigns included stellar observations interspersed with those of the planets and Moon. The subsequent adjustment of the stellar observations onto a fundamental reference catalogue was applied as well to the planetary and lunar observations. As such, the planetary and lunar observations may be considered to be differential in nature with respect to the fundamental stellar system. Systematic errors, in both the reduction processes and in the fundamental system itself, were inherited directly by the observed positions of the Moon and planets and were therefore transmitted directly into the positions and motions of the derived ephemerides.

In retrospect, it is now possible to present estimates of the errors in previous optically-based ephemerides. This may be done in two ways: first, by using the more recently determined corrections to the previous fundamental reference system along with formulae which relate how these affect the derived results, and second, by actually comparing the previous ephemerides with more modern ones.

Up to the present time, the standard reference system upon which optical observations have been reduced has been the fundamental stellar catalogue, FK4 (Fricke and Kopff, 1963). Recent determinations (Fricke, 1971, 1982) of the global errors of this reference system are 1".10/cty, 0".525 and 1".275/cty for the corrections to the general precession in longitude, the 1950 equinox offset and the equinox motion, respectively, and where cty denotes a Julian century. These parameters enter into the reductions of the observations in different ways, depending upon what type of observation (apparent, astrometric, etc.) is being considered. However, a general rule of thumb would be to expect ephemerides based upon this system to exhibit errors throughout the present century on the order of 1".

It is indeed the case that, when compared with more modern ephemerides, the positions listed in the national almanacs published prior to 1984 (and, therefore, based upon optically based ephemerides) show differences on the order of 1", sometimes even larger. A comprehensive study by Stumpff and Lieske (1984) has shown a difference in the mean motion of the earth between Newcomb's (1898) Theory of the Sun and the JPL ephemerides amounting to 1"/cty. These authors have noted that this discrepancy must be attributed to one (or more) of the three following factors: Fricke's correction to precession, the inertial mean motion of the earth in the modern ephemerides and Newcomb's of-date mean motion. However, recent evidence seems overwhelming that the major portion of the discrepancy must lie in the third factor: earth orientation studies using lunar laser-ranging (Newhall *et al.*, 1987) and VLBI (Sovers *et al.*, 1987; Herring *et al.*, 1986; Herring, 1987; Ma, 1987) have confirmed that Fricke's correction to the precession value is substantially valid, and assessments of the expected error in the mean motions of modern ephemerides (as discussed later in this chapter) indicate accuracies two orders of magnitude below that in question. Indeed, a comparison (Standish, 1987) between independently created ephemerides at JPL and at the Harvard-Smithsonian Center for Astrophysics shows agreement even closer than 0".01/cty for the mean motion of the earth.

3. EPHEMERIDES BASED UPON MODERN OBSERVATIONS

In contrast to their optically-based predecessors, modern ephemerides of the four inner planets and the Moon are determined almost entirely from the highly accurate ranging observations to which they are fit: radar echos from planetary surfaces, radio ranging to spacecraft transponders and laser ranging to the lunar reflectors. The relative positions, inertial motions, and even the orientation of the reference frame onto the dynamical equinox are all virtually independent of any other astronomical reference frame. While optical observations are also included in the data set, their effect is significant only for the outer five planets. Even for these, other new data types of greater accuracy are now becoming available.

Modern-day ephemerides of the Moon and planets are produced by numerically integrating the equations of motion which govern the motions of the relevant bodies of the solar system. The ephemerides, then, are strictly functions of only three factors: the equations of motion, the numerical integration program and the set of initial conditions & related constants (e.g., planetary masses). The equations of motion, expressed with respect to inertial space, represent our understanding of the physical laws of nature, at least to the presently observable accuracy. The numerical integration programs have been tested and have been found to be sufficiently accurate. Therefore, one can state that the numerically integrated ephemerides represent a dynamical system which is actually possible in inertial space. How well these ephemerides represent the true solar system, then, is a function of only the initial conditions and the associated dynamical constants. In turn, the initial conditions and some of the constants result from the least-squares adjustments to the observational data; some of the constants are determined from other sources.

The initial conditions play a second important role - that of the orientation of the reference frame. Certainly, any (time-independent) rotation of the initial conditions of the planets would produce an equally valid inertial ephemeris. However, in practice, the orientation of the initial conditions is determined by the observations to which the ephemerides are adjusted and by the formulae used in the reductions of the observations. With the wide variety of data types which are now fit by the ephemerides and the complexities of the reduction processes, the orientation of the frame of the integration becomes quite obscure. However, the resultant coordinates of the bodies may be used to determine the orientation of the reference frame; the accuracy of the coordinates may be used to infer the accuracy of the frame.

This section describes briefly, how the adjustment of the ephemerides to fit the ranging data automatically determines all of the facets of the planetary and lunar ephemerides. It is also shown, if so desired, how the ephemerides may be left oriented onto the reference frame of a stellar catalogue, by use of the optical data.

3.1. RELATIVE POSITIONS AND VELOCITIES

It is not difficult to visualize how the inter-body ranging can be used to determine relative positions and motions. Especially after a decade or so, during which the inner planets and the Moon have completed a number of orbits, the relative system can be determined with accuracies comparable to those of the observational ranging data.

3.2. INERTIALLY-BASED MOTIONS

What is not generally realized, on the other hand, is that the motions with respect to inertial space are also determined just from the ranging data alone. This can be visualized by a succession of three, progressively more complex examples:

1) Ranging between two mutually-orbiting point masses in a Keplerian ellipse. For this Newtonian example the ellipse does not rotate with respect to inertial space. The inertial mean motion (or inertial period) and mean anomaly at epoch are simply the period and phase of the purely periodic values of the range throughout time. The semi-major axis and eccentricity are determined from the maximum value of the range, $a(1+e)$, and from the minimum value, $a(1-e)$.

2) Ranging between two massless points in circular planar Newtonian motion about a known central mass. Over a synodic period, the values of the range vary between a maximum equal to the sum of the two radii (semi-major axes) and a minimum equal to the difference of the two radii. These yield, directly, the ratio of the two radii, which, in turn, through Kepler's third law, gives the ratio of the inertial mean motions. The synodic period gives the difference in the mean motions. Together, each individual inertial mean motion is determined.

3) Actual ranging from the earth. While the lunar or planetary orbits are not truly Keplerian, they are approximately so. The relative departures from Keplerian motion are small for the planets and larger for the Moon; they are computed to significant accuracy through the integration of the equations of motion.

The first example may be envisioned as a simplified lunar case; the second as a simplified planetary case; the third as a case closer to reality. For each, the combination of the geometry and the dynamics prevents the two from being consistent in anything but an inertial frame. Section 5 presents an analysis of the characteristic signatures of the ranging data, showing how the various orbital elements are determined and how the sensitivities of the elements may be estimated.

3.3. DYNAMICAL EQUINOX AND OBLIQUITY

Lunar laser ranges measure not only the motion of the Moon about the earth, but two other major features as well. Since they are taken from the surface of the spinning earth, the ranges are sensitive to the direction of the pole of the earth's rotation, and, therefore, to the of-date celestial equator. Furthermore, from the strong perturbations of the sun upon the lunar orbit, the ranges are also sensitive to the orientation of the of-date ecliptic. The combination of the equator and the ecliptic provide the of-date dynamical equinox and obliquity.

The (time-dependent) osculating ecliptic is a quantity inherent in a given set of ephemerides, from which one may extract the intersection of the mean ecliptic with the mean equator in a relatively straightforward manner to the level of 0".001 or better (see Standish, 1982; Chapront-Touzé and Chapront, 1983). With the mean equator coincident with the x-y plane, the ephemerides may be adjusted by a rotation about the z-axis in order to align the x-axis onto this point, the dynamical equinox. This, in fact, was done when the JPL ephemeris, DE118, was re-oriented onto the dynamical equinox (forming DE119) and precessed to the epoch of J2000 in order to form the ephemeris, DE200.

3.4. OPTICALLY-BASED ALIGNMENTS

Exported JPL Ephemerides previous to DE200 have not been so aligned to the dynamical equinox. Instead, the orientation about the z-axis had been determined by the relatively weak optical data to which they had been fit. Since the optical data were based on the FK4 catalogue, the ephemerides were allowed to become aligned to the zero-point of that system and therefore shared the same equinox offset. In particular, the z-axis rotation for DE118 (equinox offset) was found to be 0".531 (Standish, 1982), agreeing remarkably well with Fricke's (1982) determination of 0".525 for the 1950 equinox offset of the FK4.

4. EPHEMERIS DATA AND ADOPTED CONSTANTS

4.1. THE AMOUNT AND ACCURACY OF THE OBSERVATIONAL DATA

There is now a wide variety of observational data to which modern ephemerides are adjusted. Almost every major member of the solar system is unique with respect to the various types of data by which it has been observed as well as to the special considerations with which the data must be reduced. It is not possible to present a full description of the data in the present chapter; this is being done by Standish (1988). Here, a general summary is given.

Table 1 presents a list of the different data types to which the more recent JPL ephemerides are being adjusted. Similar ephemeris work at the Harvard- Smithsonian Center for Astrophysics includes similar sets of data. The table gives the approximate number of observations for each individual data type as well as approximate values for the accuracy of each. For comparison purposes, the accuracies are listed in both kilometers and arcseconds.

4.2. THE MAJOR SOURCES OF UNCERTAINTIES IN THE OBSERVATIONAL DATA

Radar ranges consist simply of round-trip times of a signal reflected from the surface of a planet; their errors are dominated by the topography on the planet's surface. For Mercury and Venus, the topographical variations introduce an rms scatter of about 1.5 km; for Mars, the scatter of 2.2 km may be reduced to 1.8 km when the surface is approximated by a tri-axial ellipsoid.

Radar closure is derived from paired radar points, separated in time, where the two members were reflected off nearly the same topographical feature. The topographical uncertainties are reduced by over an order of magnitude and the resulting difference yields an ephemeris drift with an uncertainty of about 150 meters over the time-span separating the two points of the pair.

The lunar ranges result from laser pulses bounced off of the four retroreflectors on the Moon. The range uncertainties depend on the pulse width of the laser, the instrumental effects, and the signal-to-noise ratio. Technical advances have improved the accuracy of the ranges by more than an order of magnitude since the first returns were received in 1969; the major steps are indicated in Table 1.

For the spacecraft data, the Mariner 9 range errors arise from the uncertainties of the spacecraft's orbit with respect to the center of mass of Mars and also from the uncertainties in the density of the electrons in the solar corona (especially near solar conjunction). These corona delays were modeled by the empirical formulae of Muhleman and Anderson (1981).

For the single-frequency Viking Lander range data, dual-frequency ranges, nearby in time, from the Viking Orbiters were used to calibrate and remove most of the delays from the corona. The remaining residuals show a consistency during a single day of 2-3 meters, but a day-to-day scatter of about 7 meters. Presumably, these day-to-day variations come from errors in the processes of calibrating both the corona corrections and the electronic equipment. After the orbiters ceased operating in 1980, the corona delays were estimated by the same model as that used for Mariner 9 and showed a scatter of about 12 meters away from solar conjunction.

Range and Doppler measurements from the Pioneer and Voyager encounters contain unmodeled accelerations upon the spacecraft from the solar radiation pressure and from attitude control thrusters as well as uncertainties in the spacecraft's planetocentric trajectories.

Radio measurements of the thermal emission from the Galilean satellites of Jupiter, Saturn's satellite Titan, Uranus and Neptune have assumed that the center of the body's radio emission corresponds to the body's center of mass. The accuracy of 0".03 may possibly be improved in the future with a more refined model of the brightness distribution. The viewing conditions are also affected, at times, by poor weather at the receiver.

All of the optical observations inherit the uncertainties of the stellar catalogue upon which they are based. The observations of the finite disks of the planets require additional considerations:

(i)- when the limb of an object is measured, corrections must be applied both for the object's semi-diameter and for any phase effect of non-uniform illumination;

(ii)- the sun, Mercury and Venus are observed during daylight hours when most stars are not visible;

(iii)- daytime observations also are affected by temperature variations;

(iv)- Mars has a large phase effect as well as problems due to irregular and varying surface brightness; and

(v)- the rings of Saturn present a problem with north-south limb definitions; the ring occultations of Uranus also rely upon the dynamical model of the rings.

TABLE 1.

Data types to which modern ephemerides are adjusted. The (post-fit) rms residuals indicate the accuracy of the data. The values listed without brackets are the units of the original observations; those within brackets give the comparable values for comparison purposes.

Type of observation	time span	post-fit rms residuals km	post-fit rms residuals ″	number of observations
Radar ranging				
Mercury	1966-	1.5	[0.″002]	500
Venus	1965-	1.5	[0.″002]	1000
Mars	1967-	2.2	[0.″003]	40000
Mars Closure	1969-1982	0.15	[0.″0002]	200
Spacecraft Ranging				
Ma9 Orbtr (Mars)	1972-1973	0.040	[0.″0002]	600
Vkng Lndr (Mars)	1976-1980	0.007	[0.″000003]	900
	1980-1982	0.012	[0.″000006]	400
Spacecraft Tracking (range,Doppler)				
Pio&Voy (Jup,Sat)	1973-1980	[200,400]	[0.″05]	20000
Lunar laser ranging	1969-1970	0.00100	[0.″0005]	10
	1970-1975	0.00030	[0.″00016]	1700
	1976-1985	0.00015	[0.″00008]	3000
	1985-	0.00006	[0.″00003]	600
Radio astrometry				
Jup,...,Nep	1983-	[100,...,600]	0.″03	10
Ring Occultation				
Uranus	1978-	[1500]	0.″1	14
Optical transits (manual)				
Sun,Mer,Ven	1911-	[700]	1.″0	37000
Mars,...,Nep	1911-	[150,...,10000]	0.″5	18000
Optical transits (photoelectric)				
Mars,...Nep	1982-	[100,...,6000]	0.″3	1000
Astrolabe				
Mars,...,Ura	1961-	[100,...,4000]	0.″3	1500
Astrometry				
Pluto	1914-	[15000]	0.″5	1600

TABLE 2.

Values and uncertainties of various dynamical constants used in recent ephemerides. Those determined from the ephemeris least-squares solutions are marked with an asterisk (*). The major sources of the determinations are given in the last column; the sources in brackets are external to the ephemeris determinations.

Constant	Value	Uncertainty	Determination Source
* Scale [km/au]	149597870.66	0.02	planetary ranging
* E-M mass ratio	81.3006	0.0001	[s/c tracking]
planetary masses [sun/planet]			
Mercury	6023600	400	[Mariner 10]
Venus	408523.5	0.5	[Mariner 5]
* Earth+Moon	328900.55	0.01	lunar ranging
Mars	3098710	10	[Mariner 4]
* Jupiter	1047.3492	0.0005	Pio & Voy s/c
* Saturn	3497.91	0.02	Pio & Voy s/c
* Uranus	22905	10	Voy
* Neptune	19350	50	[Triton, Viking]
Pluto	130,000,000	30,000,000	[Charon]
asteroid masses [asteroid/sun]			
0001 Ceres	5.9×10^{-10}	5 %	| [perturbations
0002 Pallas	1.08×10^{-10}	20 %	| on other
0004 Vesta	1.38×10^{-10}	10 %	| asteroids]
0007 Iris	5.4×10^{-12}	50 %	[density=Ceres]
0324 Bamberga	8.8×10^{-12}	50 %	[density=Ceres]
mean asteroid densities			
* C-types	2.0	1.0	[radius]
* S-types	2.7	1.0	[radius]
lunar tidal acceleration			
	-24".9/cty^2	1".0/cty^2	lunar ranging

4.3. THE DYNAMICAL CONSTANTS IN THE EPHEMERIDES

The ephemerides which result from the numerical integrations are a function, not only of the initial conditions, but also of the constants which enter into the dynamics through the equations of motion. Table 2 presents a list of these constants along with their associated uncertainties. As opposed to being the formal standard deviations which result directly from the least-squares solutions, these values are intended to be realistic ones, estimated by considering the possible unknown or unmodeled errors in the observational data.

5. PARAMETER DETERMINATIONS AND UNCERTAINTIES

As discussed in Section 3, the establishment of the reference frame of the ephemerides depends upon the initial conditions with which the equations of motion are numerically integrated. The initial conditions (along with a variety of other parameters) are adjusted in a least-squares sense to fit the observational data. In practice, the complete process of ephemeris creation uses cartesian coordinates exclusively; even the partial derivatives are numerically integrated. No explicit use is made of orbital elements.

It is helpful, in order to understand the accuracies of the numerical fitting procedure, to discuss an analytical model using Keplerian-type elements, even though these elements never appear explicity un the whole process of ephemeris creation. From an analysis of the model, one is able to estimate the accuracies with which various specific combinations of elements are (implicitly) determined in the fitting process.

We present here analyses of the lunar ranging data and then that of planetary ranging. At first, we shall consider the structure of only the instantaneous distances between the centers of mass of the earth and of the body being observed. Next, we discuss the choice of units; in particular, the scale of the solar system, a free parameter. Finally, we shall consider the signatures in the ranging data due to perturbations upon the orbits as well as the positioning of both the antenna (station location, precession, nutation, etc.) and the "bounce point" (planet orientation, lander or retro-reflector location, etc.). We ignore the rotational and orbital motions of the bodies during the time of flight of the signal. In actual practice, of course, these motions are highly significant and are modeled with utmost precision during the data reduction processes.

We present the low-degree expansion of the range - a Fourier series which is seen to contain a constant term and a number of periodic terms, each with a specific amplitude, phase and period (or phase rate). Using the values given in Table 1 for the amount, time-span and accuracy of the observational data, one may derive estimates of how well each of the quantities in the series is determined. With such an expansion it is easiest to estimate a sensitivity to some parameter: the size of an amplitude or phase change necessary to match the range accuracy. Estimating accuracies is harder than sensitivities since it involves knowledge of the observing geometry of a number of points and correlations with other parameters. Thus the main use of the analytical model will be for sensitivities. In Section 6 these will be compared with the accuracies resulting from two covariance analyses which account for the complexities.

5.1. LUNAR RANGING

The center-to-center distance of the Moon about the earth is given by the expansion (to first degree in eccentricity),

$$r = a - ae\cos l \qquad \{1\}$$

where, for the semi-major axis, a = 384400 km, for the eccentricity, e =0.0549, for the product, ae = 21100 km and where the period of l, the mean anomaly, is one month. Fitting the above equation to lunar ranging data would be the equivalent of determining four quantities: a constant term, the amplitude of cos l, the phase of l at some epoch and the period (mean motion). Note that the product ae is closer to an observable quantity than is e itself, so that even if a good distribution of observations causes a and ae to be uncorrelated, e will always be correlated with a.

Equation {1} may be differentiated with respect to l in order to calculate the sensitivity of the range data to a change in the mean anomaly. A maximum change of 6 cm in range (present observational accuracy) would be caused by a change of 0".0006 in the mean anomaly. A span of data improves the accuracy for the mean anomaly, and permits the rate (mean motion) and tidal acceleration to be determined. A rate error estimate is complicated by the nonuniform distribution of lunar range accuracies in Table 1, but 0".02/cty is a reasonable estimate at some favored mid-time. The rate and longitude are much degraded by the secular acceleration error of 1".0/cty^2 (Newhall et al., 1987), the determination of which is complicated by a long-period term (Williams et al., 1978), near or beyond the ends of the data span.

The center-to-center distance from the Earth to the Moon does not depend on the orientation of the reference frame. However, the mean longitude rate can be determined with respect to inertial space if one knows the precession of the Moon's orbital plane. The plane effectively precesses along the ecliptic (the Laplacian plane is tilted about 8" out of the ecliptic) and this rate depends on the semi-major axes, eccentricities, and mutual inclination of the geocentric lunar orbit and the heliocentric orbit of the Earth-Moon system. The rate also depends on the gravitational harmonics of the earth and Moon and the ratio of their masses. The zonal harmonics of the earth are very well known from the tracking of artificial satellites; a consideration of the various errors shows that it is the error in the inclination of the Moon's orbit with respect to the ecliptic which is the limiting error source. The present uncertainty in the inclination is about 0".002 which causes an uncertainty of 0".037/cty in the rate of the longitude of perigee. An additional error of 0".013/cty arises from the 0.6% uncertainty in the lunar harmonic, J_2. These two errors give a combined uncertainty of 0".039/cty for the perigee motion along the ecliptic, and this is our estimate of the drift error of the lunar orbit with respect to inertial space. This is larger than the estimate for the mean anomaly rate error, so the error in the rate of the mean longitude (mean anomaly rate plus longitude of perigee rate) is about 0".04/cty. However, this estimate applies to times near the middle of the span of observations and the uncertainty in the lunar tidal acceleration will cause this rate uncertainty to grow at more distant times. Future lunar ranges will cause both the inclination and tidal acceleration errors to decrease.

5.2. PLANETARY RANGING

For planetary ranging, it is easier to use the square of the center-to-center distance,

$$r_c^2 = r^2 + r'^2 - 2\,\mathbf{r}\cdot\mathbf{r'} \qquad \{2\}$$

since the expansion for r_c can be slowly convergent. Expanding to first power in the two eccentricities, e and e', and in the square of the inclination, gives

$$\begin{aligned} r_c^2 = \; & a^2 + a'^2 - 2\,aa'\cos(\lambda-\lambda') \\ & - 2a^2 e\cos l - 2\,a'^2 e'\cos l' \\ & + aa'\,[\,3e'\cos(\lambda-\omega') + 3e\cos(\lambda'-\omega) \\ & \qquad - e\cos(\lambda+l-\lambda') - e'\cos(\lambda'+l'-\lambda) \\ & + (aa'\sin^2 i')/2\,[\,\cos(\lambda-\lambda') - \cos(\lambda+\lambda'-2\Omega')\,] \end{aligned} \qquad \{3\}$$

where a and a' are the two semi-major axes, l and l' are the mean anomalies, λ and λ' are the mean longitudes, ω and ω' are the longitudes of perihelion, Ω' and i' are the node and inclination of the ranged planet with respect to the ecliptic. There are only six independent angles: l, l', ω, ω', Ω' and i', since use

use has been made of the relations, $\lambda = l + \omega$ and $\lambda' = l' + \omega'$, in order to shorten the expressions.

Table 3 gives the sizes of the coefficients and the periods of the arguments in equation {3} for the planets which have been measured by ranging: Mercury, Venus and Mars. As above in the case of the Moon, one may differentiate each term of the equation in order to determine the sensitivity of the range to a given change in the angular argument of the term.

TABLE 3.

The coefficients of the expansion of the squared center-to-center range between the earth (unprimed quantities) and Mercury, Venus and Mars (primed quantities).

	Mercury		Venus		Mars	
coefficient	coeff au^2	period (days)	coeff au^2	period (days)	coeff au^2	period (days)
circular terms						
$(a^2 + a'^2)$	1.15	--	1.52	--	3.31	--
$2aa'\cos(\lambda-\lambda')$	0.78	116	1.44	584	3.04	780
eccentricity terms						
$2a^2 e \cos(l)$	0.034	365	0.034	365	0.034	365
$2a'^2 e' \cos(l')$	0.063	88	0.007	225	0.430	687
$3aa'e'\cos(\lambda - \omega')$	0.241	365	0.015	365	0.424	365
$3aa'e \cos(\lambda' - \omega)$	0.020	88	0.037	225	0.078	687
$aa'e \cos(\lambda + l - \lambda')$	0.007	170	0.012	975	0.026	249
$aa'e'\cos(\lambda' + l' - \lambda)$	0.080	50	0.005	162	0.141	5764
inclination term						
$(aa'\sin^2 i')/2$						
$\times \cos(l + l' - 2\Omega')$	0.0029	71	0.0013	139	0.0008	238

5.2.1. The Circular Terms

For the first three terms, as with the Moon, there is a constant, an amplitude of the cosine term, a phase at some epoch and a phase rate; however, the correspondence with the orbital parameters is different from the lunar case. The constant term $(a^2 + a'^2)$ and the amplitude (2aa') do uniquely determine a and a', though they will be correlated; the angles in $\lambda-\lambda'$ and their rates are determined only as differences. The maximum sensitivity of the range to $\lambda-\lambda'$ in radians is just equal to the smaller of the two semi-major axes. As seen from Table 3 for Mars, a change of 7m (accuracy of the Viking data) would be caused by an angular change of about 0".00001 in $\lambda-\lambda'$, implying a remarkable sensitivity in both the earth-Mars differential longitude and, given 4 years of calibrated Viking data, in its rate. For Mercury and Venus, radar ranging of 1.5 km implies angular sensitivities of 0".005 and 0".003, respectively. Further, given a good distribution of range data, these differential mean longitudes of Mercury or Venus with respect to the earth should be known at the milliarcsecond level, while the rates of these should be known to 0".2/cty or better; for Mars, these will be known at least two orders of magnitude better. For each planet, these differential longitudes and their rates will be the most accurately measured angles and rates.

5.2.2. The Terms Containing the Eccentricities

The six eccentricity-dependent terms in equation {3} are generally smaller than the circular terms, so that the semi-major axes may be considered as already determined. Since the perihelion rates are so small, only four of the six periods are distinguishable with only a decade or so of ranging data. However, since there are only two independent angles, in addition to $\lambda-\lambda'$, it is theoretically possible to determine the associated parameters, though there will be correlations among them.

As seen in Table 3, which of the six terms is the dominating one depends on which planet is being observed. For Mars, the amplitudes of the terms with arguments l' and $\lambda-\omega'$ are both 14% of the size of the periodic circular term with argument $\lambda-\lambda'$. The sensitivity of l' and $\lambda-\omega'$ is improved by both the other terms and the geometry, so we have used 25% of the sensitivity of $\lambda-\lambda'$. Note that since $l'=\lambda'-\omega'$, both Earth and Mars have their mean longitudes and rates well-determined with respect to Mars' perihelion direction. Since the Venus-Earth and Mercury-Earth differential longitudes and rates are known from the circular terms, the mean longitudes of the two inner planets are also well determined with respect to Mars' perihelion. For each of the four innermost planets, the mean longitude and its rate, measured from Mars' perihelion, is the second best known angle and rate.

The sensitivity of the earth's mean anomaly will be a factor of 5 worse than that of Mars. The mean anomalies and rates for Mercury and Venus will be much less well-determined than their longitudes and rates referred to Mars' perihelion direction.

5.2.3. The Term Containing the Mutual Inclination

The inclination enters into terms of equation {3} with only even powers sin i'. An equivalent statement is that a mirror image of the solar system reflected about the ecliptic plane would leave the ranges unaltered. Since the leading terms with eccentricities are linear, but those with inclinations are quadratic, it follows that the accuracies of the inclinations are one to two orders of magnitude larger than for the eccentricities. For Mars (i'=1.85 degrees), a 7m change in range can be caused by a change of 0".0002 in the inclination. For Mercury and Venus, the inclination accuracies are two orders of magnitude worse. Note that the ecliptic is the natural reference plane for the planets with ranging data.

5.2.4. Inertial Mean Motions

The center-to-center distance from the earth to the Moon or a planet does not depend on the orientation of the reference system. We have seen above that the ranges cause only the differences of longitudes (mean, perihelion or nodal) to be determined so that the origin (equinox) always cancels out. However, the absolute rates of the longitudes can also be determined with respect to inertial space. For the planets and Moon, the above equations show that the mean anomaly rate is directly measurable. The longitude of perihelion rate (perigee rate for the Moon) is small and calculated to significant accuracy during the integration, since the uncertainties in these rates are many orders of magnitude smaller than the rates themselves. For the least well-determined rate, that of Mars' perihelion, the mass uncertainties of Table 2 for the outer planets and for the asteroids each contribute 0".001/cty, (see Williams, 1984, for an earlier discussion) so that the total uncertainty in Mars' perihelion rate is about 0".002/cty (DE118, DE200 and earlier ephemerides had uncertainties of about 0".006/cty, degraded by worse masses). Since it was estimated above that the longitude rates of both earth and Mars with respect to Mars' perihelion are no worse than 0".002/cty, it may be concluded that state-of-the-art ephemerides for earth and Mars should have total drift errors (of inertial mean longitudes) less than 0".003/cty at a time near the center of the Viking data span (1979). For Mercury and Venus, the rate uncertainties are about 0".2/cty, dominated by the rate uncertainties of their differential longitudes with respect to the earth, as given above.

The error estimate for the mean longitude rate of Mars degrades with extrapolation away from the time span of the Viking data. The limiting error source here is the uncertainties in the masses of certain asteroids with orbital periods which are nearly commensurate with Mars' period. These near commensurabilities cause perturbations on Mars' longitude which have periods from decades to centuries. For extrapolation over centuries, the Martian longitude rate uncertainty for a modern JPL ephemeris is 0".025/cty; modeling further asteroids (using estimates for the radii and densities) would reduce this to 0".015/cty. The above estimates for the rate errors of Mercury, Venus and earth would be little affected by extrapolation. The errors given refer to drifts along the ecliptic plane with respect to inertial space. The accuracy for the longitude rate for Mars has been discussed by Williams (1984).

The natural origin for the Moon and the four inner planets is a fixed equinox, not a moving equinox (as was appropriate for ephemerides of past decades based upon only astrometric observations). Since the orbital planes of the four inner planets tilt very slowly with time (a few arc seconds/century), the uncertainty of these tilt rates is very small and therefore the drift errors in the eclpitic latitudes of these bodies are negligible.

5.3. SCALE OF THE SOLAR SYSTEM - CONVENTION FOR UNITS

The solar system calculations follow the conventions of the IAU. Primary constants are the speed of light (299792.458 km/s) and k^2, the Gaussian constant squared, the gravitational constant times the mass of the sun (k^2=0.01720209895^2 [au^3/day^2]). The time scale is dynamical time. The ranges to the planets and Moon are actually measured as time delays, but they may be converted directly to metric units by using the speed of light. Since k^2 is defined with units of au^3/day^2 and since Kepler's third law relates semi-major axes (in au) and mean motions to k^2 and planetary masses (in solar masses), then the length of the astronomical unit in kilometers must be one of the parameters of the fits to the range data. Kepler's third law can then be written as

$$n^2 (a/A)^3 = k^2 (1+m) \qquad \{4\}$$

where n is the mean motion (in radians/day), a is the semi-major axis (in kilometers), A is the astronomical unit (in kilometers),and m is the planet's mass (in solar masses). The discussion of fits to ranges has treated semi-major axes and mean motions as though they were separately determined; in principle, doing so is the equivalent to determining the length of the au for each planet. Actually there is only one free parameter for the length of the au in the general solutions, and it is the determination of the size of the orbits of Mars and the earth from the Viking data which sets the scale of the solar system, and the error in the au is on the order of the error in those ranges. If they were separately determined, the relative error of the mean motions would be smaller than the relative error of the lengths of the semi-major axes so that the semi-major axes are known more accurately when expresssed in au's than in kilometers. Note that determining the length of the au is equivalent to determining k^2A^3, the gravitational constant times the solar mass, in km^3/day^2. Note also that the semi-major axis of the earth's orbit is not defined as one au, but is a determined quantity just like the other planets. The determination of the length of the semi-major axis of the lunar orbit can be thought of as separate from the determination of the mean motion since this is equivalent to solving for GM of the Earth-Moon system. Because of the way dynamical time is defined in terms of atomic time, there are known inconsistencies between geocentric and solar system barycentric formulations at the level of relativistic terms which affect lengths and GM values expressed in metric units (see Chapters 5 and 15).

5.4. PERTURBATIONS, STATION/REFLECTOR/LANDER COORDINATES

The analytical range model can be extended to include orbital perturbations and the finite sizes of the planets and Moon. We shall only summarize those results here.

The sun causes 3000 and 3700 km perturbations in the lunar range with distinguishable periods. Both perturbation terms involve the difference in the geocentric longitude of the Moon and the heliocentric longitude of the earth. Given the range accuracy this differential longitude is sensitive to 0".001. The earlier discussed tight tie between the longitudes of the four inner planets means the longitude of the Moon is also well known with respect to these planetary longitudes and to the perihelion directions of the earth and Mars as well.

An 80 km perturbation term in the lunar range is sensitive to the inclination of the lunar orbit to the ecliptic (0".007 peak sensitivity) and the angular position of the Moon measured from its node on the ecliptic. Thus the lunar orbital plane is also known with respect to the ecliptic and hence the planes of all of the four inner planets. The earth, Moon and planets have finite sizes which affect the ranges since the tracking stations and targets are on their surfaces. The rotation of the earth causes a nearly daily modulation of a range with a typical amplitude of thousands of kilometers. The phase of this daily variation depends on the terrestrial station longitude, UT1, and the right ascension of the ranged body. The accuracy of the best LLR data permits UT (actually UT0 for a single station) accuracies of 0.0001 seconds of time from a single night's measurements and 0".001 accuracies for station longitudes from more extended spans of data. The station longitudes are directly measured with respect to the Moon's right ascension so the terrestrial and celestial coordinate frames are tied together (once certain conventions such as the zero points for earth rotation measurements are adopted). Sensitivity to the other two station coordinates and the declination of the target comes from the remainder of the modification of the range due to the finite size of the earth.

We have previously stated that orbit perturbations permit the lunar orbit plane to be oriented with respect to the ecliptic plane; the declination dependence permits the lunar orbit plane to be oriented with respect to the equator as well. The mutual orientations of the ecliptic, lunar orbit, and equator planes are known to 0".002 accuracy. Since the intersection of the equator and ecliptic planes is the dynamical equinox, it is possible to use this point as our reference point for zero right ascension. This has been done in DE200. The accuracy with which the system of the Moon and planets and terrestrial longitudes can be aligned with respect to the dynamical equinox depends on the orientation error and is currently 0".002/sin ε = 0".005, where ε is the obliquity of the ecliptic. The advantage of using the dynamical equinox is that it coincides with the theoretical origin of the FK5. Export ephemerides before DE200 were aligned to the FK4 equinox through the use of the optical data included in the solutions. The mutual orientations of the three planes is known best near the middle of the span of data and, for example, the error in the constant of precession of the earth's equator (now less than 0".2/cty) corrupts knowledge of the position of the equator plane at other times; e.g., J2000 or B1950.0. See Newhall *et al.* (1987) for a determination of precession and nutation from the LLR data.

The coordinates of the reflectors on the Moon or the Viking landers on Mars also result from analyses of the range data. For Mercury and Venus the planetary radii result from the solution. As with the earth, a dynamically consistent Mars-centered coordinate system results once conventions for Martian precession, nutation, and sidereal time are chosen. The Martian obliquity and equinox are solved for. Because of its synchronous rotation, the Moon is different from the earth and Mars. For the numerical integrations of the Euler angles which describe the orientation of the Moon, the three principal axes are used and hence establish a Cartesian coordinate system. There are, however, small (arc minute sized), calculable orientation differences between the principal axis coordinates and coordinates lined up with the mean earth direction and mean rotation axis of the Moon. Selenocentric coordinates of the retroreflectors in the latter system can be determined with accuracies of a few meters. See Williams *et al.* (1987) for coordinates from a recent solution.

6. NUMERICAL COVARIANCE STUDY

This section presents the results of two numerical covariance studies and compares them with the estimates for the parameters obtained from the preceding section. In each case, actual observational data was used in a least-squares solution and the resultant covariance matrix was mapped into a covariance matrix representing the parameters discussed in the preceding section.

For the first case, which closely approximates the discussion of the previous section, only the Viking Lander ranging data was used in the least-squares solution and the solution parameters were only those for the orbits of earth and Mars along with the scale factor. For the second case, the full set of ephemeris data (see Table 1) was used along with the full set of solution parameters normally used in the complete ephemeris adjustment process. The results (formal errors) are shown in Table 4 along with corresponding estimated sensitivities from the analytical discussion in the preceding section.

There are a number of reasons why the three sets of values do not show complete agreement. Each of these reasons has an effect on the present study.

(i)- The analytical discussion did not account for the fact that with a substantial number of observations, the statistics improve when using more than just a single observation. The uncertainties decrease rather like the square root of the number of observations.

(ii)- There are correlations among the various parameters, sometimes large. When highly correlated, a pair of parameters, or combination of parameters, will not separate well from each other and will be poorly determined. Other combinations may be well-determined and we have tried to point out these.

(iii)- In both examples, there is incomplete modeling; when a relevant parameter is either neglected or omitted from the solution, that parameter will be assumed to be perfectly known. This will lead to an overly optimistic estimate of the uncertainties of the other parameters, but case #2 is more realistic than case #1.

(iv)- The first example, case #1, contains only a subset of the observations: additional observations, whether of the same type or of another type, will strengthen the determinations of various parameters, thereby allowing better separation of others.

For case #1, there were nearly 1300 observations, from which one might expect a statistical improvement of an order of magnitude or so from the corresponding analytical estimates. However, one sees ratios here between 2 and 5 for the angular argument uncertainties, since there are some very high correlations. For the mean motions, n, n' and n'-n, the ratio is close to 10; however, the values from the analytical discussion and those from the numerical examples are not truly equivalent: the latter are derived strictly from the uncertainties in the semi-major axes while the former were estimated from sampling the angles over the six-year time span in question.

The equatorially-referenced pairs of nodes and pairs of perihelion longitudes of both Mars and the Earth are correlated at the level of 0.9985, indicating that the data in this case has sensitivity to the differences in these quantities, but not to the orientation of the reference system itself. As such, it was necessary to define the longitudes of each body, λ and λ', with respect to the intersection of the orbit of the ranged planet upon the true ecliptic, a reference point to which the data is sensitive. Similarly, the only well-determined inclination is that of the planetary orbit to the ecliptic.

In actuality, since the partial derivatives of the real data were used, there was some degree of sensitivity to the reference frame orientation due to the signatures both of the station locations and of the perturbations. As such, both of the absolute longitudes measured from the equator show uncertainties of 0".006, the same value as that for the uncertainty of the longitude of the node of the planet's orbit, measured along ecliptic. In a strictly center-to-center Keplerian formulation, this uncertainty of 0".006 would become completely indeterminate.

The most important feature of case #1, however, is the fact that one does see approximately the same relative strengths of determination among the various elements as were predicted by the analytical discussion in the previous section.

Case #2 differs in two ways. Additional data has allowed the independent determination of the orientation of the reference frame; the presence in the solution of a number of parameters which have an effect upon the orbits of Earth and Mars introduces previously unaccounted for uncertainties. As discussed in Sections 3 and 5, the lunar laser-ranging data orients the ecliptic onto the equator, thereby locating the intersection point of the Earth-Mars orbits also with respect to the equator. The perturbations of the outer planets and the asteroids, especially upon the orbit of Mars, introduce additional signatures into the ranging data which add uncertainty to the basic orbital parameters. As shown in Table 2, we have thus far accounted for only five asteroids, those that produce the greatest perturbations for the time of the Viking data. There is a number of others, however, which do have influence; their presence would dilute the accuracies of case #2 even further. The influence of the asteroid mass uncertainties has been studied in detail by Williams (1984), who estimates that when the forces of the neglected asteroids are included in the equations of motion, the remaining mass uncertainties will increase the uncertainties of the inertial mean motions of Mars and of the Earth to a level of about 0".003/cty near the middle of the Viking data span; for times far away from this epoch, the uncertainties for Mars rise to as much as 0".015/cty.

TABLE 4.

Comparison of the analytical estimates of angular uncertainties with those computed numerically. Case #1 considered only the set of Viking Lander range data and solved for only the orbital parameters of Earth and Mars along with the scale factor. Case #2 considered the full set of planetary and lunar observational data and solved for the full set of ephemeris parameters. The values represent uncertainites near the middle of the Viking data span (1979.6); for those that grow in time, values for about 1960 or 2000 are given in parentheses. The uncertainties for the Earth are improved relative to those of Mars by the influence of other observational data, while those of Mars are degraded more strongly by the uncertainties of the masses of the outer planets and of the asteroids.

ANGLE	ANALYTICAL SENSITIVITY	NUMERICAL CASE #1	NUMERICAL CASE #2
$\lambda'-\lambda$	0".00001	0".000003 (0".00001)	0".00002 (0".0007)
$\lambda'-\omega'$ [=l']	0".00004	0".00001 (0".00004)	0".0001 (0".0005)
$\lambda-\omega'$	0".00004	0".00001 (0".00005)	0".0001 (0".0003)
l	0".0002	0".00008 (0".0001)	0".0002 (0".0003)
$\lambda'-\omega'$	0".0002	0".00008 (0".0001)	0".0002 (0".0007)
λ'	--	0".006	0".001
λ	--	0".006	0".001
i'	0".0002	0".00002	0".00005
n'-n	0".0005/cty	0".00006/cty	0".003/cty
n'	0".002/cty	0".0002/cty	0".002/cty
n	0".002/cty	0".0002/cty	0".001/cty
AU	0.007 km/au	0.013 km/au	0.007 km/au

Case #2 shows how truly accurate the ephemerides have become. Even providing for a further degradation in accuracy from that shown in the table, these uncertainties show a vast improvement of ephemerides based upon modern ranging observations over their optically-based predecessors.

7. SUMMARY

Earlier ephemerides of the Moon and planets, based upon optical observations, have inherited errors directly from the catalogues upon which they have been based. These errors amount to a number of tenths of an arcsecond in angular position and a number of tenths of an arcsecond per century in angluar motion; i.e., errors comparable to those that are known to exist in the FK4 fundamental reference system.

Modern ephemerides based upon ranging observations show at least an order of magnitude improvement over their optically-based predecessors. The major portion of this chapter has shown how the various elements of the ephemerides are determined from fitting to the ranging observations and has provided estimates for their uncertainties.

The basis for modern ephemerides is the set of observational data to which they are adjusted. We have discussed this set of data briefly. We then have selected the most important data types and have calculated how sensitive these data are to changes in certain ephemeris elements. The sensitivities, in turn, indicate how well each of these elements may be determined through the data fitting, keeping in mind that the statistics of the actual determinations are improved due to the large number of observations but also that there are correlations among the various parameters.

The lunar laser ranging data is sensitive to a change in the lunar mean anomaly and its rate at levels of 0".0006 and 0".02/cty, respectively. The data is also sensitive to the rate of the lunar longitude with respect to inertial space at a level of 0".04/cty. This rate error is dominated by the uncertainties in the precessional rates of the lunar perigee; the precessional rates themselves are due to the perturbations which depend on the orbital elements and gravitational harmonics of the Earth and Moon. At times away from the data span, the uncertainty (1".0/cty^2) in the tidally-induced acceleration in longitude becomes predominant.

For the planets, the most important data are the ranges to the Viking landers on Mars. We show that these ranges have a remarkable sensitivity to a number of differential angles: the difference in heliocentric longitudes between Earth and Mars at a level of 0".00001, each longitude with respect to the perihelion of Mars at a level of 0".00004 and each longitude with respect to the perihelion of the Earth at a level of 0".0002. Further, the corresponding level for the inclination of Mars' orbit upon the ecliptic is about 0".0002.

Radar-ranging to Mercury and Venus determines the longitudes of these planets with respect to the longitude of the Earth (and therefore to Mars). These sensitivities are on the order of 0".005 and 0".003, respectively, since the data are accurate to the level of 1.5 kilometers. The sensitivites to the inclinations upon the ecliptic are two orders of magnitude worse than that for Mars.

Solar perturbations upon the lunar orbit provide sensitivity to both the differential longitude between the heliocentric Earth and the geocentric Moon and to the inclination of the lunar orbit to the ecliptic; 0".001 and 0".007, respectively.

Since the lunar ranges are taken from the spinning Earth, sensitivites to

the Earth's orientation, coupled with the terrestrial coordinates of the observing station, allow determinations of
(i)- the mutual inclinations of the equator, the ecliptic and the lunar orbital plane (0".002);
(ii)- the longitude of the Earth and Moon with respect to the dynamical equinox (0".005); and
(iii)- a tie between the ephemeris frame and the terrestrial reference system (0".001 in longitude, comparable to 0.0001 seconds in UT0).

Finally, the fact that the lunar retroreflectors and the Viking landers are situated on the surfaces of the bodies, the ranges are sensitive to the physical orientations of the bodies themselves. The lunar librations affect the LLR data; the spin rate, obliquity and equinox of Mars influence the Viking ranges.

The analytical sensitivity analyses have been substantiated by numerical examples though the correspondance is not exact because of differences in numbers of observations, correlations, additional data and other perturbating forces. However, even when all of these factors are considered, it is seen that the dynamical reference system may be determined better than 0".01 in position with respect to the dynamical equinox. Further, the mean motions of Earth and Mars with respect to inertial space may be determined as well as 0".003/cty during the times of the highly accurate ranging data; the uncertainty for Mars will grow to about 0".015/cty over the course of many decades away from the present data.

ACKNOWLEDGMENT

Much of the analytical development of the sensitivities of the lunar ranges was first done by P. L. Bender and related to one of us (JGW) in numerous discussions. Those ideas have been used here and have inspired a similar development for the planetary ranging. The work described in this paper was carried out by the Jet Propulsion Laboratory, California Institute of Technology, under contract with the National Aeronautics and Space Administration.

BIBLIOGRAPHY

Chapront-Touzé, M. and Chapront, J., 1983, '*The lunar ephemeris ELP2000*', *Astron. Astrophys.*, **124,** p. 50-62.

Dickey, J. O., Newhall, X X and Williams, J. G., 1985, '*Earth Orientation from Lunar Laser Ranging and an Error Analysis of Polar Motion Services*', *J. Geophys. Res.*, **90**, p. 9353-9362.

Fricke, W., 1971, '*A Rediscussion of Newcomb's Determination of Precession*', *Astron. Astrophys.*, **13**, p. 298-308.

Fricke, W., 1982, *'Determination of the Equinox and Equator of the FK5', Astron. Astrophys.*, **107**, p. L13-L16.

Fricke, W. and Kopff, A., 1963, *'Fourth Fundamental Catalogue (FK4)', Veroffentlichungen #10*, Astronomisches Rechen-Institut, Heidelberg.

Herring, T. A., Gwinn, C. R. and Shapiro, I. I., 1986, *'Geodesy by Radio Interferometry: Studies of the Forced Nutations of the Earth. 1. Data Analysis', J. Geophys. Res.*, **91**, p. 4755-4765.

Herring, T. A., 1987, private communication.

Ma, C., 1987, private communication.

Muhleman, D. O. and Anderson, J. D. , 1981, *'Solar Wind Electron Densities from Viking Dual-Frequency Radio Measurements', Astrophys. J.*, **247**, p. 1093-1101.

Newcomb, S., 1898, *'Tables of the Motion of the Earth on its Axis and Around the Sun', Astronomical Papers,* U. S. Naval Observatory, Washington, DC., **vol. VI**

Newhall, X X, Preston, R. A. and Esposito, P. B., 1986, *'Relating the JPL VLBI Reference Frame and the Planetary Ephemerides'*, H. K.Eichhorn and R. J. Leacock, (eds.), *Astrometric Techniques,* D. Reidel Publ. Co., p. 789-794

Newhall, X X, Standish, E. M. and Williams, J. G., 1983, *'DE102: a numerically integrated ephemeris of the Moon and planets spanning forty-four centuries', Astron. Astrophys.*, **125**, p. 150-167.

Newhall, X X, Williams, J. G. and Dickey, J. O., 1987, *'Earth Rotation from Lunar Laser Ranging',* G. Wilkins and A. Babcock (eds.), to be published in the Proceedings of the IAU Symposium No. 129, *'The Earth's Rotation and Reference Frames for Geodesy and Geodynamics'* (Coolfont, West Virginia, October 20-24, 1986), D. Reidel, Boston.

Sovers, O. J., Edwards, C. D., Jacobs, C. S., Lanyi, G. E. and Treuhaft, R. N., 1987, *'The JPL 1986-3 Extragalactic Reference Frame'*, IAU Symposium #133, *'Mapping the Sky'*, Paris, June 1987.

Standish, E. M., 1982, *'Orientation of the JPL Ephemerides, DE200/LE200, to the Dynamical Equinox of J2000', Astron. Astrophys.*, **114**, p. 297-302.

Standish, E. M., 1987, *'Ephemeris Comparison - PEP740 vs. DE118',* Interoffice Memorandum **314.6-799**, Jet Propulsion Laboratory, Pasadena.

Standish, E. M., 1988, *'Observational Data in the Planetary and Lunar Ephemerides, DE200/LE200'*, to be submitted to *Astron. Astrophys.*

Stumpff, P. and Lieske, J. H., 1984, *'The Motion of the Earth-Moon System in Modern Tabular Ephemerides', Astron. Astrophys.*, **130**, p, 211-226.

Williams, J. G., 1984, *'Determining Asteroid Masses from Perturbations on Mars', Icarus,* **57**, p. 1-13.

Williams, J. G., Sinclair, W. S. and Yoder, C. F., 1978, *'Tidal Acceleration of the Moon', Geophysical Research Letters,* **5**, p 943-946.

Williams, J. G., Newhall, X X and Dickey, J. O., 1987, *'Lunar Gravitational Harmonics and Reflector Coordinates',* P. Holotato (ed.), in the Proceedings of the International Symposium, *'Figure and Dynamics of the Earth, Moon and Planets'*, special issue of the *Monograph Series of the Research Institute of Geodesy, Topography and Cartography*, (held September 15-20, 1986 in Prague, Czechoslovakia), p. 643-648.

REFERENCE FRAMES FOR ARTIFICIAL SATELLITES OF THE EARTH

CH. REIGBER
DGFI, München, FRG

1. INTRODUCTION

Near-Earth orbiting satellites have become one of the basic tools for solid-Earth and ocean physic studies, that is investigation of processes and forces which act upon the solid, viscous and fluid parts of the planet Earth and the related physical and geometric properties of our planet.

Manifestations of such forces and processes which are derivable from near Earth satellite observations are irregular variations in the orientation of the Earth in space (including both changes of the direction of the rotation axis and variations in the rotational speed), vertical and horizontal changes in the Earth's surface geometry (continents and oceans), the anomalous gravity field, the solid Earth and ocean tides, and the anomalous magnetic field. These phenomena cannot be regarded as single features in isolation: they have rather to be considered in the light of how they react with one another and with the atmosphere and hydrosphere, in order to obtain a significant overall picture both of the processes taking place on and inside the Earth and of the forces which drive these processes.

Satellite geodesy with its broad spectrum of existing observing systems or systems under development for: (i) target tracking from ground to space, satellite to ground or satellite to satellite, (ii) remote sensing of the ocean and ice surfaces with satellite-borne altimeters and (iii) measuring gravity gradient tensor components with satellite-borne gradiometer sensors, has in the last 10 years already contributed significantly to the improved understanding of the Earth and ocean dynamics and will continue to contribute to the explosive increase in our understanding of these phenomena with improved observational techniques and satellite missions.

Those aspects of the overall complex solid Earth and ocean physics which have profited most from the rapid development in satellite geodesy are

(i) - the global Earth gravity field;

(ii) - precise positions of terrestrial points;

(iii) - the orientation of the Earth.

All these topics are more or less non-static and therefore have to be treated as functions of time.

To make meaningful contributions to the many questions which are so far unresolved in solid Earth and ocean physics, extremely high accuracies are required for the various parameter subgroups:

J. Kovalevsky et al. (eds.), Reference Frames, 91–114.

(i) - high-resolution global gravity model parameters with such an accuracy that the resulting geoid has an accuracy of better than 10 centimetres and the resulting gravity anomalies over blocks with 100 km side length are accurate to a few milligals,

(ii) - relative positions of surface points over distances ranging from a few tens to many thousands of kilometres with an accuracy of 1- 2 centimetres at sampling intervals of a few months, so that the velocity of motion can be derived with an accuracy of less than one centimetre per year,

(iii) - Earth orientation parameters at intervals of about one day with accuracies of 1 mas for the pole coordinates and 0.1 ms for the rotational component about the polar axis (e.g. $UT1 - UTC$).

To reach this 10^{-8} to 10^{-9} relative precision from satellite data analysis, more accurate and frequent tracking along with much more sophisticated algorithms are required to be derived from better physical and mathematical theories. One element of the overall model are the reference frames which must be defined at least to an accuracy comparable to that required to express and analyze the quantities under investigation. This means, according to the above statement, with a 10^{-9} precision. Reference frame definitions with this precision are an important research subject for the near future. The following sections will give a description of the basic needs for reference frames in satellite data analysis and a description of the "state of the art" situation.

2. REFERENCE FRAMES AND TRANSFORMATIONS

The determination of near Earth satellite orbits from tracking data, the modelling of such data in terms of Earth kinematic and dynamic parameters and the interpretation of estimated parameters requires two basic reference frames: a space-fixed frame in which the satellite's motion takes place (see chapters 1 to 3) and an Earth body fixed frame in which the observer's position is defined (see chapter 7). When dealing with sensor measurements on board a spacecraft as is the case for e.g. satellite gradiometry, or with the rotational motion of the spacecraft for correcting tracking data, computing torques etc., a frame fixed to the spacecraft body is needed. Several additional intermediate reference frames may be used in practice.

2.1. REFERENCE FRAMES

2.1.1. The Quasi-Inertial Reference Frame

Let us consider two frames S and $\bar{S}$, moving with respect to each other, as shown in figure 1 and being represented by the orthonormal base vectors $\mathbf{e}_i$ and $\bar{\mathbf{e}}_i$ and their origins O and O'. If the motion of a satellite with constant mass m_S satisfies in the frame S the motion equation

$$m_S d^2\mathbf{r}/dt^2 = \mathbf{F}(t; \mathbf{r}, d\mathbf{r}/dt), \quad (1)$$

with

$\mathbf{r}, d\mathbf{r}/dt$ ... position and velocity vectors in S

$\mathbf{F}$ ... force function represented in S,

then classically the motion equation in the frame $\bar{S}$ is given by the expression (Goldstein, 1970)

$$\begin{aligned} m_S D^2\bar{\mathbf{r}}/Dt^2 = \bar{\mathbf{F}}(t, \bar{\mathbf{r}}, D\bar{\mathbf{r}}/Dt) - m_S d^2\mathbf{R}/dt^2 - \\ m_S\vec{\omega} \times (\vec{\omega} \times \bar{\mathbf{r}}) - m_S D\vec{\omega}/Dt \times \bar{\mathbf{r}} \\ -2m_S\vec{\omega} \times D\bar{\mathbf{r}}/Dt, \end{aligned} \quad (2)$$

where $\times$ denotes a cross-product and

$\frac{d}{dt}, \frac{D}{Dt}$ ... time derivatives with respect to S and $\bar{S}$, respectively
$\bar{\mathbf{r}}$ ... position vector in $\bar{S}$
$\mathbf{R}$ ... translation vector of O' in frame S
$\vec{\omega}$ ... rotation vector of $\bar{S}$ with respect to the frame S
$\bar{\mathbf{F}}$ ... vector of forces acting upon satellite, identical with $\mathbf{F}$ (but here being expressed as a function of position and velocity within $\bar{S}$)
$\bar{\mathbf{C}} = -2m_S\vec{\omega} \times D\bar{\mathbf{r}}/Dt$ ... Coriolis force
$\left.\begin{aligned} \bar{\mathbf{Z}} &= -m_S\vec{\omega} \times (\vec{\omega} \times \bar{\mathbf{r}}) \\ \bar{\mathbf{S}} &= -m_S D\vec{\omega}/Dt \times \bar{\mathbf{r}} \end{aligned}\right\}$ ... Centrifugal forces

As can be seen from the above, to the force $\mathbf{F}$ in S additional apparent forces arise in $\bar{S}$, resulting from the translational acceleration and rotational motion of the reference frame $\bar{S}$ with respect to the frame S. The privileged class of reference frames $S, S' \ldots$, characterized by $\ddot{\mathbf{R}} \equiv \mathbf{o}$ and $\vec{\omega} \equiv \mathbf{o}$ and thus by the non-existence of mass-proportional apparent forces, is called inertial reference frame. In classical mechanics inertial reference frames are postulated to exist.

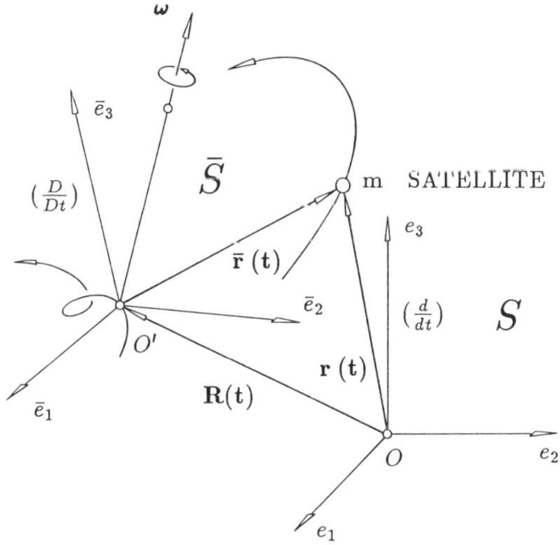

Figure 1: Orbit in frames S and $\bar{S}$, moving with respect to each other

Thus when assuming Newtonian mechanics, the motion of a satellite in an inertial reference frame is described by equation (1). Corrections which have to be applied to this formulation in the case of post-Newtonian mechanics are discussed in chapter 5.

A very good and convenient approximation of inertiality with respect to translation is to place the centre O into the barycentre of the solar system. For the description of near Earth satellite orbits, moving around the Earth as the central body, the geocentre is generally more appropriate. A quasi-inertial system centred at the geocentre will naturally show a translatoric acceleration because of the Earth's motion around the barycentre of Earth and Moon and about the barycentre of the solar system. This effect can on the other hand be adequately taken into account as part of the third body perturbations.

Inertiality with respect to rotation is best achieved by an extragalactic radio-source catalogue system and somewhat less accurate by a star catalogue system such as the FK5. The tabulated right ascensions and declinations and, in the case of a star catalogue, the proper motions define the reference axes of such a conventional celestial reference system (CCRS), here called S_I, relative to the directions of the respective celestial objects. The axes are chosen in such a way that at a basic epoch (e.g. J2000.0) they coincide in optimal approximation with the mean equatorial frame defined by the mean celestial pole and the mean dynamical equinox. More detailed explanations of quasi-inertial reference frames are given in chapters 1, 2 and 3.

2.1.2. Slowly Moving Celestial Reference Frames

Slowly moving celestial frames are such frames which show small rotational motions with respect to the quasi-inertial reference frame axes. Reference frames of this type are (cf. figure 2):

(i) - the mean equatorial reference frame S_M at date t. Its motion with respect to the fixed catalogue frame S_I, called precession, is for numerical use approximated by a model, adopted by the IAU in 1976. The Z_M and X_M axes are directed towards the mean celestial pole at date and the mean vernal equinox at date γ_M, respectively.

(ii) - the true equatorial reference frame S_t at date t. It is defined by the true celestial pole and the true vernal equinox γ_t and differs from the mean equatorial frame S_M by the effect of nutation, which is also numerically approximated by a model (currently by the model of Wahr (1981)) adopted by the IAU in 1981 (see chapter 8). This model neglects diurnal motions of the Earth rotation vector and thus defines the celestial ephemeris pole relative to the mean equatorial system or, together with the precession model, relative to the space-fixed CCRS. The Z_t axis is thus in the direction of the celestial ephemeris pole (CEP) and the X_t axis towards the true vernal equinox γ_t as described by the nutation model.

(iii) - the orbital reference frame S_O. This frame which was used by some orbit analysis groups in the past (Veis, 1963), is an intermediate frame with the third axis oriented towards the celestial ephemeris pole and the first axis in the true equatorial plane separated from the true vernal equinox by precession and nutation in right ascension since the basic epoch. This frame is space-fixed with respect to rotations about the Z_O axis, whereas the $X_O - Y_O$ plane moves due to precession and nutation.

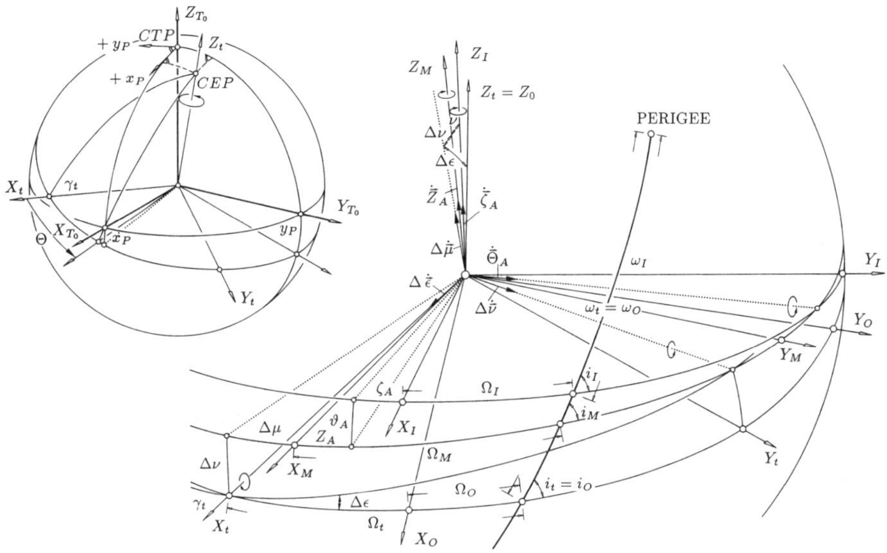

Figure 2: Relations between reference frames

2.1.3. Terrestrial Reference Frame

The frame of the conventional terrestrial reference system (CTRS), denoted as S_{T_O}, is materialized by a set of coordinates of station positions and motions, as explained in detail in chapter 7. Its Earth-fixed axes are conveniently chosen in such a way that Z_{T_O} at some epoch T_O (e.g. J1985.0) coincides with the mean position of the celestial ephemeris pole (CEP), the mean being defined over several Chandlerian periods, and X_{T_O} is close to the Greenwich meridian.

The time-variable connection between S_{T_O} and S_t is parametrized by the small polar motion angles x_p, y_p, defining the angular separation between Z_{T_O} and the Earth rotation axis, and Greenwich apparent sidereal time Θ. In numerical computations Θ is often substituted by $UT1$, which has a strict mathematical relation to Greenwich mean sidereal time Θ_O, and Θ_O is related to Θ by the equation of equinoxes according to the adopted nutation model.

2.1.4. Spacecraft Fixed Reference Frame

For the correction of tracking measurements to the centre of mass of a non-spherical spacecraft and for the transformation of onboard sensor measurements (e.g. gradiometer observations) to a terrestrial or inertial frame a vehicle-fixed rectangular coordinate frame is required. Most commonly the origin of this frame is the centre of mass of the spacecraft. Normally the axes are defined by the spacecraft designers as

(i) x_b-longitudinal (roll) axis directed towards the front part of the S/C,

(ii) y_b-lateral (pitch) axis and

(iii) z_b-vertical (yaw) axis.

2.2. TRANSFORMATIONS BETWEEN THE FRAMES

In the satellite orbit and parameter estimation process transformations between various coordinate frames are required for each observation instant and at each integration step. Position vectors, velocities and accelerations have to be expressed in more than one coordinate frame. To perform these transformations, a few geometric and kinematic relations are needed. The transition from a coordinate frame A with the orthonormal base $\mathbf{e}^A$ to a coordinate frame B with the orthonormal base $\mathbf{e}^B$ can be accomplished by three elementary rotations. Describing the relative orientation of the bases by Cardan angles α, β, γ as shown in figure 3, the transformation is realized by (1) a rotation about the e_3^A-axis with angle γ, (2) a rotation about the new e_2'-axis with angle β and (3) a rotation about the again new e_1''-axis with angle α. Mathematically each elementary rotation is represented by a rotation matrix $R_i(\alpha)$ about the i-axis for which, because of their orthogonality, holds

$$R_i^{-1}(\alpha) = R_i^T(\alpha) = R_i(-\alpha). \tag{3}$$

The transformation of the $\mathbf{e}^A$-base into the $\mathbf{e}^B$-base is thus described by

$$\begin{aligned} \mathbf{e}^B &= R_1(\alpha)R_2(\beta)R_3(\gamma)\mathbf{e}^A \\ &= R(\alpha,\beta,\gamma)\mathbf{e}^A. \end{aligned} \tag{4}$$

The rotations R_i are combined in this equation to the matrix $R(\alpha, \beta, \gamma)$, which when applied to the coordinates of any vector in the frame A, e.g. a position vector $\mathbf{r}^A$, performs the rotation into the frame B.

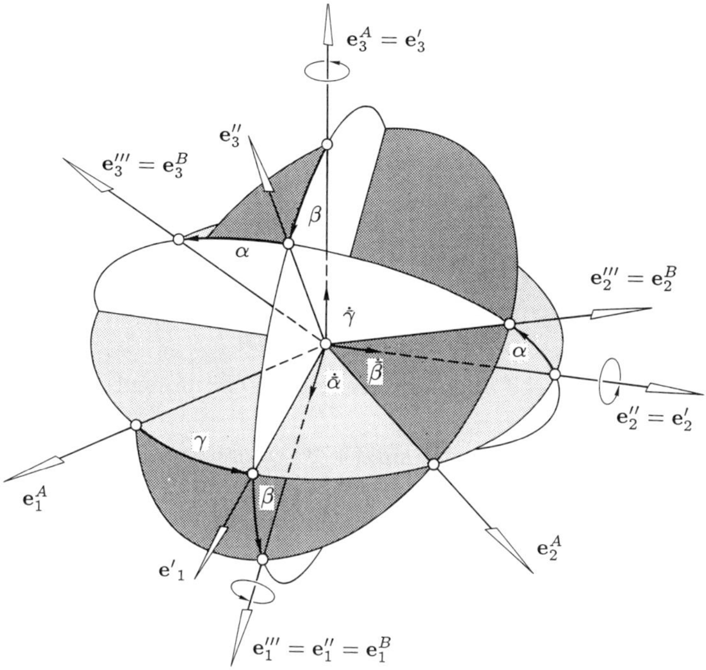

Figure 3: Definition of Cardan angles

For the further derivations in section 3 the angular velocity of a frame B with respect to a frame A is needed. According to figure 3 the angular velocity vector $\vec{\omega}(B/A)$ can be considered as being composed of a sequence of rotations with angular velocities $\dot{\vec{\alpha}}, \dot{\vec{\beta}}, \dot{\vec{\gamma}}$, i.e.

$$\vec{\omega}(B/A) = \dot{\vec{\alpha}} + \dot{\vec{\beta}} + \dot{\vec{\gamma}}. \quad (5)$$

It has to be noted that the individual angular velocity vectors are aligned along different axes and have therefore to be transformed by the rotation matrices R_i into the frame, in which $\vec{\omega}(B/A)$ is supposed to be represented. Supposing $\vec{\omega}(B/A)$ should be given in the system B, it is easy to verify from figure 3 that in an explicit form

$$\vec{\omega}^B(B/A) = \dot{\alpha}\mathbf{e}_1''' + \dot{\beta}R_1(\alpha)\mathbf{e}_2'' + \dot{\gamma}R_1(\alpha)R_2(\beta)\mathbf{e}_3'. \quad (6)$$

For the components of the rotation vector $\vec{\omega}(B/A)$ in the frame B one thus gets

$$\begin{aligned} \omega_1^B &= \dot{\alpha} - \dot{\gamma}\sin\beta \\ \omega_2^B &= \dot{\gamma}\sin\alpha\cos\beta + \dot{\beta}\cos\alpha \\ \omega_3^B &= \dot{\gamma}\cos\alpha\cos\beta - \dot{\beta}\sin\alpha. \end{aligned} \quad (7)$$

Introducing the operators

$$\begin{aligned} d/dt &: \text{ time derivative in } A \\ D/Dt &: \text{ time derivative in } B \end{aligned}$$

and, taking into account the following relation between the operators (Goldstein, 1970),

$$\frac{d}{dt} = \frac{D}{Dt} + \vec{\omega}(B/A)\times \tag{8}$$

and thus

$$\frac{d\vec{\omega}(B/A)}{dt} = \frac{D\vec{\omega}(B/A)}{Dt} \tag{9}$$

we finally obtain for the angular acceleration components

$$\begin{aligned} \frac{D\omega_1^B}{Dt} &= \dot{\omega}_1^B = \ddot{\alpha} - (\ddot{\gamma}\sin\beta + \dot{\beta}\dot{\gamma}\cos\beta) \\ \frac{D\omega_2^B}{Dt} &= \dot{\omega}_2^B = \dot{\alpha}\omega_3^B + (\ddot{\gamma}\cos\beta - \dot{\beta}\dot{\gamma}\sin\beta)\sin\alpha + \ddot{\beta}\cos\alpha \\ \frac{D\omega_3^B}{Dt} &= \dot{\omega}_3^B = -\dot{\alpha}\omega_2^B + (\ddot{\gamma}\cos\beta - \dot{\beta}\dot{\gamma}\sin\beta)\cos\alpha - \ddot{\beta}\sin\alpha \end{aligned} \tag{10}$$

With the relations (4) - (10) it is now easy to perform transformations between the various reference frames discussed in section 2.1 and to derive the angular velocity and angular acceleration vector components. What are needed are the rotation angles for precession, nutation and the orientation of the Earth fixed frame with respect to the true equatorial frame and for the orientation of the satellite body fixed frame with respect to any of the other frames plus their first and second derivative with respect to time.

(i) - The precession is given by the three angles ζ_A, ϑ_A and z_A, defined by the IAU (1976) system of constants. The values of these angles may be found, for example, in (Lieske 1979).

(ii) - The nutation can be described by the three rotation angles $\Delta\epsilon, \Delta\nu = \Delta\psi\sin\epsilon, \Delta\mu = \Delta\psi\cos\epsilon$, where $\Delta\epsilon, \Delta\nu, \Delta\mu$ are the nutation components in obliquity, declination and right ascension respectively. Expressions for the mean obliquity of date ϵ, the nutation parameters $\Delta\psi$ and $\Delta\epsilon$, using the IAU (1976) system of constants and the 1980 IAU theory of nutation, can be found in (Kaplan, 1981).

(iii) - The angle Θ, that is the Greenwich apparent sideral time, and the pole coordinates x_p, y_p connect the terrestrial reference frame S_{T_o} with the true equatorial frame S_t. The computation of Θ involves a mixture of conventional expressions to account for the mean rate of the Earth's rotation (including precession and nutation effects) and an empirical term to account for irregular variations. Both the irregular variations (e.g. $UT1 - UTC$) and the pole coordinates are derived from measurements and are disseminated by services such as BIH, USNO and in future IERS. Expressions for the computation of the mean rotation defined by the Greenwich mean sideral time in the IAU (1976) system of constants are given in (Kaplan, 1981).

(iv) - Finally the orientation of the satellite body fixed frame with respect to any of the other frames can be described by the angles α, δ, φ where α, δ are the longitude and latitude angle of the satellite roll axis with respect to the other frame and φ is the rotation angle about the roll axis in the $y_B - z_B$ plane.

With this notation of the rotation angles the transformation of the orthonormal base vectors from one system to the next is obtained through the rotations as given in table 1.

Frame Designation	To Base	Rotation Matrix	Matrix Designation	From Base
S_I	$\mathbf{e}_I =$	I		$\mathbf{e}_I$
S_M	$\mathbf{e}_M =$	$R_3(-z_A)R_2(\vartheta_A)R_3(-\zeta_A)$	P	$\mathbf{e}_I$
S_t	$\mathbf{e}_t =$	$R_1(-\Delta\epsilon)R_2(\Delta\nu)R_3(-\Delta\mu)$	N	$\mathbf{e}_M$
S_o	$\mathbf{e}_o =$	$R_3(\zeta_A + z_A + \Delta\mu)$	O	$\mathbf{e}_t$
S_b	$\mathbf{e}_b =$	$R_1(\varphi)R_2(-\delta)R_3(\alpha)$	B	$\mathbf{e}_t$
S_{T_o}	$\mathbf{e}_{T_o} =$	$R_2(-x_p)R_1(-y_p)R_3(\theta)$	S	$\mathbf{e}_t$
	$\mathbf{e}_{T_o} =$	SNP	R	$\mathbf{e}_I$
	$\mathbf{e}_I =$	$P^T N^T S^T$	R^T	$\mathbf{e}_{T_o}$

Table 1:Rotations between various orthonormal bases $\Delta\mu = \Delta\psi \cos\epsilon$; $\Delta\nu = \Delta\psi \sin\epsilon$

With the first and second time derivatives of the rotation angles as derivable without any approximations from the expressions given in (Lieske 1979, Kaplan 1981) and the notation $R_i'(\alpha) = dR_i(\alpha)/d\alpha$ it is possible to express the angular velocity and angular acceleration vectors between some of the considered reference frames in the explicit compact form as given in table 2. These vectors are needed for the computation of the apparent forces when expressing the motion equation in one of the moving frames (cf. equation (2)). The vector components are obtained by performing the matrix multiplications on the right hand sides.

Angular velocity between frames	Represented in frame; Time derivative designation	Designation of vectors	
S_M/S_I	S_M	$\vec{\omega}_p =$	$-R_3(-z_A)R_2(\vartheta_A)\dot{\zeta}_A\mathbf{e}'_3 + R_3(-z_A)\dot{\vartheta}_A\mathbf{e}''_2 - \dot{z}_A\mathbf{e}'''_3$
	$\tilde{D}/\tilde{D}t$	$\tilde{D}\vec{\omega}_p/\tilde{D}t =$	$-R'_3(-z_A)R_2(\vartheta_A)\dot{z}_A\dot{\zeta}_A\mathbf{e}'_3 - R_3(-z_A)R'_2(\vartheta_A)\dot{\vartheta}_A\dot{\zeta}_A\mathbf{e}'_3$
			$-R_3(-z_A)R_2(\vartheta_A)\ddot{\zeta}_A\mathbf{e}'_3 + R'_3(-z_A)\dot{z}_A\dot{\vartheta}_A\mathbf{e}''_2$
			$+R_3(-z_A)\ddot{\vartheta}_A\mathbf{e}''_2 - \ddot{z}_A\mathbf{e}'''_3$
S_t/S_M	S_t	$\vec{\omega}_N =$	$-R_1(-\Delta\epsilon)R_2(\Delta\nu)\Delta\dot{\mu}\mathbf{e}'_3 + R_1(-\Delta\epsilon)\Delta\dot{\nu}\mathbf{e}''_2 - \Delta\dot{\epsilon}\mathbf{e}'''_1$
	D/Dt	$D\vec{\omega}_N/Dt =$	$-R'_1(-\Delta\epsilon)R_2(\Delta\nu)\Delta\dot{\mu}\Delta\dot{\epsilon}\mathbf{e}'_3 - R_1(-\Delta\epsilon)R'_2(\Delta\nu)\Delta\dot{\mu}\Delta\dot{\nu}\mathbf{e}'_3$
			$-R_1(-\Delta\epsilon)R_2(\Delta\nu)\Delta\ddot{\mu}\mathbf{e}'_3 + R'_1(-\Delta\epsilon)\Delta\dot{\nu}\Delta\dot{\epsilon}\mathbf{e}''_2$
			$+R_1(-\Delta\epsilon)\Delta\ddot{\nu}\mathbf{e}''_2 - \Delta\ddot{\epsilon}\mathbf{e}'''_1$
S_{T_0}/S_t	S_{T_0}	$\vec{\omega}_S =$	$R_2(-x_p)R_1(-y_p)\dot{\theta}\mathbf{e}'_3 - R_2(-x_p)\dot{y}_p\mathbf{e}''_1 - \dot{x}_p\mathbf{e}'''_2$
	$\partial/\partial t$	$\partial\vec{\omega}_S/\partial t =$	$R'_2(-x_p)R_1(-y_p)\dot{x}_p\dot{\theta}\mathbf{e}'_3 + R_2(-x_p)R'_1(-y_p)\dot{y}_p\dot{\theta}\mathbf{e}'_3$
			$+R_2(-xp)R_1(-y_p)\ddot{\theta}\mathbf{e}'_3 - R'_2(-x_p)\dot{x}_p\dot{y}_p\mathbf{e}''_1$
			$-R_2(-x_p)\ddot{y}_p\mathbf{e}''_1 - \ddot{x}_p\mathbf{e}'''_2$
S_M/S_I	S_M	$\vec{\omega}_M =$	$\vec{\omega}_P$
	$\tilde{D}/\tilde{D}t$	$\tilde{D}\vec{\omega}_M/\tilde{D}t =$	$\tilde{D}\vec{\omega}_P/\tilde{D}t$
S_t/S_I	S_t	$\vec{\omega}_t =$	$N\vec{\omega}_p + \vec{\omega}_N$
	D/Dt	$D\vec{\omega}_t/Dt =$	$(DN/Dt)\vec{\omega}_p + N(\tilde{D}\vec{\omega}_p/\tilde{D}t) + D\vec{\omega}_N/Dt$
S_{T_0}/S_I	S_{T_0}	$\vec{\omega}_{T_0} =$	$S\vec{\omega}_t + \vec{\omega}_S$
	$\partial/\partial t$	$\partial\vec{\omega}_{T_0}/\partial t =$	$(\partial S/\partial t)\vec{\omega}_t + S(D\vec{\omega}_t/Dt) + \partial\vec{\omega}_S/\partial t$

Table 2: Angular velocity and acceleration vectors between various reference frames
$\Delta\dot{\mu} = \Delta\dot{\psi}\cos\epsilon - \dot{\epsilon}\Delta\nu$; $\Delta\dot{\nu} = \Delta\dot{\psi}\sin\epsilon + \dot{\epsilon}\Delta\mu$; $\Delta\ddot{\mu} = (\Delta\ddot{\psi} + \dot{\epsilon}^2\Delta\psi)\cos\epsilon - (2\dot{\epsilon}\Delta\dot{\nu} + \ddot{\epsilon}\Delta\nu)$; $\Delta\ddot{\nu} = (\Delta\ddot{\psi} + \dot{\epsilon}^2\Delta\psi)\sin\epsilon + (2\dot{\epsilon}\Delta\dot{\mu} + \ddot{\epsilon}\Delta\mu)$

3. SATELLITE MOTION REPRESENTATION IN VARIOUS REFERENCE FRAMES

After an artificial Earth satellite has been separated from the carrier rocket, it begins orbiting the Earth. Taking the Newton-Euler formulation of the problem of motion, the satellite centre of mass motion satisfies in a quasi-inertial reference frame S_I the equation (1), i.e.

$$\frac{m_S d^2\mathbf{r}_I}{dt^2} = \mathbf{F}_I(t; \mathbf{r}_I, \frac{d\mathbf{r}_I}{dt}) \tag{11}$$

with $\mathbf{F}_I$ the composite set of forces which accelerate the satellite as seen from S_I. For a satellite orbiting in the near Earth space these forces result from interactions of the spacecraft with the primary body, other natural celestial bodies and the environmental field to which the satellite is exposed along its path. The latter non-gravitational forces, also often referred to as surface forces, are a function of the satellite size, mass and orientation. For the precise computation of geodetic satellite orbits the total acceleration vector $\mathbf{F}/m_S$ has to include the components

$$\mathbf{F}_S/m = \mathbf{a} = \mathbf{a}_K + \mathbf{a}_R + \mathbf{a}_B + \mathbf{a}_E + \mathbf{a}_O + \mathbf{a}_D + \mathbf{a}_S + \mathbf{a}_A \tag{12}$$

the designations of which are given in table 3.

Source of acceleration	Designation in equ.(12)		Magnitude of acceleration [m/s] STARLETTE	AJISAI	LAGEOS	GPS
		semimajor axis (km)	7337	7869	12266	26559
		area/mass ratio (m^2/kg)	$9,6 \cdot 10^{-4}$	$5,3 \cdot 10^{-3}$	$6,9 \cdot 10^{-4}$	$2,0 \cdot 10^{-2}$
Kepler Term	$\mid a_K \mid$		$7,4$	$6,4$	$2,6$	$0,6$
C_{20}	$\mid a_R \mid$		$8 \cdot 10^{-3}$	$6 \cdot 10^{-3}$	$2 \cdot 10^{-3}$	$5 \cdot 10^{-5}$
other harmonics			$1 \cdot 10^{-4}$	$9 \cdot 10^{-5}$	$5 \cdot 10^{-6}$	$3 \cdot 10^{-7}$
Third body perturbations	$\mid a_B \mid$		$1 \cdot 10^{-6}$	$1 - 2 \cdot 10^{-6}$	$2 \cdot 10^{-6}$	$5 \cdot 10^{-6}$
Earth tides	$\mid a_E \mid$		$2 \cdot 10^{-7}$	$1 - 2 \cdot 10^{-7}$	$3 \cdot 10^{-8}$	$1 \cdot 10^{-9}$
Ocean tides	$\mid a_O \mid$		$3 \cdot 10^{-8}$	$2 \cdot 10^{-8}$	$2 \cdot 10^{-9}$	$1 \cdot 10^{-9}$
Atmospheric drag	$\mid a_D \mid$		$1 - 2 \cdot 10^{-10}$	$1 - 2 \cdot 10^{-10}$	$3 \cdot 10^{-11}$ (empirical)	0
Solar radiation pressure	$\mid a_S \mid$		$5 \cdot 10^{-9}$	$5 \cdot 10^{-8}$	$4 \cdot 10^{-9}$	$1 \cdot 10^{-7}$
Albedo pressure	$\mid a_A \mid$		$5 \cdot 10^{-10}$	$8 \cdot 10^{-9}$	$7 \cdot 10^{-11}$	$1 \cdot 10^{-9}$

Table 3: Accelerations on some geodetic satellites in m/sec^2

This table also includes for intercomparison the magnitude of the various accelerations for those four satellites which are currently most intensively tracked to derive geodetic and geodynamic parameters and which orbit at quite different altitudes: STARLETTE (700 km), AJISAI (1200 km), LAGEOS (6000 km), GPS (20000 km). It is obvious that the Earth geopotential U, conveniently expressed in S_{T_O} by a spherical harmonic expansion as (Heiskanen and Moritz, 1967)

$$U = \frac{GM}{r}[1 + \sum_{l=2}^{\infty} \sum_{m=0}^{l} (\frac{a_E}{r})^l P_{lm}(\sin\varphi)(C_{lm}\cos m\lambda + S_{lm}\sin m\lambda)] \tag{13}$$

with G the gravitational constant, M the mass and a_E the mean equatorial radius of the Earth and the C_{lm}, S_{lm} the Stokes coefficients, produces the largest accelerations on these satellites.

$$\mathbf{a}_K + \mathbf{a}_R = \nabla U_o + \nabla(U - U_o) \tag{14}$$

with $U_o = GM/r$ the zero degree term, corresponds to the potential of a spherically symmetric Earth. Expressions to compute the other accelerations can be found in several textbooks on satellite dynamics and satellite geodesy (e.g. King-Hele, 1964, Milani et al., 1987).

Now, given initial epoch (t_o) values for the spacecraft position $\mathbf{r}_I(t_o)$ and velocity $\dot{\mathbf{r}}_I(t_o)$ and all the forces $\mathbf{F}_I$ which accelerate the satellite, all expressed in the frame S_I, the solution of the second-order differential equation (11) yields the orbital motion representation in S_I in the form

position vectors $\mathbf{r}_I(t)$
velocity vectors $\dot{\mathbf{r}}_I(t)$.

A complete analytical solution of these equations has not yet been achieved, and historically two approaches have been used. In the first approach the acceleration model is limited such that an analytical solution is possible which provides the widest possible approximation of the real solution and which can serve as a basis of a perturbation theory. In the second approach, numerical integration techniques are applied to directly integrate the differential equations (11) in rectangular coordinates. In this approach referred to in the literature as Cowell's method the entire acceleration model can be included in the equations of motion.

3.1. UNPERTURBED SATELLITE MOTION

Limiting the entire acceleration field to the part $\mathbf{a}_K$, which means to assume that the satellite is only accelerated by a spherically symmetric Earth, leads to

$$\frac{d\mathbf{r}_I^2}{dt^2} = \mathbf{a}_K = \nabla U_o = -\frac{GM}{r^3}\mathbf{r}_I. \tag{15}$$

These are the differential equations of the central force problem, which are directly integrable. The solution of equation (15) can be obtained by straightforward vector operations on the above equation, resulting in the well-known Keplerian motion. This motion is specified by six integration constants. The solution of the central force problem is characterized by the relations (see e.g. Kaplan 1976, Schneider 1979)

$$\mathbf{r} \times \dot{\mathbf{r}} = L\mathbf{L}_o \qquad \dot{\mathbf{r}} \times \mathbf{L} = (GM\mathbf{r}/r) + P\mathbf{P}_o$$

(angular momentum integral) (Laplace integral)

(16)

$$r = (L^2/GM)/(1 + (P/GM)\cos v) \qquad E - e\sin E = M_o + n(t - t_o)$$

(general polar equation of conic section) (Kepler equation)

In these equations the quantities $\mathbf{L}_o, \mathbf{P}_o, L, P$ and t_o play the role of integration constants. Being concerned here with elliptical motion, $\mathbf{L}_o$ is a unit vector normal to the orbital plane (along the angular momentum vector $\mathbf{L}$) and $\mathbf{P}_o$ is directed along the apsidal line towards the perigee ($r = r_{min}$).

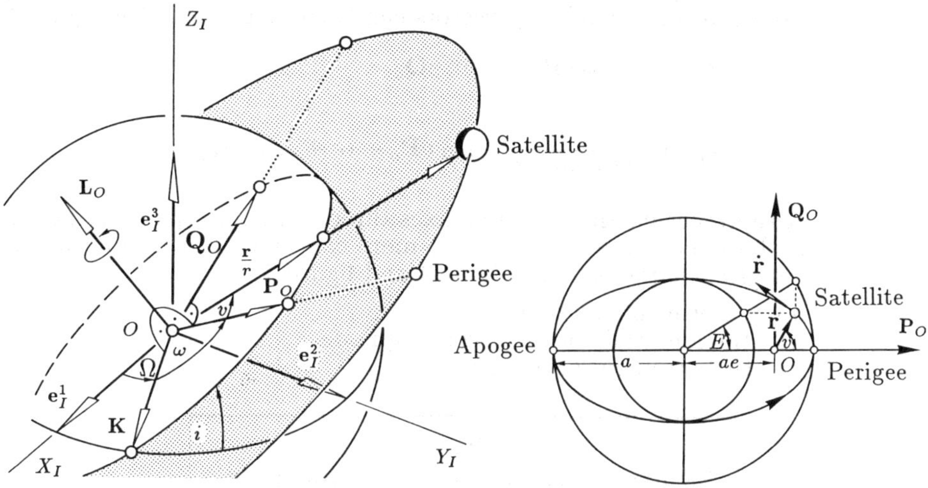

Figure 4: Specification of orbit orientation

These orthogonal vectors describe the orientation (three constants) of the orbit with respect to the $\mathbf{e}_I$-base (see figure 4). L and P are measures for the size and form of the orbit and can be replaced by the semi-major axis a and the eccentricity e, because for an elliptical orbit

$$r = \frac{a(1-e^2)}{1+e\cos v} \tag{17}$$

and thus $e = P/GM$ and $a = (L^2/GM)/(1-e^2)$. The last constant is t_o, a reference epoch to time-tag the position of the satellite in the orbit. Taking as t_o the time of perigee passage t_p, that is the point where also the true anomaly v and the eccentric anomaly E (figure 4) are zero, $M_o = 0$ and the Kepler equation reads

$$E - e\sin E = n(t - t_p) = M(t). \tag{18}$$

M is the mean anomaly and n the mean motion, which is related to the semi-major axis a through Kepler's third law

$$n^2a^3 = GM. \tag{19}$$

Finally the relation between the true anomaly v and the mean anomaly M is obtained via the eccentric anomaly E through

$$tan(v/2) = (\frac{1+e}{1-e})^{1/2} tan(E/2) \tag{20}$$

and the Kepler equation.

In the orbital plane system S_p with the orthonormal vectors $\mathbf{L}_o, \mathbf{P}_o, \mathbf{Q}_o$ the satellite position and velocity vectors can then be given as (see figure 4)

$$\begin{aligned} \mathbf{r}_p &= r \cos v \mathbf{P}_o + r \sin v \mathbf{Q}_o \\ \dot{\mathbf{r}}_p &= \frac{na}{(1-e^2)^{1/2}}(-\sin v \mathbf{P}_o + (e + \cos v)\mathbf{Q}_o). \end{aligned} \tag{21}$$

Instead of representing the S_p frame axes directly in the $S_I(\mathbf{e}_I)$ frame, the orientation between these two frames is more conveniently accomplished by the three angles, characterized in table 4 and figure 4

i - the orbit inclination
Ω - the longitude of the ascending node
ω - the argument of perigee.

Element	**Counted counterclock– wise against**	**Range in degrees**
i	$\mathbf{e}_I^3$	$0 \leq i \leq 180°$
Ω	$\mathbf{e}_I^1$ *in the* $\mathbf{e}_I^1, \mathbf{e}_I^2$ – plane	$0 \leq \Omega \leq 360°$
ω	nodal line $\mathbf{K}$ in orbital plane	$0 \leq \omega \leq 360°$

Table 4: Orientation Elements

With these angles the transformation from the $\mathbf{e}_I$-base to the $\mathbf{e}_P$-base is achieved by

$$\mathbf{e}_P = R_3(\omega)R_1(i)R_3(\Omega)\mathbf{e}_I. \tag{22}$$

The Keplerian motion (21) is thus represented in the S_I frame as

$$\begin{aligned} \mathbf{r}_I &= R_3(-\Omega)R_1(-i)R_3(-\omega)\mathbf{r}_P(a,e,M) \\ \dot{\mathbf{r}}_I &= R_3(-\Omega)R_1(-i)R_3(-\omega)\dot{\mathbf{r}}_P(a,e,M) \end{aligned} \tag{23}$$

through the six Keplerian elements $a, e, i, \omega, \Omega, M$(or v), the first five elements of which are constant in time.

3.2. PERTURBED SATELLITE MOTION

In the real world the Keplerian motion, formally representable in S_I as

$$\mathbf{r}_K(t) = \mathbf{r}_K(t; u_i); \quad \dot{\mathbf{r}}_K(t) = \dot{\mathbf{r}}_K(t; u_i); \quad i = 1 \cdots 6 \tag{24}$$

with $u_1, u_2 \cdots u_6$ the six integration constants, is affected (perturbed) by all accelerations in equation (12) which exist in addition to the central force acceleration. To solve the general motion problem

$$\frac{d^2\mathbf{r}_I}{dt^2} = \mathbf{a}_I = \mathbf{a}_{K,I} + \mathbf{A}_I \tag{25}$$

with $\mathbf{a}_I$ the entire acceleration field and $\mathbf{A}_I = \mathbf{a}_I - \mathbf{a}_{K,I}$ the vector of perturbing accelerations, the parametrization used in the solution of the central force problem can still be maintained, but the u_i parameters have now to be considered as functions of time. The solution of the perturbed motion problem then formally reads

$$\mathbf{r}(t) = \mathbf{r}(t; u_i(t)); \quad \dot{\mathbf{r}}(t) = \dot{\mathbf{r}}(t; u_i(t)); \quad i = 1 \cdots 6. \tag{26}$$

In celestial mechanics this approach is called variation of constants of the unperturbed problem.

Indirectly one makes use here of what is called an osculating Keplerian orbit. The satellite state vector $(\mathbf{r}, \dot{\mathbf{r}})$ may be obtained at a time instant $t = t_o$ from an integration of equations (25). Assuming that the perturbing acceleration $\mathbf{A}$ would vanish for instants $t > t_o$, the satellite would continue to move in an unperturbed orbit specified by the initial conditions

$$\mathbf{r}(t_o), \dot{\mathbf{r}}(t_o) \text{or} \quad a(t_o), e(t_o), i(t_o), \omega(t_o), \Omega(t_o), M(t_o).$$

This orbit is called the osculating orbit and the related elements the osculating elements. This concept can successively be used to describe the real orbit by the envelope of a successive sequence of osculating orbits with the osculating elements $a(t_i), e(t_i) \cdots M(t_i)$.

From equation (26) it is obvious that the perturbed motion is known if the time functions $u_i(t)$ are known. To determine them, the formal solution (26) is introduced into the equation of motion (25). This results in a linear system for the $\dot{u}_i$ which when solved leads to a system of six ordinary first order differential equations for the $\dot{u}_i(t)$, the so-called perturbation equations. With the Keplerian orbital elements as variational constants and the Keplerian orbit as an intermediate orbit, these perturbation equations take the form of the Lagrange perturbation equations (LPE). When expressed in the quasi-inertial system S_I, they read (see e.g. Levallois and Kovalevsky, 1971, Schneider, 1979):

$$\begin{aligned}
\frac{da_I}{dt} &= \frac{2}{na_I} A_M, \\
\frac{de_I}{dt} &= \frac{1-e_I^2}{na_I^2 e_I} A_M - \frac{(1-e^2)^{1/2}}{na_I^2 e_I} A_\omega, \\
\frac{di_I}{dt} &= \frac{\cos i_I A_\omega - A_\Omega}{na_I^2 (1-e_I^2)^{1/2} \sin i_I}, \\
\frac{d\omega_I}{dt} &= -\frac{\cos i_I A_i}{na_I^2 (1-e_I^2)^{1/2} \sin i_I} + \frac{(1-e_I^2)^{1/2}}{na_I^2 e_I} A_e,
\end{aligned} \tag{27}$$

$$\frac{d\Omega_I}{dt} = \frac{A_i}{na_I^2(1-e_I^2)^{1/2}\sin i_I},$$
$$\frac{dM_I}{dt} = n - \frac{1-e_I^2}{na_I^2 e_I}A_e - \frac{2}{na_I}A_a.$$

In these equations the A_{u_i} are the dot products

$$A_{u_i} = \mathbf{A}_I \cdot \frac{\partial \mathbf{r}_I}{\partial u_i} \tag{28}$$

with $\mathbf{A}_I$ the perturbing acceleration represented in the S_I frame and u_i the Keplerian orbital elements. The partials are easily derivable from equations (23).

In the form of equations (27) or in the Gaussian form (see e.g. Kaula, 1966) the LPE's are valid for conservative and non- conservative forces. If the perturbing acceleration $\mathbf{A}$ can be represented as a gradient of a scalar potential S, one gets for the scalar products in (28)

$$\mathbf{A}_I \cdot \frac{\partial \mathbf{r}_I}{\partial u_i} = \nabla S_I \cdot \frac{\partial \mathbf{r}_I}{\partial u_i} = \frac{\partial S_I}{\partial u_i}. \tag{29}$$

With Kaula's development of the geopotential disturbing function S in terms of Keplerian elements (Kaula, 1966), the partials can easily be obtained and introduced in the right hand sides of equations (27).

3.3. CHANGE TO MOVING FRAMES

Up to this point we have discussed the generation of satellite orbit ephemerides in terms of Cartesian coordinates and Keplerian elements in the inertial reference frame S_I (basic epoch J2000.0) by integrating the equivalent differential equations (25) and (27). In both sets of differential equations the effective acceleration field needs to be given in the S_I frame. Designating R_m the rotation matrix which transforms a vector given in S_I into one of the moving reference frames in table 1, with $\vec{\omega}_m$ the corresponding angular velocity vector represented in the moving frame (table 2) and with D_m/Dt the time derivative in the same moving frame, one easily obtains with equations (4) and (8) the orbit ephemerides in rectangular coordinates in the moving frame as

$$\begin{aligned} \mathbf{r}_m &= R_m \mathbf{r}_I \\ \frac{D_m \mathbf{r}_m}{Dt} &= \frac{d\mathbf{r}_m}{dt} - \vec{\omega}_m \times \mathbf{r}_m \\ &= R_m\left(\frac{d\mathbf{r}_I}{dt} - \vec{\omega}_m \times \mathbf{r}_I\right). \end{aligned} \tag{30}$$

With equations (16) to (23) the satellite state vector $(\mathbf{r}_m(t_i), \dot{\mathbf{r}}_m(t_i))$ at time $t = t_i$ can be transferred into the equivalent set of osculating orbital elements $u_{i,m}(t_i) = (a(t_i), e(t_i), \cdots M(t_i))_m$. No simple accurate expression exists to directly relate the $u_{i,m}$ to the $u_{i,I}$ (see 3.3.2).

The orbit ephemerides in a moving frame are obtained directly when integrating the Newton-Euler motion equations or the Lagrange perturbation equations expressed in this moving frame. When the satellite data analysis is performed in one of the moving frames, this is the computationally more efficient approach because less transformation steps are required.

3.3.1. Dynamic Transformations

Using the notation of the previous section and applying the relation (8) again on the velocity (30), the motion equations in an arbitrary non-uniformly moving reference frame S_m read (see also equation (2))

$$\begin{aligned} D_m^2\mathbf{r}_m/Dt^2 &= \mathbf{a}_m - \vec{\omega}_m \times (\vec{\omega}_m \times \mathbf{r}_m) \\ &\quad -(D_m\vec{\omega}_m/Dt) \times \mathbf{r}_m - 2\vec{\omega}_m \times D_m\mathbf{r}_m/Dt. \end{aligned} \tag{31}$$

In this equation $\mathbf{a}_m$ is the acceleration field now expressed in the moving frame

$$\mathbf{a}_m(t;\mathbf{r}_m, D_m\mathbf{r}_m/Dt) = R_m\mathbf{a}_I(t;\mathbf{r}_I, dr_I/dt) = R_m d^2\mathbf{r}_I/dt^2 \tag{32}$$

and the additional accelerations on the right-hand side are the apparent forces accelerations. In equation (2) these forces were denoted as $\mathbf{C}_m, \mathbf{Z}_m, \mathbf{S}_m$. Having tabulated $\vec{\omega}_m$ and $\dot{\vec{\omega}}_m$ in the moving frame in advance, these additional accelerations can be added to the main acceleration field without much computational effort.

The main additional transformation which has to be performed is for the geopotential. In the Earth body fixed frame S_{T_O} the spherical harmonics expansion of the geopotential U is given by equation (13). If the Z_{T_O} axis and the principal axis of maximum inertia coincide, the $C_{2,1}, S_{2,1}$ coefficients vanish in S_{T_O}. Changing into any frame differentially rotated against S_{T_O}, one can still keep the original development (13) and express the potential coefficients in the rotated frame. In closed form this relation is given for any unnormalized (l,m)-term by (Ilk, 1983) as

$$\begin{aligned} C_{lm}^{new} &= C_{lm}^{old} + \Delta C_{lm} \\ S_{lm}^{new} &= S_{lm}^{old} + \Delta S_{lm} \end{aligned} \tag{33}$$

with

$$\begin{aligned} \Delta C_{lm} &= \frac{1}{2}\{[-(l+m+1)(l-m)S_{l,m+1} - (1-\delta_{Om})(1-\delta_{1m})S_{l,m-1}]\alpha_1 \\ &\quad + [(l+m+1)(l-m)C_{l,m+1} - (1-\delta_{Om})(1+\delta_{1m})C_{l,m-1}]\alpha_2 \\ &\quad + 2m\, S_{lm}\,\alpha_3\} \\ \Delta S_{lm} &= \frac{1-\delta_{om}}{2}\{[(l+m+1)(l-m)C_{l,m+1} + (1+\delta_{1m})C_{l,m-1}]\alpha_1 \\ &\quad + [(l+m+1)(l-m)S_{l,m+1} - (1-\delta_{1m})S_{l,m-1}]\alpha_2 \\ &\quad - 2m\, C_{lm}\,\alpha_3\} \end{aligned} \tag{34}$$

and α_i the infinitesimal rotations about the i-th axis. Applying (34) to transform from Z_{T_0} to $Z_t \equiv CEP$ ($\alpha_1 = y_P$, $\alpha_2 = x_P$) one gets for example for second degree coefficients

$$\begin{aligned} C_{2,0}^t &= C_{2,0}^{T_0} - 3y_P\, S_{2,1}^{T_0} + 3x_P\, C_{2,1}^{T_0} \\ C_{2,1}^t &= C_{2,1}^{T_0} - 2y_P\, S_{2,2}^{T_0} - x_P\, C_{2,0}^{T_0} + 2x_P\, C_{2,2}^{T_0} \\ S_{2,1}^t &= S_{2,1}^{T_0} + 2x_P\, S_{2,2}^{T_0} + y_P\, C_{2,0}^{T_0} + 2y_P\, C_{2,2}^{T_0} \\ C_{2,2}^t &= C_{2,2}^{T_0} - \frac{1}{2}x_P\, C_{2,1}^{T_0} - \frac{1}{2}y_P\, S_{2,1}^{T_0} \\ S_{2,2}^t &= S_{2,2}^{T_0} - \frac{1}{2}y_P\, C_{2,1}^{T_0} - \frac{1}{2}x_P\, S_{2,1}^{T_0}. \end{aligned} \tag{35}$$

Not considering rotational deformation effects in $C_{2,1}$, $S_{2,1}$ (Wahr, 1987) $C_{2,1}^{T_0}$, $S_{2,1}^{T_0}$ can be assumed to vanish and of the second degree terms only the $C_{2,1}^t$, $S_{2,1}^t$ become time-dependent in S_t. When transforming to the other moving frames and into the inertial frame S_I the precession and nutation angles will make all terms time-dependent. For high-accuracy requirements a number of other potential coefficients need to be transformed too.

3.3.2. Kinematic Transformations

Because of the invariance of a, e and M against system transformations, the variations in these elements are the same in the inertial System S_I and in any of the moving frames.
Thus

$$\frac{da}{dt} = \frac{D_m a}{Dt}; \ \frac{de}{dt} = \frac{D_m e}{Dt}; \ \frac{dM}{dt} = \frac{D_m M}{Dt}. \tag{36}$$

This is not the case for the angular quantities i, ω, Ω because of the motion of the equatorial plane and the equinox when changing to a moving frame. Various ways exist to derive this kinematic effect (Kozai and Kinoshita, 1973, Lambeck, 1973, Reigber, 1981). With the derivations given in (Reigber, 1981) one finds in an arbitrary moving frame S_m the relations

$$\begin{aligned} \frac{D_m i_m}{Dt} &= \frac{di_m}{dt} - \omega_m^1 \cos\Omega_m - \omega_m^2 \sin\Omega_m \\ \frac{D_m \Omega_m}{Dt} &= \frac{d\Omega_m}{dt} - \omega_m^3 + \operatorname{ctg} i_m(\omega_m^1 \sin\Omega_m - \omega_m^2 \cos\Omega_m) \\ \frac{D_m \omega_m}{Dt} &= \frac{d\omega_m}{dt} + \operatorname{cosec} i_m(\omega_m^2 \cos\Omega_m - \omega_m^1 \sin\Omega_m). \end{aligned} \tag{37}$$

In these equations the argument of perigee ω_m referred to the intersection of the orbit plane with the S_m system equator should not be mixed with the components ω_m^i of the rotation vector $\vec{\omega}_m$. With equations (36) and (37) one can directly write down the Lagrange equations which hold for an arbitrary moving reference frame S_m

$$
\begin{aligned}
\frac{D_m a_m}{Dt} &= \frac{da_m}{dt} \\
\frac{D_m e}{Dt} &= \frac{de_m}{dt} \\
\frac{D_m i_m}{Dt} &= \frac{di_m}{dt} - \omega_m^1 \cos\Omega_m - \omega_m^2 \sin\Omega_m \\
\frac{D_m \Omega_m}{Dt} &= \frac{d\Omega_m}{dt} - \omega_m^3 + \operatorname{ctg} i_m(\omega_m^1 \sin\Omega_m - \omega_m^2 \cos\Omega_m) \\
\frac{D_m \omega_m}{Dt} &= \frac{d\omega_m}{dt} + \operatorname{cosec} i_m(\omega_m^2 \cos\Omega_m - \omega_m^1 \sin\Omega_m) \\
\frac{D_m M}{Dt} &= \frac{dM_m}{dt}
\end{aligned}
\tag{38}
$$

in which the time derivates $du_{i,m}/dt$ with respect to S_I are introduced from equations (27) with a change of indices.

The geometric meaning of the kinematic part of equations (38) becomes clear when integrating the kinematic part only, which is the same as assuming unperturbed motion $(du_{i,m}/dt = 0)$.

Considering precession and nutation (index m replaced by t) one obtains from table 2 for the rotation vector components when assuming infinitesimal rotations

$$
\begin{aligned}
\omega_t^1 &\approx \vartheta_A\dot{\zeta}_A - z_A\dot{\vartheta}_A - \Delta\mu\dot{\vartheta}_A + \Delta\nu\dot{\zeta}_A + \Delta\nu\dot{z}_A + \Delta\nu\Delta\dot{\mu} - \Delta\dot{\epsilon} \\
\omega_t^2 &\approx \Delta\mu\vartheta_A\dot{\zeta}_A - \Delta\mu z_A\dot{\vartheta}_A + \vartheta_A z_A\dot{\zeta}_A + \dot{\vartheta}_A + \Delta\epsilon\dot{\zeta}_A \\
&\quad -\Delta\epsilon\dot{z}_A + \Delta\epsilon\Delta\dot{\mu} + \Delta\dot{\nu} \\
\omega_t^3 &\approx \Delta\nu\vartheta_A\dot{\zeta}_A - \Delta\nu z_A\dot{\vartheta}_A + \Delta\epsilon\dot{\vartheta}_A - \dot{\zeta}_A - \dot{z}_A - \Delta\dot{\mu} + \Delta\epsilon\Delta\dot{\nu}.
\end{aligned}
\tag{39}
$$

Integration over a time interval Δt (assuming $\dot{\vec{\omega}}_t = \mathbf{0}$)

$$
\int_0^{\Delta t} \vec{\omega}_t dt \cong \vec{\omega}_t\Delta t \Rightarrow \vec{\omega}_t\Delta t \cong \begin{pmatrix} -\Delta\epsilon \\ \Delta\nu + \vartheta_A \\ -\zeta_A - z_A - \Delta\mu \end{pmatrix}
\tag{40}
$$

leads with equations (37) to the following linear first order expressions for the kinematic changes for the true system of date S_t

$$
\begin{aligned}
\Delta i = i_t - i_I &= -(\omega_t^1\Delta t\cos\Omega_t + \omega_t^2\Delta t\sin\Omega_t) \\
&= \Delta\epsilon\cos\Omega_t - (\Delta\nu + \vartheta_A)\sin\Omega_t \\
\Delta\Omega = \Omega_t - \Omega_I &= +(\zeta_A + z_A + \Delta\mu) - \operatorname{ctg} i_t[\Delta\epsilon\sin\Omega_t + (\Delta\nu + \vartheta_A)\cos\Omega_t] \\
\Delta\omega = \omega_t - \omega_I &= \operatorname{cosec} i_t[(\Delta\nu + \vartheta_A)\cos\Omega_t + \Delta\epsilon\sin\Omega_t].
\end{aligned}
\tag{41}
$$

With $\vec{\omega}_S$ (table 2), polar motion and small variations in Θ can be similarly taken into account.

3.4. REFERENCE FRAMES USED IN PRACTICE

In principle there is no reason to give preference to one or the other reference frame as long as all transformations are properly taken into account and no approximations are introduced in the dynamic and kinematic transformations. This also shows up in the fact that different orbit analysis groups work with different reference frames, depending on their original philosophy used to set up the software systems.

Most groups work nowadays either (i) in the slowly moving true equatorial reference frame S_t, (ii) in an inertial reference frame S_R (very often called the true of reference date frame) whose axes at a reference epoch t_R at the beginning or in the middle of a satellite arc coincide with the true equatorial frame, (iii) in the quasi-inertial frame S_I whose axes at the basic epoch J2000.0 coincide with the mean equatorial frame. The advantage of solutions (i) and (ii) is that these frames are always near to the terrestrial frame rotated by θ. Thus the transformation of the geopotential is as simple as shown by equations (35). In case (i) the motion equations are given by (31) with $m = t$ and at each integration step the apparent forces accelerations have to be computed. With $\vec{\omega}_t$, $D\vec{\omega}_t/Dt$ being tabulated in advance this can be done efficiently and precisely (Balmino, 1974). In case (ii) all transformations are related to the reference date t_R. With (30) we get

$$\begin{aligned} \mathbf{r}(t_R) &= R(t_R, 2000.0)\mathbf{r}_I \\ &= N(t_R)\, P(t_R, 2000.0)\mathbf{r}_I \end{aligned} \quad (42)$$

and thus for the coordinate transformation from the S_R to the S_t frame

$$\mathbf{r}(t) = R(t, 2000.0)R^T(t_R, 2000.0)\mathbf{r}(t_R) \quad (43)$$

with $R(t_R, 2000.0)$ coming from tabulated files. At $t = t_R$, $RR^T = I$. Since the reference epoch for the alignment of the axes of S_R has been chosen close to the observation epochs, the time-dependent rotation between S_R and S_t will be small during the observed arc. In order to compute the corresponding angular velocity vector of S_t and its derivative with respect to time, infinitesimal rotations for precession and nutation (relative to S_R) can be applied if the time interval does not exceed a few weeks and if no extreme precision is required.

In the case of (iii) the geopotential is transformed with expressions of the form (34) directly into the inertial frame S_I and the motion equations being integrated are equations (25).

4. SATELLITE EPHEMERIS ESTIMATION

4.1. SATELLITE EPHEMERIS AND ACCESS TO A CCRS

Let us start with the following thoughts. Assume that the state vector $\mathbf{X}_S(t_o; \mathbf{r}_S(t_o); \dot{\mathbf{r}}_S(t_o))$ of a satellite at an epoch t_o has been obtained without errors in a space-fixed frame S_I from error-free observations around the epoch t_o and assume in addition that all accelerations acting on the satellite are perfectly known. Then, the integration (assuming no numerical errors) of the second order differential equations (25) in Cartesian coordinates or integration of the first order differential equations (27), with initial conditions $a(t_o) \cdots M(t_o)$ derived from $\mathbf{X}_S$, would provide, at any moment $t > t_o$, the real state of motion of the spacecraft represented in a non-rotating quasi-inertial reference frame. The computed orbit would exactly match the real perturbed motion of the satellite in space if Newtonian mechanics are valid. To check whether the orbits really match, one could from time to time perform perfect range and direction observations from a geocentrically perfectly known location on the Earth and compare observed with computed observations from the integrated orbits. If all assumptions made are valid, the vector equation (44) (relating the satellite's and observer's geocentric position through the measured topocentric position of the spacecraft) would be completely satisfied at any time. Thus, by assuming that all physical and mathematical models are perfect, the satellite ephemerides generated either in the form of Cartesian coordinates or Keplerian elements establish direct access to an inertial reference system even for a perturbed orbit and without using observations for times $t > t_o$. One osculating orbital plane or a fictitious mean plane could be selected as the reference plane of such a dynamically defined inertial reference system and all other osculating planes rotating relative to this plane could be precisely referred to it because all orbital perturbations are assumed to be known. In this ideal situation observations from the Earth to the target in space would primarily serve to observe the motion of the Earth in space and deformations of the Earth's surface.

Changing into the real world it is appropriate to take the geodynamics satellite LAGEOS as an example, which because of its altitude, size and mass is presently the least perturbed geodetic satellite (cf. table 3). Starting from a best determined LAGEOS state vector at epoch t_o and integrating the orbit with the best acceleration model which is presently available over a one-year time span, the difference between a laser-observed range and the corresponding computed range can be as large as 5 kilometres after one year. For LAGEOS this is mainly due to present limitations in our capability to model some of the non- conservative forces accelerations (Barlier et al., 1986), but also due to a mismodelling in the conservative forces accelerations producing together orbit position errors primarily in along-track direction. A best fit of the integrated orbit to the real orbit (represented e.g. by precise laser range observations) can only be achieved by using tracking data along the orbit and adjusting in a least squares estimation process a number of model and/or empirical parameters to compensate for the large deviations mentioned before. For such long arcs, orbital fits of the order of one metre and better are nowadays obtainable (Tapley et al., 1985). For monthly arcs *rms* orbital fits are not larger than 5 to 10 centimetres for LAGEOS. Such impressive orbital fit results of course do not necessarily imply that the adjusted orbit ephemerides are related with the same precision to the axes of the CCRS, defined at a basic epoch which differs from t_o at present by more than 10 years. This would only be the case if the acceleration model were known to a very high accuracy even for its time-dependent parts. For the yearly and monthly arcs mentioned it only means that with the a-priori geometric, dynamic and measurement models introduced into the orbit determination process and by adjusting a few model parameters the relation between the observer's geocentric position and the satellite's geocentric position in space can be maintained with the small differences mentioned before for the time period of the orbital arc length. Further explanations on this point are given in chapter 9.

In any case the former remarks show that with the present knowledge of the dynamic models one quickly loses the precise connection of the orbit ephemerides to the CCRS system axes. Tracking data for a satellite therefore not only serve to compute the initial state vector and kinematic parameters of the Earth but also to correct those acceleration model components to which the satellite is specifically sensitive. This is done in a combined orbit and parameter estimation process, the principles of which are described in the sequel.

4.2. PRINCIPLES OF ORBIT AND PARAMETER ESTIMATION

Dynamic methods for estimating orbital, kinematic and dynamic model parameters from satellite tracking data start from the basic vector equation

$$\vec{\varrho} = \mathbf{r}_S - \mathbf{R}_T \tag{44}$$

expressed in a reference frame $S(X_1, X_2, X_3)$ which the analyser considers to be most appropriate for the data reduction process. This equation connects (figure 5) the motion $\mathbf{R}_T(t)$ of the topocentre T (which can be a tracking station on the solid parts of the Earth surface, an altimeter footprint in the oceans or even another object in space) with the geocentric motion $\mathbf{r}_s(t)$ of the satellite through the topocentric position vector $\vec{\varrho}(t)$.

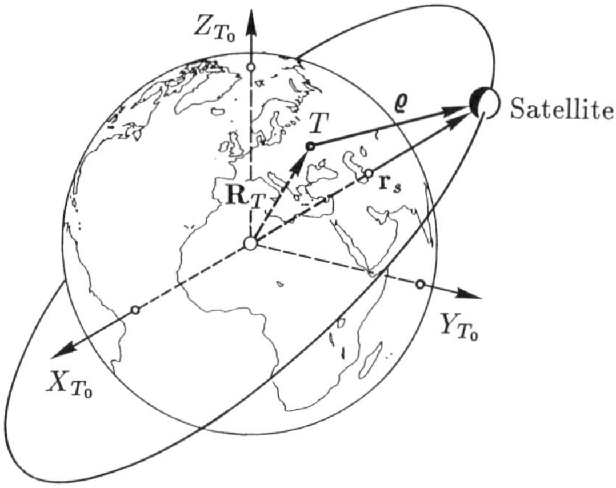

Figure 5: Station – satellite geometry

Single elements of the vector $\vec{\varrho}$ result from the measurement and data preprocessing process (e.g. ranges).

The vector $\mathbf{r}_S$ results from the theory of the satellite's orbital motion in space and the vector $\mathbf{R}_T$ from the theory of the motion of the topocentre on the deformable Earth and the motion of the Earth in space. These theories are based on models which are described by a great number of parameters.

Equation (44) can therefore be regarded as a generalized observation equation. The instantaneous position of the satellite is derived from an integration of the motion equations and is therefore a function of the state vector components at some initial epoch and the parameters describing the various force field components. The instantaneous position of the topocentre, when described in the same frame as the geocentric satellite position, is primarily a function of the three-dimensional coordinates in the conventional terrestrial reference frame and the Earth orientation parameters between these two frames. So we can write the observation equation in the form

$$q(t, \mathbf{P}) = q\Big(\mathbf{X}_S(t; \mathbf{Y}_S^o, \beta, \epsilon), \mathbf{X}_T(t; \mathbf{S}_T, \mathbf{o})\Big), \tag{45}$$

where

q: the observed quantity (e.g. range)

t: time

$\mathbf{P}$: the set of all parameters to be estimated

$\mathbf{X}_S/\mathbf{X}_T$: the satellite/the topocentre position coordinates at the observation epoch t in the frame S

$\mathbf{S}_T$: the station position coordinates in the terrestrial frame S_{T_O}

$\mathbf{Y}_S^o$: the satellite state vector containing the coordinates of position and velocity at epoch in the frame S

β: the set of parameters representing the geopotential

ϵ: the set of surface force model parameters (drag and solar radiation pressure)

$\mathbf{o}$: the set of Earth orientation parameters $(x_p, y_p, \Delta UT1)$ at the observation epoch t

Since this model is considerably underdetermined due to the number of orientation parameters being three times the number of observation epochs and an infinite number of geopotential parameters, it has to be approximated by a simpler model. This is achieved by neglecting higher order geopotential parameters, by expressing the Earth orientation parameters by a simple function of time (e.g. a polygon function using only values at finite time intervals) and by replacing unknown parameters by a priori known information, according to the respective purpose of the computation.

Thus the problem is converted into an overdetermined one and can be solved by least squares adjustment. Because of the non-linearity, the observation equations have to be linearized by using a priori approximate values $\mathbf{P}_R$ for all parameters to be determined:

$$\Delta q = q(t, \mathbf{P}) - q_c(t, \mathbf{P}_R) = \Big[\frac{\partial q}{\partial \mathbf{X}_S}, \frac{\partial q}{\partial \mathbf{X}_T}\Big]_{\mathbf{P}_R} \sum_i \Big(\frac{\partial}{\partial P_i}\Big)_{\mathbf{P}_R} \begin{bmatrix}\mathbf{X}_S \\ \mathbf{X}_T\end{bmatrix} \Delta P_i. \tag{46}$$

The parameters $\mathbf{Y}_S^o, \beta, \epsilon$ influence the motion of the satellite. The partial derivatives of $\mathbf{X}_S$ with respect to them are obtained by numerical integration of the so-called variational equations together with the orbit integration, which is necessary to get the approximate value $q_c(t, \mathbf{P}_R)$. The other partial derivatives in equation (46) can be computed directly.

REFERENCES

Balmino, G.,1974,*Celestial Mechanics* **10**, p.423

Barlier, F., Carpino, M., Farinella, P., Mignard, F., Milani, A., and Nobili, A.M.,1986,*Annales Geophysicae*,**4**, A, 3, p.193

Goldstein, H.,1970,*Classical Mechanics*, Addison-Wesley, Reading, Mass.

Heiskanen, W.A. and Moritz, H.,1967,*'Physical Geodesy'*, Freeman, San Francisco

Ilk, K.H.,1983,*Deutsche Geod. Komm.*, C, 288

Kaplan, G.H.,1981,*U.S. Nav. Obs.*, Circ. 163

Kaplan, M.H.,1976,*'Modern Spacecraft Dynamics & Control'*, John Wiley & Sons, New York

Kaula, W.M.,1966,*'Theory of Satellite Geodesy'*, Blaisdell, Waltham, Mass.

King-Hele, D.G.,1964,*'Theory of Satellite Orbits in an Atmosphere'*, Butterworks, London

Kozai, Y. and Kinoshita, H.,1973,*Celestial Mechanics* **7**, p.356

Lambeck, K.,1973,*Celestial Mechanics* **7**, p.139

Levallois, J.J. and Kovalevsky, J.,1971,*'Géodésie Générale'*, Tome IV, Géodésie Spatiale, Eyrolles, Paris

Lieske, J.H.,1979,*Astron. Astrophys.*,**73**, p.282

Milani, A., Nobili, A.M. and Farinella, P.,1987,*'Non-Gravitational Perturbations and Satellite Geodesy'*, Adam Hilger, Bristol

Reigber, Ch.,1981,*Bull. Geod.*,**55**, p.111

Schneider, M.,1979,*'Himmelsmechanik'*, Bibliographisches Institut, Mannheim

Tapley, B.D., Schutz, B.E. and Eanes, R.J.,1985,*J.Geophys.Res.*, **90**, p.9235

Veis, G.,1963,*SAO Spec. Rep.* No. 123

Wahr, J.,1981,*Geophys. J.R. Astron. Soc.* **64**, p.705

Wahr, J.,1987,*Geophys. J.R. Astron. Soc.* **88**, p.265

RELATIVISTIC THEORY OF CELESTIAL REFERENCE FRAMES

V. A. BRUMBERG
Institute of Applied Astronomy, 197042 Leningrad – USSR
S. M. KOPEJKIN
Sternberg State Astronomical Institute, 119899 Moscow – USSR

1. INTRODUCTION

At present, the general theory of relativity (GRT) should be considered as the necessary framework for the description of the gravitational field and the construction of astronomical reference frames. In contrast with Newtonian mechanics one cannot introduce in GRT the global Galilean (inertial) coordinates. The coordinates of GRT are in general not unique and equally admissible. This results in the intrusion of coordinate-dependent, unmeasurable quantities into astronomical ephemeris. For example, in ephemeris astronomy, in order to use the well-known numerical planetary and lunar theories of motion DE-200/LE-200 referred to the barycentric coordinate time it is necessary to take into account the type of the space-time coordinates inherent to these theories.

In this paper we suggest to use everywhere in astronomical applications the harmonic coordinates of GRT. Harmonic coordinates have no physical privileges but are convenient in the mathematical sense. A lot of problems of relativistic celestial mechanics were solved in these coordinates. Needless to say, one can use any coordinates appropriate for a specific problem. But to apply the relativistic theories in astronomical practice it is suitable that a certain type of GRT coordinates be recommended by the IAU. Such a recommendation will facilitate the comparison of various results and may help to avoid ambiguities in dealing with coordinate-dependent quantities.

The concept of reference frame is often differently used in physics and astronomy. This paper is addressed to astronomers and we shall follow the operational definitions given in this book and in Kovalevsky and Mueller (1981); Kovalevsky (1985). Therefore, a reference (coordinate) system is the primary mathematical construction whereas a reference frame results from the materialization of a reference system by means of some astronomical objects. A reference system is given in GRT by setting up the

J. Kovalevsky et al. (eds.), Reference Frames, 115–141.

appropriate metric form. To construct a reference frame is to prescribe some definite values of coordinates for reference astronomical objects. The reference frame is called dynamical if it is constructed using observations of the solar system bodies. If the reference objects represent distant fixed sources (quasars) the associated reference frame is called kinematical. Some researchers call this latter frame the geometrical one preserving the term kinematical for a frame based on the observations of stars (Podobed and Nesterov, 1982). There is no intrinsic distinction between these frames provided that all necessary relationships are taken into account (light propagation, motion of the solar system bodies, rotation of the Earth, stellar proper motions, reduction of observations, etc...).

So far, the most popular approach in the relativistic theory of astronomical reference frames is to construct the proper reference frame of a single observer. This approach had been applied in astrometry as early as in Mast and Strathdee (1959) and was further developed in text books such as Synge (1960), Moeller (1972) and Misner et al. (1973). The proper reference frame of a moving observer represents a tetrad of unit orthogonal vectors propagated along the wordline of the observer by Fermi-Walker transport. The time axis of this system is the wordline of the observer. Three space axes of the system are spacelike geodesics. Such coordinates are called Fermi normal coordinates. The most detailed mathematical description of this system taking into account possible acceleration and space rotation is given in Ni and Zimmermann (1978). In the absence of accelerations and space rotation such a reference frame transferred parallelly along the observer's geodesic represents a generalization of the inertial frame of the special theory of relativity. Application of such an approach to the geocentric reference frame involves difficulties because one cannot consider the Earth as a massless body and neglect its own gravitational field. To overcome these difficulties the metric tensor of the whole solar system is separated into a "local" part describing the gravitational field of the Earth and an "external" one due to all other bodies of the solar system. The Earth is supposed to move in a fictitious space-time defined by the "external" part of the metric tensor. In the vicinity of the Earth one may construct the Fermi normal system. The corresponding coordinate transformation is substituted into the full metric resulting in a generalization of the proper reference frame for the massive Earth. This approach is developed in Fukushima et al. (1986a), Bertotti (1986), Boucher (1986), Fujimoto and Grafarend (1986) and Ashby and Bertotti (1986). The splitting of the metric tensor into two parts is not free of ambiguities (the 'external' part may contain the terms depending on the Earth but not prevent from introducing the Fermi normal coordinates). The generalized Fermi normal coordinates as any other GRT coordinates are unmeasurable quantities. If one ignores tidal terms due to the external masses they become the coordinates of the Schwarzschild problem.

Another approach based on a finite linear transformation from barycentric to geocentric coordinates is exposed in Pavlov (1984a, b).

There is a different approach to relativistic discussion of observations. This approach is based on the possibility of local splitting of the space-time at the point of observation into the time axis (proper time) and the three dimensional space. This enables us to construct a local inertial frame in the infinitesimal vicinity of the point of observation.The different versions of such an approach are presented in Murray (1983), Brumberg (1986), Hellings (1986).

Considering that one may use, in general, any coordinates in constructing the reference frames, it is important to bring the solution of the dynamical problem to expressions for observable quantities.This involves, as a rule, the solution of the problem of light propagation in the same coordinates and the operational description of the observation technique. If a reference system is not adequate then both the solution of the dynamical problem and the transformation to the observational data will contain a number of extra terms caused only by the unsatisfactory choice of the reference system.These terms cancel out in the expressions of measurable quantities and the resulting relativistic effects turn out to be much smaller than the perturbations in the coordinate description of the dynamical problem.This is the case, for example, of recent solutions for the relativistic motions of the Moon (Lestrade and Chapront-Touzé, 1982) and of an Earth satellite (Martin *et al.*, 1985) elaborated actually in barycentric coordinates (the geocentric coordinate of the Moon or of a satellite is meant here as the difference of the barycentric coordinates of the corresponding body and the Earth). On the contrary, if a reference system is physically adequate, the solution of the dynamical problem may be presented in more compact form with insignificant changes when converting to measurable quantities. For instance, the geocentric reference system is regarded as physically adequate if the influence of the external masses manifests in such system not directly but only in the form of tidal terms.

In developing the relativistic theory of celestial reference frames we try to construct coordinate systems in an unambiguous and dynamically reliable manner. Our all basic reference systems satisfy two conditions :

(i)- they are constructed in the harmonic coordinates,

(ii)- their metric tensors represent the physically adequate solutions of the Einstein field equations. Our theory is based on the solutions of the GRT field equations developed in Anderson and Decanio (1975), Ehlers (1980), Thorne and Hartle (1985) and Kopejkin (1987).

The solutions of the Einstein equations for the whole solar system

(barycentric metric), for the vicinity of the centre of mass of the Earth (geocentric metric), for a terrestrial observatory (topocentric metric) and for an Earth orbiting observatory (satellite metric) determine the hierarchy of corresponding reference systems. In order to use these systems it is necessary to set up the relationships between them. This is achieved by asymptotic matching the global and local metrics (D'Eath, 1975; Kates, 1981; Damour, 1983, 1987; Kopejkin, 1987; see also Ashby and Bertotti, 1986). This matching enables us to :
(i)- set up the relationships between global and local coordinates,
(ii)- determine the coefficients of the local metric expansions,
(iii)- obtain in the global reference system the equations of motion of the body placed at the origin of the local system (barycentric equations of the Earth motion, geocentric equations of an Earth satellite motion).

All these reference systems, excluding the barycentric one, are elaborated in two versions, i.e. with or without space rotation. Harmonic coordinate conditions are imposed only on non-rotating systems. Space rotation is introduced in a pure kinematical way as in (Fukushima *et al.*, 1986a). We would preserve harmonic conditions in the case of rotating systems as well (Suen, 1986). But any metric in the harmonic rotation coordinates is rather complicated for using in application.

Principles of construction and mathematical formulation of the reference systems are given in Section 2 and 3. Section 4 is devoted to the transformations from one system to another. Section 5 deals with relativistic reduction of astrometric observations.

In this paper we use notations as in Misner *et al.* (1973). In particular, the greek indices run from 0 to 3; the latin indices run from 1 to 3; the capital latin indices A, B, C,... specify the bodies of the solar system (the major planets, the Moon, the Sun) :

$\eta_{\alpha\beta}=\eta^{\alpha\beta}$ is the Minkowski tensor ($\eta_{00}=-1$, $\eta_{oi}=0$, $\eta_{ij}=\delta_{ij}$);

δ_{ij} is the Kronecker symbol ($\delta_{ij}=0$ for $i\neq j$ and 1 for $i=j$);

e_{ijk} is the fully antisymmetric Levi-Civita symbol ($e_{123}=+1$);

$g_{\alpha\beta}$ is the metric tensor; $g^{\alpha\beta}$ are the elements of the inverse matrix of ($g_{\alpha\beta}$);

c is the light velocity;

G is the gravitational constant.

In our formulae one may not distinguish between upper and lower latin indices. The Einstein summation rule is used everywhere. This means the summation over every index occuring twice in any expression irrespective of the places of this repeated index.

2. MATHEMATICAL PRINCIPLES OF CONSTRUCTION OF THE COORDINATE SYSTEMS

The GRT space-time may be mathematically represented as the real, connected four-dimensional Riemanian manifold. At every point of the manifold the metric tensor is of hyperbolic type with the signature (-, +, +, +) and determines the metric relations with the aid of the metric :

$$ds^2 = g_{\alpha\beta} dx^\alpha dx^\beta \qquad (2.1)$$

Any given metric defines the gravitational field and a specific coordinate system. Our technique of constructing coordinate systems results from the determination of the metric coefficients from the Einstein field equations. These equations are solved in harmonic coordinates. Harmonic coordinates are defined by the explicit conditions :

$$\frac{\partial}{\partial x^\beta}\left(\sqrt{-g}\, g^{\alpha\beta}\right) = 0 \;,\; g = \det\left(g_{\alpha\beta}\right) \qquad (2.2)$$

The Einstein field equations take in these coordinates a rather simple form :

$$\eta^{\mu\nu} \frac{\partial^2 \gamma^{\alpha\beta}}{\partial x^\mu \partial x^\nu} = -\frac{16\pi G}{c^4} W^{\alpha\beta} \qquad (2.3)$$

$$W^{\alpha\beta} \equiv \Theta^{\alpha\beta} + \frac{c^4}{16\pi G}\left(\frac{\partial \gamma^{\alpha\mu}}{\partial x^\nu}\frac{\partial \gamma^{\beta\nu}}{\partial x^\mu} - \gamma^{\mu\nu}\frac{\partial^2 \gamma^{\alpha\beta}}{\partial x^\mu \partial x^\nu}\right) \qquad (2.4)$$

$$\gamma^{\alpha\beta} \equiv \eta^{\alpha\beta} - \sqrt{-g}\, g^{\alpha\beta} ; \; \Theta^{\alpha\beta} \equiv (-g)\left(T^{\alpha\beta} + t^{\alpha\beta}\right) \qquad (2.5)$$

Here, $T^{\alpha\beta}$ is the stress-energy tensor and $t^{\alpha\beta}$ is the Landau-Lifshitz pseudotensor.

Coordinate conditions (2.2) do not fix completely the coordinate system. In particular, there remains the arbitrariness in the choice of ;
(i)- the origin of the system,
(ii)- the wordline of the origin,
(iii)- the angular velocity of space rotation.
This arbitrariness is used in specifying the reference system.
The construction of relativistic reference systems should be based in our mind on the following principles :
(i)- Consistent and unambiguous mathematical formulation of reference

systems,

(ii)- Universality of systems, i.e. their validity for solving various dynamical and kinematical problems of celestial mechanics, astrometry, geodesy, etc.,

(iii)- Conversion of one system into another by means of simple and rigorous transformations (series or finite sum of analytical functions).

The aim of this paper is to elaborate reference systems for the solar system bodies. The following features of the solar system should be taken into account :

(i)- Slowness of the orbital motion: $v << c$, v being the characteristic velocity of the bodies,

(ii)- Weakness of the gravitational field everywhere (including the interior of the bodies): $U << c^2$, U being the Newtonian gravitational potential,

(iii)- Quasi-point structure of the bodies: $L << R$, L is the characteristic size of the bodies, R is the separation between them.

This enables us to introduce three small parameters :

(i)- $\varepsilon \sim v/c \sim (U/c^2)^{1/2} \sim (m/R)^{1/2} << 1$,

(ii)- $\eta \sim (m/L)^{1/2} << 1$,

(iii)- $\delta \sim L/R << 1$, M being the characteristic mass of the bodies, $m=GM/c^2$. These parameters are used in solving eqs (2.3) by successive approximations.

The solar system moves in the gravitational field of the Galaxy. For our purposes this field may be ignored and the solar system is presumed to be isolated.

Each body of the solar system has its own gravitational field characterized by the multipole moments. The gravitational field of other bodies may be regarded as external with respect to the body under consideration. The external field near the body has three characteristic length scales: inhomogeneity scale $\mathcal{L}_e$, radius of curvature $\mathcal{R}_e$ and time scale $\mathcal{T}_e$ for changes of curvature (Thorne and Hartle, 1985). Each body of the solar system is isolated in the sense :

$$L << \mathcal{L}_e, \quad L << \mathcal{R}_e, \quad L << c\mathcal{T}_e \tag{2.6}$$

For example, the external gravitational field for the Earth is caused mainly by the Sun and the Moon. It is easy to verify the validity of (2.6) for the Earth.

For an isolated body the space-time may be split up into three regions (Thorne and Hartle, 1985): the internal region $L < r < r_i$, a buffer region $r_i < r < r_o << \mathcal{L}_e, \mathcal{R}_e, c\mathcal{T}_e$ and the external region $r_o < r$. In the body's internal region its own gravitational field dominates, in the external region the gravitational field of other bodies dominates and in the buffer region both fields have comparable effects.

The Einstein field equations (2.3) are solved using the post-Newtonian approximation method and the multipole formalism (Ehlers, 1980; Damour, 1987; Thorne and Hartle, 1985). The occuring series result from the superposition of the series of three types. The series of the first type expanded in powers of ε, η, δ are due to the gravitational field of the body. The series of the second type being caused by the external bodies represent an expansion in powers of $r/\mathcal{R}_e$, $r/\mathcal{L}_e$, $r/(c.\mathcal{T}_e)$. The series of the third type describe the gravitational interaction between the body and the external sources and include different powers of both sets of these parameters.

Thus, the solution of Eqs (2.3) is looked up in the form :

$$\gamma^{\alpha\beta} = \varepsilon \underset{1}{\gamma}^{\alpha\beta} + \varepsilon^2 \underset{2}{\gamma}^{\alpha\beta} + \varepsilon^3 \underset{3}{\gamma}^{\alpha\beta} + \ldots \tag{2.7}$$

The odd powers are due to the rotation of the reference system or else to the gravitational radiation (Damour, 1983; Futamase and Schutz, 1983; Kopejkin, 1985; Grishchuk and Kopejkin, 1986). The right-hand side of (2.3) is also expanded in a similar way. Substituting these series into (2.3) and (2.2) yields :

$$\frac{\partial^2}{\partial x^s \partial x^s} \underset{n}{\gamma}^{\alpha\beta} = -\frac{16\pi G}{c^4} \underset{n-2}{W}^{\alpha\beta} + \frac{1}{c^2} \frac{\partial^2}{\partial t^2} \underset{n-2}{\gamma}^{\alpha\beta} \tag{2.8}$$

$$\frac{1}{c} \frac{\partial}{\partial t} \underset{n-1}{\gamma}^{\alpha 0} + \frac{\partial}{\partial x^i} \underset{n}{\gamma}^{\alpha i} = 0 \tag{2.9}$$

The solution of Eq. (2.8) may be presented in form :

$$\underset{n}{\gamma}^{\alpha\beta} = \underset{n}{p}^{\alpha\beta} + \underset{n}{f}^{\alpha\beta} \tag{2.10}$$

where the first term is a particular solution of the inhomogeneous equation (2.8) and the second term is a general solution of the homogeneous equation. The detailled exposition of the solving technique may be found in Anderson and Decanio, 1975; Ehlers (1980); Thorne and Hartle (1985); Suen (1986); Zhang (1986).

3. MATHEMATICAL DESCRIPTION OF THE ASTRONOMICAL REFERENCE SYSTEMS

3.1. BARYCENTRIC REFERENCE SYSTEM (BRS)

- Coordinates : $x^0 = ct$, $\underline{x} = (x^1, x^2, x^3)$
- Metric tensor : $g_{\alpha\beta}(t, \underline{x})$

In the internal region where the influence of the Galaxy may be ignored the solar system metric has the well known form (Fock, 1959) :

$$g_{00}\ (t,\underline{x}) = -1 + c^{-2}\, 2U + c^{-4}\, 2W + O\,(\varepsilon^5) \tag{3.1}$$

$$g_{0i}\ (t,\underline{x}) = -c^{-3}\, 4U^i + O\,(\varepsilon^5) \tag{3.2}$$

$$g_{ij}\ (t,\underline{x}) = \delta_{ij} + c^{-2}\, 2U\, \delta_{ij} + O\,(\varepsilon^4) \tag{3.3}$$

In the general case, the metric describes the solar system gravitational field both outside and inside the bodies. Outside the bodies the Newtonian potential U, the vector-potential U^i and the additional potential W have the values :

$$U = \sum_A \frac{GM_A}{r_A} + \frac{1}{2}\sum_A \frac{GI_{ij}^{(A)}}{r_A^3}\left(-\delta_{ij} + 3\frac{r_A^i\, r_A^j}{r_A^2}\right) + \ldots, \tag{3.4}$$

$$U^i = \sum_A \frac{GM_A}{r_A} v_A^i + \sum_A G\varepsilon_{ijk}\, \omega_A^j\, I_{km}^{(A)} \frac{r_A^m}{r_A^3} + \frac{1}{2}\sum_A GI_{km}^{(A)} v_A^i \frac{\partial^2 r_A^{-1}}{\partial r_A^k \partial r_A^m} + \ldots, \tag{3.5}$$

$$W = -U^2 + \frac{3}{2}\sum_A \frac{GM_A}{r_A} v_A^2 - \sum_A \sum_{B\neq A} \frac{G^2 M_A M_B}{r_A r_{AB}} + \frac{1}{2}\frac{\partial^2}{\partial t^2}\sum_A GM_A r_A + 3\sum_A G\varepsilon_{ijk}\, \omega_A^i v_A^k I_{jm}^{(A)} \frac{r_A^m}{r_A^3} +$$

$$+ \frac{1}{4}\frac{\partial^2}{\partial t^2}\sum_A GI_{km}^{(A)} \frac{\partial^2 r_A}{\partial r_A^k \partial r_A^m} + \frac{3}{4}\sum_A GI_{km}^{(A)} v_A^2 \frac{\partial^2 r_A^{-1}}{\partial r_A^k \partial r_A^m} +$$

$$+\sum_{A}\sum_{B\neq A} G^2 M_B I^{(A)}_{km}\left[\frac{r_A^m r_{AB}^k}{r_A^3 r_{AB}^3} - \frac{1}{2r_{AB}}\frac{\partial^2 r_A^{-1}}{\partial r_A^k \partial r_A^m} - \frac{1}{2}\left(\frac{1}{r_A}+\frac{1}{r_B}\right)\frac{\partial^2 r_{AB}^{-1}}{\partial r_A^k \partial r_A^m}\right] + \ldots \tag{3.6}$$

Here M_A is the mass of the body A, $x_A^i(t)$ and $v_A^i(t)$ are coordinates and components of velocity of its centre of mass, ω_A^i are the components of the angular velocity and $I_{ij}^{(A)}$ are quadrupole moments defined by :

$$I_{ij}^{(A)} = \int_A \rho_A (x^i - x_A^i)(x^j - x_A^j)\, d^3x \tag{3.7}$$

ρ_A being the mass density. In deriving (3.5) and (3.6) it is assumed that the velocity inside the bodies approximately satisfy the kinematical relation :

$$v^i = v_A^i + \varepsilon_{ijk}\,\omega_A^j (x^k - x_A^k) + \ldots \tag{3.8}$$

Besides this :

$$r_A^i = x^i - x_A^i,\quad r_{AB}^i = x_A^i - x_B^i,\quad r_A = (r_A^i r_A^i)^{1/2},\quad r_{AB} = (r_{AB}^i r_{AB}^i)^{1/2} \tag{3.9}$$

Metric (3.1)-(3.6) represents the well-known EIH (Einstein-Infeld-Hoffman) metric (in the post-Newtonian approximation their coordinate conditions coincide with harmonic ones) with addition (in the linear approximation) of quadrupole and spin terms. Adding such terms to the EIH metric, Thorne and Hartle (1985) used a different (inconsistent with EIH) mass density and neglected the symmetrical part of the spin terms. This introduces a slight discrepancy between their metric and (3.4)-(3.6). The origin of the coordinate system (3.1)-(3.6) is not fixed yet. The coordinates of the centre of mass of the solar system are defined by :

$$MX^i = c^{-2}\int \Theta^{00}(t,\underline{x})\,x^i d^3x \tag{3.10}$$

M being the total mass of the solar system bodies and the gravitational field :

$$M = c^{-2}\int \Theta^{00}(t,\underline{x})\,d^3x \tag{3.11}$$

Integration in (3.10), (3.11) is performed over the whole three-dimensional space. In neglecting the gravitational radiation, differentiation of (3.10), (3.11) yields (Fock, 1959) :

$$\frac{dM}{dt} = 0 \,, \quad M \frac{d^2 X^i}{dt^2} = 0 \tag{3.12}$$

Thus, the coordinates of the solar system barycenter are linear functions of time. They vanish, provided that :

$$\sum_A M_A x_A^i \left(1 + \frac{1}{2c^2} v_A^2 - \frac{1}{2c^2} \sum_{B \neq A} \frac{GM_B}{r_{AB}}\right) + \frac{1}{c^2} \sum_A \varepsilon_{ijk}\, \omega_A^\kappa\, I_{jm}^{(A)}\, v_A^m = 0 \tag{3.13}$$

where the pure quadrupole terms are omitted.

This relation where the pure quadrupole terms are omitted may be derived also by determining from (3.1)-(3.6) the Lagrangian of the equations of motion and obtaining the integrals of momentum and centre of mass (Brumberg, 1972).

The reference system determined by (3.1)-(3.6) with relation (3.13) is called barycentric reference system (BRS). Its origin coincides with the solar system barycenter and moves along the worldline with zero acceleration, i.e. along the geodesic. This system is non-rotating as there is no term of the form $c^{-1}\varepsilon_{ijk}\omega^j x^k$ in g_{oi}.
Such a term would describe the rotation of the coordinate vectors with respect to the vectors submitted to the Fermi-Walker transport along the worldline of the origin of the coordinate system.

3.2. GEOCENTRIC NON-ROTATING REFERENCE SYSTEM (GRS)

Coordinates : $w^o = cu$, $\underline{w} = (w^1, w^2, w^3)$
Metric tensor : $\hat{g}_{ab}(u,\underline{w})$

The geocentric metric includes the terms due to the gravitational field of the Earth and the external bodies (the Sun, the Moon, the major planets). For our purposes it is possible to neglect the mixed terms caused by the interaction of the terrestrial and external fields (but this cannot be done in solving dynamical problems such as lunar or satellite motion). Under this condition the metric may be presented in the form (Thorne and Hartle, 1985):

$$\hat{g}_{00}(u,\underline{w}) = -1 + c^{-2}(2\hat{U}_E + 2Q_i w^i + 3Q_{ij} w^i w^j + 5Q_{ijk} w^i w^j w^k + \ldots) + c^{-4}(2\hat{W}_E + \ldots) + \ldots \tag{3.14}$$

$$\hat{g}_{0i}(u,\underline{w}) = -c^{-3} 4\hat{U}_E^i - c^{-3}(4\varepsilon_{ijk} C_{jm} w^k w^m + \ldots) + \ldots \tag{3.15}$$

$$\hat{g}_{ij}(u,\underline{w}) = \delta_{ij} + c^{-2}(2\hat{U}_E + 2Q_k w^k + 3Q_{km} w^k w^m + 5Q_{kmn} w^k w^m w^n + \ldots)\,\delta_{ij} + \ldots \tag{3.16}$$

Letter E marks quantities relating to the Earth. Functions $\hat{U}_E$, $\hat{U}_E^i$, $\hat{W}_E$ as all other coefficients (3.14)-(3.16) may be derived from matching to the BRS (3.1)-(3.6). Their values are also known from the problem of a slow rotating, non spherical body (Fock, 1959; Brumberg, 1972) :

$$\hat{U}_E = \frac{G\hat{M}_E}{\rho} + \frac{1}{2\rho^3} G\hat{I}^{(E)}_{ij}\left(-\delta_{ij} + \frac{3}{\rho^2} w^i w^j\right) + \ldots \tag{3.17}$$

$$\hat{U}^i_E = G\varepsilon_{ijk}\,\hat{\omega}^j_E\,\hat{I}^{(E)}_{km}\frac{w^m}{\rho^3} + \ldots \tag{3.18}$$

$$\hat{W}_E = -\hat{U}^2_E + \ldots \tag{3.19}$$

where $\rho = |\underline{w}|$.

We have omitted the dipole term in (3.17) since it is possible to retain the centre of mass of the Earth in the origin of the coordinate system for all times (see below). This is done with the help of a special choice of the function Q_i in (3.14), (3.16), which characterizes the acceleration of the worldline of the origin of the chosen coordinate system with respect to the local inertial system of the external background space-time due to the external bodies.

By analogy with (3.10), (3.11) the mass and the coordinates of the centre of mass of the Earth are defined by :

$$\hat{M}_E = c^{-2}\int_E \hat{\theta}^{00}_E\, d^3w \tag{3.20}$$

$$\hat{M}_E w^i_E = c^{-2}\int_E \hat{\theta}^{00}_E\, w^i d^3w \tag{3.21}$$

with θ^{00}_E relating only to the gravitational field of the Earth. $\hat{M}_E$ being actually constant, differentiation of (3.21) enables to find the coordinate acceleration $\hat{a}^i_E$ of the centre of mass of the Earth :

$$\hat{M}_E\,\hat{a}^i_E = \hat{M}_E Q_i + \frac{15}{2}\left(\hat{I}^{(E)}_{jk} - \frac{1}{3}\delta_{jk}\hat{I}^{(E)}_{mm}\right)Q_{ijk} + \ldots \tag{3.22}$$

This acceleration should vanish if the centre of the Earth is to coincide with the origin of GRS for all times. Vanishing the right-hand side of (3.22) results in the estimation $Q_i \sim 5.10^{-12}$ cm/s^2. This quantity may be regarded at present as negligibly small leading to the conclusion that the centre of mass of the Earth moves on geodesic. It may be noted that in our theory the geodetic motion of the Earth is not an assumption but a theoretical consequence valid within the present level of accuracy.

The reference system (3.14)-(3.19) with the value of Q_i calculated from Eq. (3.22) under the condition $\hat{a}^i_E = 0$ is called geocentric reference system (GRS). All coefficients in (3.14)-(3.16) will be determined in Section 4 by matching to the BRS.

3.3. GEOCENTRIC ROTATING REFERENCE SYSTEM (GRS$^+$)

Coordinates : $y^0 = cu$, $\underline{y} = (y^1, y^2, y^3)$
Metric tensor : $g'_{\alpha\beta}(u, \underline{y})$

GRS$^+$ is obtained from GRS by application of a pure kinematical relation, describing a rigid three-dimensional rotation (Fukushima et al., 1986a) :

$$y^0 = w^0 \ , \ y^i = \mathcal{P}_{ik} w^k \tag{3.23}$$

The orthogonal matrix $\mathcal{P}$=SNP includes the matrices P of rotation for precession, N for nutation and S for Earth rotation taking into account polar motion (Mueller, 1981). Not going into details it is sufficient to keep in mind that :

$$\frac{d\mathcal{P}_{ik}}{du} = \varepsilon_{inj} \, \hat{\omega}^j_E \, \mathcal{P}_{nk} \tag{3.24}$$

GRS$^+$ metric has the form :

$$g'_{00}(u, \underline{y}) = -1 + c^{-2}\left[2U'_E + (\underline{\hat{\omega}}_E \wedge \underline{y})^2 + 3Q'_{ij} y^i y^j + 5Q'_{ijk} y^i y^j y^k + \ldots\right] + c^{-4}(2W'_E + \ldots) + \ldots, \tag{3.25}$$

$$g'_{0i}(u, \underline{y}) = c^{-1}\varepsilon_{ijk}\hat{\omega}^j_E y^k + c^{-3}(2\varepsilon_{ijk}\hat{\omega}^j_E y^k U'_E - 4U'^i_E - 4\varepsilon_{ijk} C'_{jm} y^k y^m + \ldots, \tag{3.26}$$

$$g'_{ij}(u, \underline{y}) = \delta_{ij} + c^{-2}(2U'_E + 3Q'_{km} y^k y^m + 5Q'_{kmn} y^k y^m y^n + \ldots)\,\delta_{ij} + \ldots \tag{3.27}$$

where U'_E, W'_E stand for $\hat{U}_E$, $\hat{W}_E$ expressed in new variables :

$$U'^i_E = \mathcal{P}_{ij} \hat{U}^j_E \ , \ Q'_{ij} = \mathcal{P}_{ik}\mathcal{P}_{jm} Q_{km} \ , \ Q'_{ijk} = \mathcal{P}_{im}\mathcal{P}_{jn}\mathcal{P}_{kl} Q_{mnl} \tag{3.28}$$

and similarly for C'^{ij}.

3.4. TOPOCENTRIC REFERENCE SYSTEM (TRS)

Coordinates : $z^0 = ct$, $\underline{z} = (z^1, z^2, z^3)$

Metric tensor : $\tilde{g}_{\alpha\beta}(\tau, \underline{z})$

TRS is constructed in the vicinity of an observer on the Earth surface. Thus, the TRS metric is due only to the external bodies (Earth, Sun, Moon, etc...) and has the form (Zhang, 1986) :

$$\tilde{g}_{00}(\tau,\underline{z}) = -1 + c^{-2}(2E_i z^i + 3E_{ij} z^i z^j + \ldots) + \ldots \tag{3.29}$$

$$\tilde{g}_{0i}(\tau,\underline{z}) = -c^{-3}\, 4\, \varepsilon_{ijk} H_{jm} z^k z^m + \ldots \tag{3.30}$$

$$\tilde{g}_{ij}(\tau,\underline{z}) = \delta_{ij} + c^{-2}(2E_k Z^k + 3E_{km} z^k z^m + \ldots) + \ldots \tag{3.31}$$

The point on the surface of the Earth does not move on the geodesic. Therefore, the term with acceleration E_i is significant. The coefficients of the TRS may be determined by matching to the GRS.

3.5. TOPOCENTRIC ROTATING REFERENCE SYSTEM (TRS+)

Coordinates : $\xi^0 = c\tau$, $\underline{\xi} = (\xi^1, \xi^2, \xi^3)$

Metric tensor : $g^*_{\alpha\beta}(\tau, \underline{\xi})$

TRS+ is obtained from TRS just in the same manner with the aid of the similar matrix $\mathcal{P}$ as GRS+ is obtained from GRS. Thus, resulting from the transformation :

$$\xi^0 = z^0,\ \xi^i = \tilde{\mathcal{P}}_{ik} z^k,\ \frac{d\tilde{\mathcal{P}}_{ik}}{d\tau} = \varepsilon_{inj}\, \tilde{\omega}^j_E\, \tilde{\mathcal{P}}_{nk}$$
$$\tilde{\mathcal{P}}_{ik}(\tau) = \tilde{\mathcal{P}}_{ik}(u) + \ldots,\quad \tilde{\omega}^k_E = \hat{\omega}^k_E + \ldots \tag{3.32}$$

the TRS+ metric has the form :

$$\overset{*}{g}_{00}(\tau,\xi) = -1 + c^{-2}\left[(\underline{\tilde{\omega}}_E \wedge \xi)^2 + 2\overset{*}{E}_i \xi^i + 3\overset{*}{E}_{ij} \xi^i \xi^j + \dots \right] + \dots \tag{3.33}$$

$$\overset{*}{g}_{0i}(\tau,\xi) = c^{-1} \varepsilon_{ijk} \tilde{\omega}^j_E \xi^k + c^{-3}(-4\varepsilon_{ijk} \overset{*}{H}_{jm} \xi^k \xi^m + 2\varepsilon_{ijk} \tilde{\omega}^j_E \overset{*}{E}_n \xi^k \xi^n + \dots) \dots \tag{3.34}$$

$$\overset{*}{g}_{ij}(\tau,\xi) = \delta_{ij} + c^{-2}(2\overset{*}{E}_k \xi^k + 3E^*_{km} \xi^k \xi^m + \dots)\delta_{ij} + \dots \tag{3.35}$$

Therewith :

$$\overset{*}{E}_i = \tilde{\mathcal{P}}_{ij} E_j \ , \ \overset{*}{E}_{ij} = \tilde{\mathcal{P}}_{ik} \tilde{\mathcal{P}}_{jm} E_{km} \tag{3.36}$$

and similar for H^*_{ij}.

3.6. SATELLITE REFERENCE SYSTEM (SRS)

SRS is essentially the same as TRS with the important difference that the satellite moves on geodesic around the Earth. Therefore, retaining the notations of Section 3.4 the SRS may be described by (3.29)-(3.31) with $E_i=0$.

3.7. SATELLITE ROTATING REFERENCE SYSTEM (SRS$^+$)

Retaining the notation of Section 3.5 the SRS$^+$ results from SRS by the transformation :

$$\xi^0 = z^0 \ , \ \xi^i = \tilde{\mathcal{P}}_{ik} z^k \ , \ \frac{d\tilde{\mathcal{P}}_{ik}}{d\tau} = \varepsilon_{inj} \tilde{\omega}^j_S \tilde{\mathcal{P}}_{nk} \tag{3.37}$$

where $\mathcal{P}_{ik}$ is an orthogonal matrix of space rotation of the satellite with angular velocity $\tilde{\omega}^1_S$. The SRS$^+$ metric is described by (3.33)-(3.36) with $E^*_i=0$ and replacing $\tilde{\omega}^i_E$ by $\tilde{\omega}^i_S$.

Evidently, one should use different notations for TRS and SRS if it is necessary to relate these systems directly.

4. RELATIONSHIPS BETWEEN REFERENCE SYSTEMS

Starting from the solution (3.1)-(3.6) in BRS it is possible to relate all systems introduced above. We consider in details three such relationships.

4.1. TRANSFORMATION FROM BRS TO GRS

This transformation is of the form :

$$u=t+c^{-2}\left(\frac{1}{2}v_E^2 t+A-v_E^k x^k\right)+c^{-4}\left[\begin{array}{l}\frac{3}{8}v_E^4 t-\frac{1}{2}v_E^2 v_E^k x^k+B+B^k(x^k-x_E^k)+\\ +B^{km}(x^k-x_E^k)(x^m-x_E^m)+\ldots\end{array}\right]+\ldots \tag{4.1}$$

$$w^i=x^i-x_E^i(t)+c^{-2}\left[\left(\frac{1}{2}v_E^i v_E^k+F^{ik}+D^{ik}\right)(x^k-x_E^k)+D^{ikm}(x^k-x_E^k)(x^m-x_E^m)\right]+\ldots \tag{4.2}$$

where functions A, B, B^i, B^{ij}, $F^{ij}(=-F^{ji})$, $D^{ij}(=D^{ji})$, $D^{ijk}(=D^{ikj})$ depend only on barycentric time t. These functions, the coefficients Q_{ij}, Q_{ijk}, C_{ij} of the GRS metric (3.14)-(3.16) and the acceleration of the geocentre $\underline{a}_E=d\underline{v}_E/dt$ in the BRS are derived by matching the BRS and the GRS metrics on the basis of the tensor relation :

$$g_{\alpha\beta}(t,\underline{x})=\hat{g}_{\mu\nu}(u,\underline{w})\frac{\partial w^\mu}{\partial x^\alpha}\frac{\partial w^\nu}{\partial x^\beta} \tag{4.3}$$

In matching (3.1)-(3.3) and (3.14)-(3.16) the potentials (3.4)-(3.6) are unambiguously separated into an internal part due to the Earth and an external one caused by the remaining bodies. For example :

$$U_E=U_E+\bar{U},\quad U^i=U_E^i+\bar{U}^i \tag{4.4}$$

where U_E, U^i_E are determined by (3.4), (3.5) for one term A=E and $\overline{U}$, $\overline{U^i}$ are determined by the sums in (3.4), (3.5) with A≠E. In the vicinity of the Earth functions $\overline{U}$, $\overline{U^i}$ are expanded in series of the type :

$$\overline{U}(\underline{x})=\overline{U}(\underline{x}_E)+\frac{\partial\overline{U}(\underline{x}_E)}{\partial x^i}(x^i-x_E^i)+\frac{1}{2}\frac{\partial^2\overline{U}(\underline{x}_E)}{\partial x^i\partial x^j}(x^i-x_E^i)(x^j-x_E^i)+\ldots$$

By equating in (4.3) the terms of the same form one has :

$$a_E^i=\frac{\partial\bar{U}(\underline{x}_E)}{\partial x^i}=-\sum_{A\neq E}\frac{GM_A}{r_{EA}^3}r_{EA}^i \tag{4.5}$$

$$\frac{dA}{dt} = a_E^k x_E^k - (a_E^k v_E^k)\, t - \bar{U}(\underline{x}_E) = \sum_{A\neq E} \frac{GM_A}{r_{EA}^3}\left[(v_E^k t - x_E^k) r_{EA}^k - r_{EA}^2 \right] \tag{4.6}$$

$$Q_{ij} = \frac{1}{3} \frac{\partial^2 \bar{U}(\underline{x}_E)}{\partial x^i \partial x^j} = \sum_{A\neq E} \frac{GM_A}{r_{EA}^5} (r_{EA}^i r_{EA}^j - \frac{1}{3} r_{EA}^2 \delta_{ij}) \tag{4.7}$$

$$Q_{ijk} = \frac{1}{15} \frac{\partial^3 \bar{U}(\underline{x}_E)}{\partial x^i \partial x^j \partial x^k} = \sum_{A\neq E} \frac{GM_A}{r_{EA}^5} \left(\frac{1}{5}\delta_{jk} r_{EA}^i + \frac{1}{5}\delta_{ki} r_{EA}^j + \frac{1}{5}\delta_{ij} r_{EA}^k - \frac{1}{r_{EA}^2} r_{EA}^i r_{EA}^j r_{EA}^k \right) \tag{4.8}$$

$$D^{ij} = \delta_{ij} \bar{U}(\underline{x}_E) = \delta_{ij} \sum_{A\neq E} \frac{GM_A}{r_{EA}} \tag{4.9}$$

$$D^{ijk} = \frac{1}{2}\left[\delta_{ij} \frac{\partial \bar{U}(\underline{x}_E)}{\partial x^k} + \delta_{ik} \frac{\partial \bar{U}(\underline{x}_E)}{\partial x^j} - \delta_{jk} \frac{\partial \bar{U}(\underline{x}_E)}{\partial x^i} \right] =$$

$$= \frac{1}{2} \sum_{A\neq E} \frac{GM_A}{r_{EA}^3} \left(\delta_{jk} r_{EA}^i - \delta_{ij} r_{EA}^k - \delta_{ik} r_{EA}^j \right) \tag{4.10}$$

$$B^i = 4 \bar{U}^i(\underline{x}_E) - 3 v_E^i \bar{U}(\underline{x}_E) = \sum_{A\neq E} \frac{GM_A}{r_{EA}} (4 v_A^i - 3 v_E^i) \tag{4.11}$$

$$B^{ij} = \frac{1}{2} \frac{dD^{ij}}{dt} + \frac{\partial \bar{U}^i(\underline{x}_E)}{\partial x^j} + \frac{\partial \bar{U}^j(\underline{x}_E)}{\partial x^i} - \frac{1}{2} v_E^i \frac{\partial \bar{U}(\underline{x}_E)}{\partial x^j} - \frac{1}{2} v_E^j \frac{\partial \bar{U}(\underline{x}_E)}{\partial x^i} =$$

$$= \sum_{A\neq E} \frac{GM_A}{r_{EA}^3}\left[(\frac{1}{2} v_E^i - v_A^i) r_{EA}^j + (\frac{1}{2} v_E^j - v_A^j) r_{EA}^i - \frac{1}{2}\delta_{ij} (v_E^k - v_A^k) r_{EA}^k \right] \tag{4.12}$$

$$\frac{dF^{ij}}{dt} = \frac{3}{2}\left[v_E^i \frac{\partial \bar{U}(\underline{x}_E)}{\partial x^j} - v_E^j \frac{\partial \bar{U}(\underline{x}_E)}{\partial x^i}\right] - 2\left[\frac{\partial \bar{U}^i(\underline{x}_E)}{\partial x^j} - \frac{\partial \bar{U}^j(\underline{x}_E)}{\partial x^i}\right] =$$

$$= \sum_{A \neq E} \frac{GM_A}{r_{EA}^3}\left[\frac{3}{2}\left(v_E^j r_{EA}^i - v_E^i r_{EA}^i\right) + 2\left(v_A^j r_{EA}^i - v_A^i r_{EA}^j\right)\right] \tag{4.13}$$

$$C_{ij} = \frac{1}{6}\varepsilon_{ikm}\left[v_E^k \frac{\partial^2 \bar{U}(\underline{x}_E)}{\partial x^j \partial x^m} - \frac{\partial^2 \bar{U}^k(\underline{x}_E)}{\partial x^j \partial x^m}\right] + \frac{1}{6}\varepsilon_{jkm}\left[v_E^k \frac{\partial^2 \bar{U}(\underline{x}_E)}{\partial x^i \partial x^m} - \frac{\partial^2 \bar{U}^k(\underline{x}_E)}{\partial x^i \partial x^m}\right] =$$

$$= \frac{1}{2}\sum_{A \neq E} \frac{GM_A}{r_{EA}^5}\left(\varepsilon_{ikm} r_{EA}^j + \varepsilon_{jkm} r_{EA}^j\right) r_{EA}^m \left(v_E^k - v_A^k\right) \tag{4.14}$$

and neglecting the terms of the order ε^2 :

$$\hat{M}_E = M_E \ , \quad \hat{I}_{ij}^{(E)} = I_{ij}^{(E)} \ , \quad \hat{\omega}_E^i = \omega_E^i \tag{4.15}$$

Thus the expansions (3.14)-(3.19) of the GRS metric are completely defined.

Matching of BRS and GRS was performed here up to the order ε^3 inclusively. Hence, function B in (4.1) remains undetermined and the barycentric acceleration (4.5) of the Earth includes only the Newtonian terms. To obtain B and the relativistic equations of motion of the Earth it is necessary to match the terms of the order ε^4 due to the function W in (3.6) (D'Eath, 1975). But this lies beyond the scope of the present chapter.

Let us consider the transformations (4.1)-(4.2) in more details. The coordinate timescales t for BRS and u for GRS stand in our theory for TDB and TDT respectively. Because we use the same type of GRT coordinates (harmonic ones) these timescales become unambiguously defined being related by (4.1). The transformation (4.2) of space coordinates is remarkable for being rigorous with respect to the differences $x^i - x^i_E$ (neglecting terms of the order ε^4). This is also a consequence of using the same type of GRT coordinates. Functions F^{ij} in (4.2) include the effect of the geodetic precession. Functions A and F^{ij} are determined only by their derivatives. Expressions (4.6) and (4.13) are still to be integrated in analytical or numerical form.

4.2. TRANSFORMATION FROM GRS+ TO TRS+

Transformation from the GRS+ metric to the TRS+ metric may be presented in the form:

$$\tau = u + c^{-2}\left(\frac{1}{2} v_T^2 u + \mathcal{A} - v_T^k w^k\right) + \ldots \tag{4.16}$$

$$\xi^i = y^i - y_T^i(u) + \ldots \tag{4.17}$$

It is assumed that the ground observatory generating the TRS+ has the coordinates $y^i{}_T$ in GRS+. With respect to GRS it has a velocity $\underline{v}_T$ and an acceleration $\underline{a}_T$:

$$\underline{v}_T = (\tilde{\underline{\omega}}_E \wedge \underline{w}_T) + \underline{v}_{TT} \tag{4.18}$$

$$\underline{a}_T = \left(\frac{d\tilde{\underline{\omega}}_E}{du} \wedge \underline{w}_T\right) + \left[\tilde{\underline{\omega}}_E \wedge \left(\tilde{\underline{\omega}}_E \wedge \underline{w}_T\right)\right] + 2\left(\tilde{\underline{\omega}}_E \wedge \underline{v}_{TT}\right) + \underline{a}_{TT} \tag{4.19}$$

Due to the deformations and other geophysical factors, the ground observatory may have in TRS+ the relative velocity $v^i{}_{TT}$ and the relative acceleration $a^i{}_{TT}$.

On the basis of the tensor relation involving $g'_{\alpha\beta}(\tilde{U},\underline{y})$ and $g^*{}_{\alpha\beta}(\tau,\underline{\xi})$ one has:

$$\frac{d\mathcal{A}}{du} = \left(a_T^k w_T^k\right) - \left(a_T^k v_T^k\right) u - U_E'\left(\underline{y}_T\right) - \frac{3}{2} Q'_{ij} y_T^i y_T^j - \frac{5}{2} Q'_{ijk} y_T^i y_T^j y_T^k + \ldots \tag{4.20}$$

$$E_i^* = -\tilde{\mathcal{P}}_{ik} a_T^k + \frac{\partial U'_E(\underline{y}_T)}{\partial y^i} + 3Q'_{ik} y_T^k + \frac{15}{2} Q'_{ijk} y_T^j y_T^k + \ldots \tag{4.21}$$

$$E_{ij}^* = \frac{1}{3} \frac{\partial^2 U'_E(\underline{y}_T)}{\partial y^i \partial y^j} + Q'_{ij} + 5Q'_{ijk} y_T^k + \ldots \tag{4.22}$$

Functions H_{ij} need not to be reproduced here in virtue of their smallness.

The timescale τ of TRS+ averaged over a number of ground observatories may serve for the determination of TAI (cf. Guinot, 1986).

4.3. TRANSFORMATION FROM GRS TO SRS

This transformation is of the form :

$$\tau = u + c^{-2}\left(\frac{1}{2}v_S^2 u + \mathcal{A} - v_S^k w^k\right) + \ldots \tag{4.23}$$

$$z^i = w^i - w_S^i(u) + c^{-2}\left[\left(\frac{1}{2}v_S^i v_S^k + \mathcal{F}^{ik} + \mathcal{D}^{ik}\right)\left(w^k - w_S^k\right) + \mathcal{D}^{ikm}\left(w^k - w_S^k\right)\left(w^m - w_S^m\right)\right] + \ldots \tag{4.24}$$

where $v^i_S = dw^i_S/du$ is the geocentric velocity of the satellite. The coefficients of these relations and the basic coefficient E_{ij} of the SRS metric are as follows :

$$a_S^i = \frac{\partial \hat{U}_E(w_S)}{\partial w^i} + 3\mathcal{Q}_{ik} w_S^k + \frac{15}{2}\mathcal{Q}_{ijk} w_S^j w_S^k + \ldots \tag{4.25}$$

$$\frac{d\mathcal{A}}{du} = (a_S^k w_S^k) - (a_S^k v_S^k)\, u - \hat{U}_E(w_S) - \frac{3}{2}\mathcal{Q}_{ij} w_S^i w_S^j + \ldots \tag{4.26}$$

$$E_{ij} = \frac{G\hat{M}_E}{\rho_S^3}\left(\frac{w_S^i w_S^j}{\rho_S^2} - \frac{1}{3}\delta_{ij}\right) + \mathcal{Q}_{ij} + 5\mathcal{Q}_{ijk} w_S^k + \ldots \tag{4.27}$$

$$\mathcal{D}^{ij} = \delta_{ij}\left(\frac{G\hat{M}_E}{\rho_S} + \frac{3}{2}\mathcal{Q}_{km} w_S^k w_S^m + \frac{5}{2}\mathcal{Q}_{kmn} w_S^k w_S^m w_S^n + \ldots\right) \tag{4.28}$$

$$\mathcal{D}^{ijk} = \frac{1}{2}\frac{G\hat{M}_E}{\rho_S^3}\left(\delta_{jk} w_S^i - \delta_{ij} w_S^k - \delta_{ik} w_S^j\right) + \frac{3}{2}\left(\delta_{ij}\mathcal{Q}_{km} + \delta_{ik}\mathcal{Q}_{jm} - \delta_{jk}\mathcal{Q}_{im}\right) w_S^m +$$
$$+ \frac{5}{2}\left(\delta_{ij}\mathcal{Q}_{kmn} + \delta_{ik}\mathcal{Q}_{jmn} - \delta_{jk}\mathcal{Q}_{imn}\right) w_S^m w_S^n \tag{4.29}$$

$$\frac{d\mathcal{F}^i}{du} = \frac{3}{2}\left(v_S^i a_S^j - v_S^j a_S^i\right) - 6\varepsilon_{kmn}\frac{\hat{w}_E^m}{\rho_S^3}\left[\left(\frac{3}{\rho_S^2} w_S^j w_S^n - \delta_{jn}\right)\hat{I}_{ik}^{(E)} - \right. \tag{4.30}$$

$$-\left(\frac{3}{\rho_s^2} w_s^i w_s^n - \delta_{in}\right) \hat{I}_{jk}^{(E)} \Bigg] + 4\varepsilon_{ijk} C_{km} w_s^m + \ldots$$

This matching is performed here up to the terms of the order ε^3 inclusively. For this reason, the geocentric equations of satellite motion (4.25) are obtained only in the Newtonian approximation. To obtain them in the post-Newtonian approximation one should take into account in (3.14) the external and mixed terms of the order ε^4 and then apply the geodesic principle or matching technique.

5. PROPAGATION OF ELECTROMAGNETIC SIGNALS

Relativistic reduction of astrometric observations is based on the well-known propagation laws of electromagnetic (light, radio, ...) signals in the solar system gravitational field. We give here basic relations within the post-Newtonian approximation neglecting in (3.1)-(3.6) quadrupole and spin terms (Japanese Ephemeris, 1985; Brumberg, 1986).

5.1. PROPAGATION OF A PHOTON WITH A GIVEN INITIAL POSITION AND A GIVEN DIRECTION

Let the motion of a photon be given by conditions :

$$\underline{x}(t_0) = \underline{x}_0 \quad , \quad \frac{d\underline{x}(-\infty)}{dt} = c\underline{\sigma} \quad , \quad \underline{\sigma}^2 = 1 \tag{5.1}$$

that is by its position at some moment t_0 of the coordinate time and the trajectory direction at the remote past on actually infinite distances r_A from all bodies of the solar system. Retaining previous notation $\underline{r}_A = \underline{x} - \underline{x}_A$, $\underline{r}_{0A} = \underline{x}_0 - \underline{x}_A$, $m_A = GM_A/c^2$ one has the following expressions for the coordinates of the photon and its velocity at any moment t :

$$\underline{x}(t) = \underline{x}_0 + c(t-t_0)\underline{\sigma} + 2\sum_A m_A \left[\frac{\underline{\sigma}\wedge(\underline{r}_{0A}\wedge\underline{\sigma})}{r_{0A} - \underline{\sigma}\,\underline{r}_{0A}} - \frac{\underline{\sigma}\wedge(\underline{r}_A\wedge\underline{\sigma})}{r_A - \underline{\sigma}\,\underline{r}_A} - \underline{\sigma}\ln\frac{r_A + \underline{\sigma}\,\underline{r}_A}{r_{0A} + \underline{\sigma}\,\underline{r}_{0A}} \right] \tag{5.2}$$

$$\frac{1}{c}\frac{d\underline{x}(t)}{dt} = \underline{\sigma} - 2\sum_A \frac{m_A}{r_A}\left[\underline{\sigma} + \frac{\underline{\sigma}\wedge(\underline{r}_A\wedge\underline{\sigma})}{r_A - \underline{\sigma}\,\underline{r}_A} \right] \tag{5.3}$$

The coordinate direction (5.3) may be converted into the observable (coordinate-independent) direction $\underline{p}(t)$ at the point of observation (see Section 5.4 for the convertion technique) :

$$\underline{p} = \underline{\sigma} - 2\sum_A \frac{m_A}{r_A}\frac{\underline{\sigma}\wedge(\underline{r}_A\wedge\underline{\sigma})}{r_A - \underline{\sigma}\,\underline{r}_A} \tag{5.4}$$

5.2. PROPAGATION OF A PHOTON WITH TWO GIVEN BOUNDARY POSITIONS

Boundary value conditions :

$$\underline{x}(t_0) = \underline{x}_0 \;,\; \underline{x}(t) = \underline{x} \;,\; \underline{R}(t,t_0) = \underline{x} - \underline{x}_0 \; (t > t_0) \tag{5.5}$$

may be satisfied by :

$$\underline{\sigma} = \frac{\underline{R}}{R} + 2\sum_A \frac{m_A}{R}\frac{r_A - r_{0A} + R}{\left|\underline{r}_{0A}\wedge\underline{r}_A\right|^2}\left[\underline{R}\wedge(\underline{r}_{0A}\wedge\underline{r}_A)\right] \tag{5.6}$$

Substituting (5.6) into (5.3),(5.4) yields the coordinate and observable directions at moment t of the signal reception :

$$\frac{1}{c}\frac{d\underline{x}(t)}{dt} = \frac{\underline{R}}{R} - \frac{2}{R}\sum_A \frac{m_A}{r_A}\left[\underline{R} + \frac{\underline{R}\wedge(\underline{r}_{0A}\wedge\underline{r}_A)}{r_{0A}r_A + \underline{r}_{0A}\underline{r}_A}\right] \tag{5.7}$$

$$\underline{p} = \frac{\underline{R}}{R} - \frac{2}{R}\sum_A \frac{m_A}{r_A}\frac{\underline{R}\wedge(\underline{r}_{0A}\wedge\underline{r}_A)}{r_{0A}r_A + \underline{r}_{0A}\underline{r}_A} \tag{5.8}$$

The time of propagation will be :

$$c(t-t_0) = R + 2\sum_A m_A \ln\frac{r_A + r_{0A} + R}{r_A + r_{0A} - R} \tag{5.9}$$

5.3. ANGULAR DISTANCE BETWEEN TWO SOURCES

If two light rays come to an observer at moment t with observed direction $\underline{p}_1$, $\underline{p}_2$ their mutual angular distance will be :

$$\cos\psi = \underline{p}_1\underline{p}_2 \tag{5.10}$$

For the problem (5.1) there results :

$$\cos\psi = \underline{\sigma}_1\underline{\sigma}_2 + 2\sum_A \frac{m_A}{r_A}\left(\frac{\underline{r}_A \wedge \underline{\sigma}_1}{r_A - \underline{\sigma}_1\underline{r}_A} - \frac{\underline{r}_A \wedge \underline{\sigma}_2}{r_A - \underline{\sigma}_2\underline{r}_A}\right)\left(\underline{\sigma}_1 \wedge \underline{\sigma}_2\right) \tag{5.11}$$

With the aid of (5.6) one may derive the expression of the angular distance for the problem (5.5)

5.4. CONVERSION OF THE LIGHT COORDINATE DIRECTION INTO OBSERVED DIRECTION

Let the metric (2.1) be related to any of the reference systems described in Section 3. This system may admit rotation. Therefore, for the metric tensor :

$$g_{\alpha\beta} = \eta_{\alpha\beta} + h_{\alpha\beta} \;,\; \eta_{00} = -1 \;,\; \eta_{0i} = 0 \;,\; \eta_{ij} = \delta_{ij} \tag{5.12}$$

components h_{oi} may be of the order ε and components h_{oo}, h_{ij} are of the order ε^2. Local splitting of (2.1) into mesurable infinite small time and space intervals results in :

$$ds^2 = \eta_{\alpha\beta}\, dx^{(\alpha)}\, dx^{(\beta)} \tag{5.13}$$

$$dx^{(o)} = \left(1 - \frac{1}{2}h_{oo} - \frac{1}{8}h_{oo}^2\right)c\,dt - \left(h_{oi} + \frac{1}{2}h_{oo}h_{oi}\right)dx^i \tag{5.14}$$

$$dx^{(i)} = dx^i + \frac{1}{2}\left(h_{ij} + h_{oi}\,h_{oj}\right)dx^j \tag{5.15}$$

The observable direction of the photon propagation will be :

$$p^{(i)} = \frac{dx^{(i)}}{dx^{(o)}} \tag{5.16}$$

or :

$$p^{(i)} = \left[1 + \frac{1}{c}h_{ok}\frac{dx^k}{dt} + \left(\frac{1}{c}h_{ok}\frac{dx^k}{dt}\right)^2 + \frac{1}{2}h_{oo}\right]\frac{1}{c}\frac{dx^i}{dt} + \frac{1}{2}\left(h_{ij} + h_{oi}h_{oj}\right)\frac{1}{c}\frac{dx^j}{dt} \tag{5.17}$$

This formula enables to convert the coordinate direction $dx^i/(cdt)$ into observable vector $\underline{p}=p^{(1)}, p^{(2)}, p^{(3)}$ (Synge, 1960; Brumberg, 1986). If x^α are the BRS coordinates then the photon coordinate direction is given directly by (5.3) or (5.7) and (5.17) leads to (5.4) or (5.8). If x^α represent coordinates of any other reference system then one has first to find the photon coordinate direction in this system on the basis of the relationship with BRS and expression (5.3), thereupon, the expression (5.17) yields the observed direction vector.

5.5. REDUCTION OF OBSERVATIONS IN GRS

On the basis of (4.1), (4.5), (4.6) one has :

$$\frac{du}{dt} = 1 + c^{-2}\left[-v_E^k \frac{dx^k}{dt} + \frac{1}{2} v_E^2 - \bar{U}(\underline{x}_E) - a_E^k (x^k - x_E^k) \right] \tag{5.18}$$

Starting from (5.3) with the aid of (4.2) and (5.18) one finds the photon coordinate direction :

$$\frac{dw^i}{cdu} = \frac{dx^i}{cdt} + c^{-1}\left[\underline{\sigma}\wedge(\underline{\sigma}\wedge\underline{v}_E)\right]^i + c^{-2}\left\{\begin{array}{l}(\underline{\sigma}\,\underline{v}_E)\left[\underline{\sigma}\wedge(\underline{\sigma}\wedge\underline{v}_E)\right]^i - \frac{1}{2}\left[\underline{v}_E\wedge(\underline{\sigma}\wedge\underline{v}_E)\right]^i + \\ + F^{ik}\sigma^k + 2D^{ik}\sigma^k + 2D^{ijk}(x^j - x_E^j)\sigma^k + a_E^k(x^k - x_E^k)\sigma^i\end{array}\right\} \tag{5.19}$$

Then, the expression (5.17) applied to the metric (3.14)-(3.16) results in :

$$\hat{p}^{(i)} = \sigma^i + c^{-1}\left[\underline{\sigma}\wedge(\underline{\sigma}\wedge\underline{v}_E)\right]^i + c^{-2}(\underline{\sigma}\underline{v}_E)\left[\underline{\sigma}\wedge(\underline{\sigma}\wedge\underline{v}_E)\right]^i - \frac{1}{2}c^{-2}\left[\underline{v}_E\wedge(\underline{\sigma}\wedge\underline{v}_E)\right]^i +$$

$$-2\sum_A \frac{m_A}{r_A}\frac{\left[\underline{\sigma}\wedge(\underline{r}_A\wedge\underline{\sigma})\right]^i}{r_A - \underline{\sigma}\,\underline{r}_A} + c^{-2}F^{ij}\sigma^j + \sum_{A\neq E}\frac{m_A}{r_{EA}^3}\left[r_{EA}^i(x^k - x_E^k)\,\sigma^k - (x^i - x_E^i)(r_{EA}^k\sigma^k)\right] \tag{5.20}$$

This is the reduction formula from GRS to BRS including explicitly annual aberration of first order (the second term) and of second order (the second and the third terms), light deflection (the fourth term), geodesic precession (the fifth term) and relativistic contraction (the last terms in (5.20)). The corrections for planetary aberration, annual parallax and proper motions (if necessary) are contained implicitly in (5.6) and (5.9) in expressing $\underline{\sigma}$ in terms of boundary values (cf. Japanese Ephemeris, 1985; Brumberg, 1986).

5.6. REDUCTION OF OBSERVATIONS IN TRS+

On the basis of (4.16), (4.20) one has :

$$\frac{d\tau}{du} = 1 + c^{-2}\left[-v_T^k \frac{dw^k}{du} + \frac{1}{2} v_T^2 - U'_E(\underline{v}_T) - a_T^k (w^k - w_T^k) - \frac{3}{2} \mathcal{Q}'_{ij} y_T^i y_T^j - \frac{5}{2} \mathcal{Q}'_{ijk} y_T^i y_T^j y_T^k + \dots \right] \qquad (5.21)$$

where $v^k{}_T$, $a^k{}_T$ are expressed by (4.18) and (4.19). Using this relation and (4.17) one obtains the photon coordinate direction in TRS^+ :

$$\frac{d\xi^i}{cd\tau} = \frac{dy^i}{cdu} + c^{-1}\left[\left(v_T^k \frac{dw^k}{cdu} \right) \frac{dy^i}{cdu} - \mathcal{P}_{ik} v_{TT}^k \right] + \dots \qquad (5.22)$$

In accordance with (3.23) and (3.24) :

$$\frac{dy^i}{cdu} = \mathcal{P}_{ik} \frac{dw^k}{cdu} + c^{-1} \varepsilon_{inj} \hat{\omega}_E^j \mathcal{P}_{nk} w^k \qquad (5.23)$$

Combining (5.3), (5.19), (5.23) and (5.22) and using (5.17) for the metric (3.33)-(3.35) one obtains the reduction formula converting the observable light direction at the ground observatory into the observable direction in BRS. The leading terms of this formula are as follows :

$$p^{*(i)} = \mathcal{P}_{ik} \hat{p}^{(k)} + c^{-1} \mathcal{P}_{ik} \left[\underline{\sigma} \wedge (\underline{\sigma} \wedge \underline{v}_T) \right]^k + c^{-1} (\underline{\omega} \wedge \underline{\xi})^k \mathcal{P}_{im} \mathcal{P}_{kn} \sigma^m \sigma^n - c^{-1} (\underline{\omega} \wedge \underline{\xi})^i + \dots \qquad (5.24)$$

This formula contains all necessary classical and relativistic reductions. In addition to the effects taken into account in (5.20) it contains the diurnal aberration (the second term) and the terms due to the rotation of the Earth which vanish if the point of observation coincides with the origin of TRS^+ (ξ=0).

Not going into details let us note that the geodesic precession F^{ik} is combined with the classical precession P_{ik} enabling to consider both terms together.

5.7. REDUCTION OF OBSERVATIONS IN SRS

With the aid of (4.23) and (4.26) one has :

$$\frac{d\tau}{du} = 1 + c^{-2}\left[-v_S^k \frac{dw^k}{du} + \frac{1}{2} v_S^2 - \hat{U}_E(\underline{w}_S) - a_S^k (w^k - w_S^k) - \frac{3}{2} \mathcal{Q}_{ij} w_S^i w_S^j + \dots \right] \qquad (5.25)$$

where $v^i{}_S$, $a^i{}_S$ are the velocity and the acceleration of the satellite in GRS. Using this relation and (4.24) it is easy to find the photon coordinate direction in SRS :

$$\frac{dz^i}{cd\tau}=\frac{dw^i}{cdu}+c^{-1}\left[\left(v_S^k\frac{dw^k}{cdu}\right)\frac{dw^i}{cdu}-v_S^i\right]+c^{-2}\left(\underline{\sigma}\,\underline{v}_S\right)\left[\underline{\sigma}\wedge\left(\underline{\sigma}\wedge\underline{v}_S\right)\right]^i-$$

$$-\frac{1}{2}c^{-2}\left[\underline{v}_S\wedge\left(\underline{\sigma}\wedge\underline{v}_S\right)\right]^i+c^{-2}\left[\mathcal{F}^{ik}\sigma^k+2\mathcal{D}^{ik}\sigma^k+2\mathcal{D}^{ijk}\left(w^j-w_S^j\right)\sigma^k+a_S^k\left(w^k-w_S^k\right)\sigma^i\right] \tag{5.26}$$

where the coefficients should be taken from (4.17), (4.20)-(4.22). Substituting (5.19) and applying (5.17) with the values (3.29)-(3.31) for the SRS metric one obtains the photon observed direction :

$$\tilde{p}^{(i)}=\left[1+c^{-2}3E_{km}\left(w^k-w_S^k\right)\left(w^m-w_S^m\right)\right]\frac{dz^i}{cd\tau} \tag{5.27}$$

5.8. PHYSICAL REALIZATION OF REFERENCE SYSTEMS

Physical materialization of reference systems is achieved with the aid of some reference astronomical objects. Actually, having observable direction $\underline{\tilde{p}}$ in TRS+ or SRS (HIPPARCOS system) one may determine direction $\underline{\sigma}$ for BRS. The set of $\underline{\sigma}$ for various sources referred to the same moment of time and corrected for parallax will give the astronomical reference frame. It is impossible to describe here all the details of the related reduction techniques. At present, it is high time to reconsider all classical reduction methods. The first steps have been already done (Murray, 1983; Japanese ephemeris, 1985). The theory presented here may hopefully help to develop the astrometric reduction techniques in the relativistic framework.

CONCLUSION

In conclusion, let us summarize the basic statements of the present paper which may be particularly useful in elaborating the IAU recommendations concerning the astronomical reference frames.

1. All basic (non-rotating) reference systems should satisfy the specific type of coordinate conditions, for example, harmonic ones.
2. The hierarchy of the reference systems including BRS, GRS, TRS, SRS should fit the physically adequate solutions of the Einstein field equations.
3. The coordinate time scales of BRS and GRS may serve as a basis for definition of TDB and TDT respectively. The coordinate timescale of TRS averaged over various observatories may serve as TAI.
4. Asymptotic matching of global and local coordinate systems results in the unambiguously determination of the local metric expansion and its

relationship with the global metric irrespective of the presence of the local gravitational field. This matching technique offers a new way to obtain the relativistic equations of motion of celestial bodies.

5. The present theory enables us to solve dynamical problems in the most appropriate reference systems (heliocentric planetary motion, geocentric lunar or satellite motion, etc...) and to perform the rigorous reduction of astrometric observations made from the Earth, its satellites or any other solar system body. The sequence of reduction should be set by an agreement (for example, the reduction $TRS^+ \rightarrow GRS^+ \rightarrow GRS \rightarrow BRS$ is theoretically different from the reduction $TRS^+ \rightarrow TRS \rightarrow GRS \rightarrow BRS$).

Evidently,some questions of the relativistic treatment of the astronomical reference frames have not been discussed here such as, for example, the problem of astronomical constants in different frames (Fukushima *et al.*, 1986b; Hellings, 1986). These questions may be considered on the basis of the relationships between reference systems introduced here.

The authors are indebted to Dr L.P. Grishchuk for valuable discussions.

BIBLIOGRAPHY

Anderson, J.L. and Decanio, T.C., 1975, *Gen. Rel. Grav.*, **6**, 197.

Ashby, N. and Bertotti, B, 1986, *Phys. Rev.*, **D 34**, 2246.

Bertotti, B., 1986, in *'Relativity in Celestial Mechanics and Astrometry'*, J. Kovalevsky and V.A. Brumberg (eds), Reidel Publ.Co., Dordrecht, 233.

Boucher, C., 1986, in *`Relativity in Celestial Mechanics and Astrometry'*, J. Kovalevsky and V.A. Brumberg (eds), Reidel Publ.Co., Dordrecht, 241.

Brumberg, V.A., 1972, *'Relativistic Celestial Mechanics'*, Nauka, Moscow (in Russian).

Brumberg, V.A., 1986, in *'Astrometric Techniques'*, H.K. Eichhorn and R.J. Leacock (eds), Reidel Publ.Co., Dordrecht, 19.

Damour, T., 1983, in *'Gravitational Radiation'*, N. Deruelle and T. Piran (eds), North-Holland, Amsterdam, 59.

Damour, T., 1987, in *'300 Years of Gravitation'*, S.W. Hawking and W. Israel (eds), Cambridge Univ. Press, 128.

D'Eath, P.D., 1975, *Phys. Rev.*, **D 11**, 1387.

Ehlers, J., 1980, *Ann. N.Y. Acad. Sci.*, **336**, 279.

Fock, V.A., 1959, *'The Theory of Space, Time and Gravitation'*, Pergamon Press, London.

Fujimoto, M.K. and Grafarend, E., 1986, in *'Relativity in Celestial Mechanics and Astrometry'*, J. Kovalevsky and V.A. Brumberg (eds), Reidel Publ.Co., Dordrecht, 269.

Fukushima, T., Fujimoto, M.K., Kinoshita, H. and Aoki, S., 1986a, in *'Relativity in 'Celestial Mechanics and Astrometry'*, J. Kovalevsky and V.A. Brumberg (eds), Reidel Publ.Co.,

Dordrecht, 145.
Fukushima, T., Fujimoto, M.K., Kinoshita, H. and Aoki, S., 1986b, *Celestial Mechanics*, **38**, 215.
Futamase, T. and Schutz, B.F., 1983, *Phys. Rev.*, **D 28**, 2363.
Grishchuk, L.P. and Kopejkin, S.M., 1986, in *'Relativity in Celestial Mechanics and Astrometry'*, J. Kovalevsky and V.A. Brumberg (eds), Reidel Publ.Co., Dordrecht, 19.
Guinot, B., 1986, *Celestial Mechanics*, **38**, 155.
Hellings, R.W., 1986, *Astron. J.*, **91**, 650.
Japanese Ephemeris, 1985, Basis of the New Japanese Ephemeris, Tokyo.
Kates, R.E., 1981, *Ann. Phys., USA*, **132**, 1.
Kopejkin, S.M., 1985, *Astron. J. USSR*, **62**, 889 (in Russian).
Kopejkin, S.M., 1987, *Trans. Sternberg State Astron. Inst*, **59**, 53 (in Russian).
Kovalevsky, J. and Mueller, I.I., 1981, in *'Reference Coordinate Systems for Earth Dynamics'*, E.M. Gaposchkin and B. Kolaczek (eds), Reidel Publ.Co., Dordrecht, 375.
Kovalevsky, J., 1985, *Bull. Astron. Obs. Roy. Belgique*, **10**, 87.
Lestrade, J.F. and Chapront-Touzé, M., 1982, *Astron. Astrophys.*, **116**, 75.
Martin, C.F., Torrence M.H. and Misner, C.W., 1985, *J. Geophys. Research*, **90**, 9403.
Mast, C.B. and Strathdee, J., 1959, *Proc. Roy. Soc.*, **A 252**, 476.
Misner, C.W., Thorne, K.S. and Wheeler, J.A., 1973, *'Gravitation'*, Freeman, San Francisco.
Moeller, C., 1972, *'The Theory of Relativity'*, Clarendon Press, Oxford.
Mueller, I.I., 1981, in *'Reference Coordinate Systems for Earth Dynamics'*, E.M. Gaposchkin and B. Kolaczek (eds), Reidel Publ.Co., Dordrecht, 1.
Murray, C.A., 1983, *'Vectorial Astrometry'*, Adam Hilger, Bristol.
Ni, W.T. and Zimmermann, M., 1978, *Phys. Rev.*, **D 17**, 1473.
Pavlov, N.V., 1984a, *Astron. J., USSR*, **61**, 385 (in Russian).
Pavlov, N.W., 1984b, *Astron. J., USSR*, **61**, 600 (in Russian).
Podobed, V.V. and Nesterov, V.V., 1982, *'General Astrometry'*, Nauka, Moscow (in Russian).
Suen, W.M., 1986, *Phys. Rev.*, **D 34**, 3617.
Synge, J.L., 1960, *'Relativity: The General Theory'*, North-Holland.Publ.Co.
Thorne, K.S. and Hartle, J.B., 1985, *Phys. Rev.*, **D 31**, 1815.
Zhang, X.H., 1986, *Phys. Rev.*, **D 34**, 991.

PART 2

TERRESTRIAL REFERENCE FRAMES

HORIZONTAL AND VERTICAL GEODETIC DATUMS

IVAN I. MUELLER and RICHARD H. RAPP
Ohio State University, Columbus, Ohio

1. INTRODUCTION

In conventional geodetic systems, locations of points on the surface of the earth may be defined either by means of natural (astronomic) or geometric (geodetic) coordinates. The natural coordinates, the astronomic latitude (Φ), longitude (Λ), and the orthometric (mean sea level height (H), being gravity dependent, are conventionally referenced to the geoid and are determined from "natural" observations (astronomic, gravimetric and spirit leveling). The geometric coordinates, the geodetic latitude (ϕ), longitude (λ) and height (h), are referenced to a (generally) rotational ellipsoid of arbitrary size, shape and orientation, and are determined from geometric (length and/or direction) observations.

In connection with the above practice, two problems arise:

(i)- the relationship between the natural and geometric sets of coordinates, expressed through the parameters of the geodetic datum;

(ii)- the reduction of the geometric and natural observations to their appropriate reference surfaces.

This paper deals with some of the theoretical and practical aspects of handling these problems correctly and the consequences of not doing so.

2. ON THE DEFINITION OF A GEODETIC DATUM

The parameters that define the relationships between the natural and geometric coordinates referred to their respective reference surfaces are known as datum parameters. These define the size and shape of the reference ellipsoid and its position and orientation with respect to an earth (geoid) fixed terrestrial reference frame. In the latter, conventionally right-

J. Kovalevsky et al. (eds.), Reference Frames, 145–162.

handed, coordinate system, the third and the first axes are oriented respectively towards the Conventional Terrestrial Pole (CTP) and Zero Meridian as defined by the Bureau International de l'Heure (BIH) (Mueller, 1969), or since 1 January 1988, by the International Earth Rotation Service (IERS) (Wilkins and Mueller, 1986).

The reference ellipsoid, if rotational, is defined by two parameters, usually the length of its semimajor axis (a), and its flattening (f). The position of its center with respect to an earth-fixed point, usually the initial point of the triangulation system, is determined by three additional parameters, traditionally by the components of the deflection of the vertical (ξ_o, η_o) and the geoid undulation (N_o) at the fixed point, i.e., by

$$\begin{aligned} \xi_o &= \Phi_o - \phi_o \\ \eta_o &= (\Lambda_o - \lambda_o) \cos \phi_o \\ N_o &= h_o - H_o \end{aligned} \tag{1}$$

It may be shown that the measured quantities (Φ_o, Λ_o, H_o), and the defined ones (ξ_o, η_o, N_o, a, f) are equivalent to the Cartesian coordinates of the initial point with respect to the center of the Earth's mass, provided that the ellipsoidal axes are parallel to the earth-fixed axes mentioned earlier. *If this condition exists*, the datum is said to be defined by the five parameters, ξ_o, η_o, N_o, a and f.

It may also be shown that the parallelism of the axes is assured if at one point the following conditions are fulfilled (Hotine, 1969)

$$\begin{aligned} A - \alpha &= \eta \tan \phi + (\xi \sin \alpha + \eta \cos \alpha) \cot z \\ Z - z &= (\xi \cos \alpha + \eta \sin \alpha) \end{aligned} \tag{2}$$

In these equations, known as the extended Laplace equations, A and Z are the astronomic azimuth and zenith distance to another station, α and z are their geodetic (ellipsoidal) equivalents. In conventional work, largely because vertical angles cannot be measured accurately, and also because generally $z \approx 90°$, only a simplified version of the above set known as the (short) Laplace equation is enforced, i.e.,

$$A - \alpha = \eta \tan \phi \tag{3}$$

In this case the assumption is that condition (3) *throughout the network* (i.e., at several stations), from the point of view of parallelism, is equivalent with condition (2) *at a single point.*

If the datum parameters are such that the center of the reference ellipsoid coincides with the center of mass of the earth and the parameters, defining its size and shape, correspond to those of the equipotential (level) ellipsoid defining the normal gravity field, the datum is said to be 'absolute,' as opposed to 'astrogeodetic' or 'relative.' Similar terms are used for the deflections and undulations. Thus one may speak of astrogeodetic and absolute (sometimes gravimetric) deflections, undulations, etc.

The choice between astrogeodetic and absolute datum depends on how well the reference ellipsoid needs to fit the geoid within the extent of the geodetic control. This 'need to fit,' on the other hand, largely depends on how accurately the reductions, raised in the second question in Section 1, can be carried out. If the reductions discussed below can be and are applied with care, there are very few disadvantages, if any, of using an absolute datum instead of a relative one. The advantages, on the other hand, include convenience in worldwide applications, consistency with gravity reference fields and easy use in conjunction with three-dimensional (possibly time-variant) geodetic systems. There are about 100 local or astrogeodetic datums in use around the world. The most recent absolute datum is the World Geodetic System (WGS) 84. The ellipsoid of this geocentric datum is defined by the following ellipsoidal parameters:

$$a = 6\,378\,137 \text{ m}$$
$$f = 0.003\,352\,810\,664\,74\ldots$$

The orientation of this ellipsoid (rotation axis and zero meridian) is identical to the terrestrial reference frame defined by the BIH at 1984.0 (DMA, 1987).

3. ON THE REDUCTION OF OBSERVATIONS

3.1 GEOMETRIC OBSERVATIONS

The reduction of observations from the topography (where they are carried out) to the appropriate reference surfaces (where they are used in the calculations), unlike the question of the datum definition, has been more uniformly discussed in the literature and, therefore, is only summarized here.

The geometric (distance, direction) observations need to be reduced to the reference ellipsoid. The necessary corrections (and the quantities most significantly affecting them) are as follows:

1) distance correction (at the terminal points: orthometric heights, undulations, and sometime deflection of the vertical components);

2) direction corrections:

a) for the deflection of the vertical (deflection components at the point of observation),
b) for the noncoplanarity of the ellipsoidal normals at the observation and the observed points, commonly known as the correction for the 'skewness of the normals' (ellipsoidal height of the *observed* point),
c) for the angle between the normal section on the ellipsoid and the geodesic between the observation and observed points (distance on the ellipsoid between the points).

In many cases in computing the distance correction, only the orthometric heights (not the ellipsoidal heights) are considered and the influence of the deflection components is completely ignored at the direction correction 2a) as well. In this situation, the method of calculating the coordinates is called 'development' as opposed to 'projection,' when all the corrections are properly taken into consideration in accordance with the above summary. The former is considered a misnomer (not a method) and should not be used. Details of these corrections may be found in (Bomford, 1971; Heiskanen and Moritz, 1967).

3.2 NATURAL OBSERVATIONS

The natural quantities need to be reduced to the geoid. The necessary corrections (and the quantities most significantly affecting them) are

1) observed astronomic latitude, longitude and azimuth need to be corrected (Mueller, 1969):
a) for polar motion (coordinates of the true celestial pole with respect to the CTP) (see Chapter 11),
b) to UT1, for longitudes only (UTC–UT1 corrections),
c) for the curvature of the plumbline between the observation station and the geoid (average horizontal gradient of gravity along the plumbline section),
d) for the curvature of the plumbline between the observed station and the geoid, for azimuths only (same as in 1(c));

2) orthometric correction to spirit-leveled height (average gravity along the plumbline between the station and the geoid, gravity observations along the leveling lines) (Heiskanen and Moritz, 1967). See section on Vertical Datums below.

The greatest difficulties in practice arise in connection with 1(c), 1(d) and 2, largely because, for their accurate calculations, the distribution of densities in the earth's crust between the topography and the geoid needs to be considered. Since this is not readily available information, assumptions and approximations are made. Similar problems arise in connection with the reduction of observed gravity values to the geoid. To avoid this and other problems, new theories have been developed that avoid the use of the geoid (and the reductions thereto) (Molodenskii *et al.*, 1960; Hirvonen, 1960).

4. DATUM TRANSFORMATIONS AND NETWORK DISTORTIONS

For the user of the geodetic control the datum parameters are represented by the final (adjusted) coordinates of the geodetic stations within a network. Thus it is of interest to compare these with external standards, such as the satellite-determined coordinates of common stations. From such comparisons, one would expect information on the position and orientation of the coordinate system inherent to the datum with respect to the external standard and also on the distortions within the network. Such systematic distortions may be present from various sources, most of them mentioned earlier, e.g., the problems surrounding the use of the short Laplace equation throughout the network (including possible systematic errors in the astronomic azimuths therein), and/or the neglect or improper application of the observational corrections listed in the previous section.

The difficulty in making such comparisons is twofold: first , the external standard (e.g., satellite network) is not errorless and may also contain systematic internal distortions; secondly, if the external standard consists of the three-dimensional Cartesian coordinates of a set of stations, the comparison requires the knowledge of the astrogeodetic undulations throughout the net. If the astrogeodetic geoid is tilted, or any other way systematically distorted with respect to the geoid (which is often the case in practice), the separation of the datum position (transformation) parameters from the network distortions becomes difficult.

Examples are shown in Figures 1–3. Figures 1a, 2a and 3a show the differences in latitude, longitude and ellipsoidal heights between the WGS 84 absolute datum, largely based on satellite-determined information and the North American Datum (NAD) 1927. Figures 1b, 2b and 3b illustrate the same for the European Datum (ED) 1950. The differences shown are partly due to the differences in datum parameters but also to the network distortions evident from the irregularities in the contour lines. The differences in the size and shape of the NAD 1927 reference ellipsoid (Clarke 1866), the ED 1950 ellipsoid (International) and those of the WGS 84 ellipsoid, and coordinates of the centers of these ellipsoids with respect to the WGS 84 origin together with some other major datums are in Table 1 (DMA, 1987). As it is seen, no rotations or scale differences are evident between these datums.

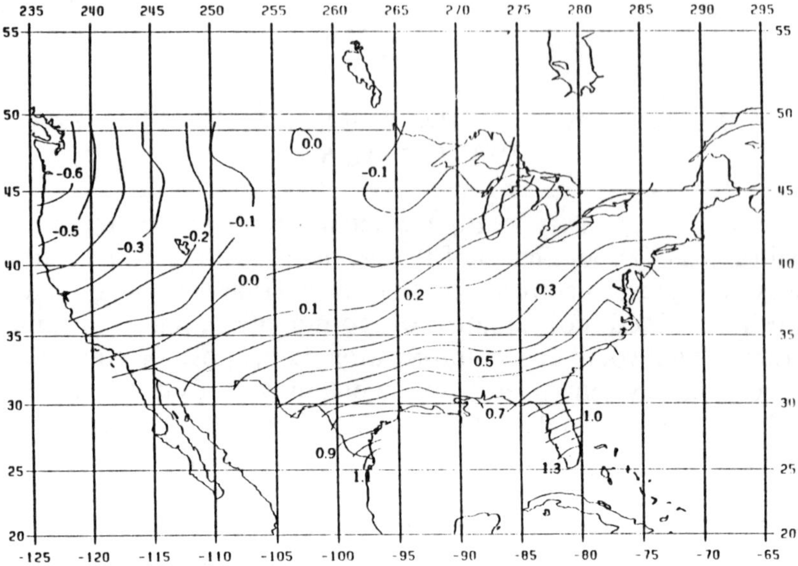

Figure 1a. Contour chart—latitude differences, WGS 84 minus NAD 27 (US; contour interval = 0.1 arcsecond)

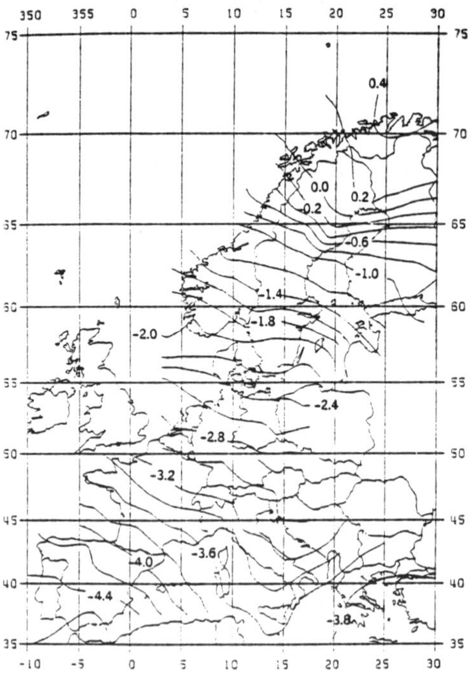

Figure 1b. Contour chart—latitude differences, WGS 84 minus ED 50 (Western Europe; contour interval = 0.2 arcsecond)

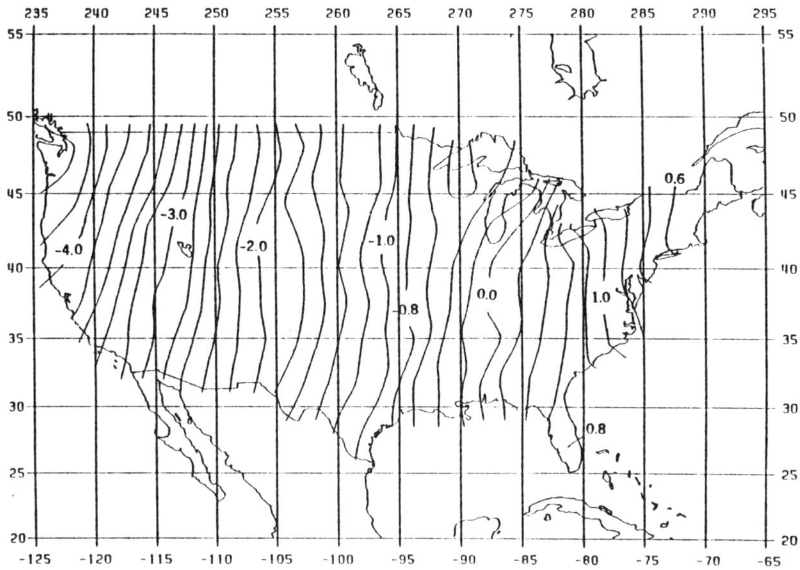

Figure 2a. Contour chart—longitude differences, WGS 84 minus NAD 27 (US; contour interval = 0.2 arcsecond)

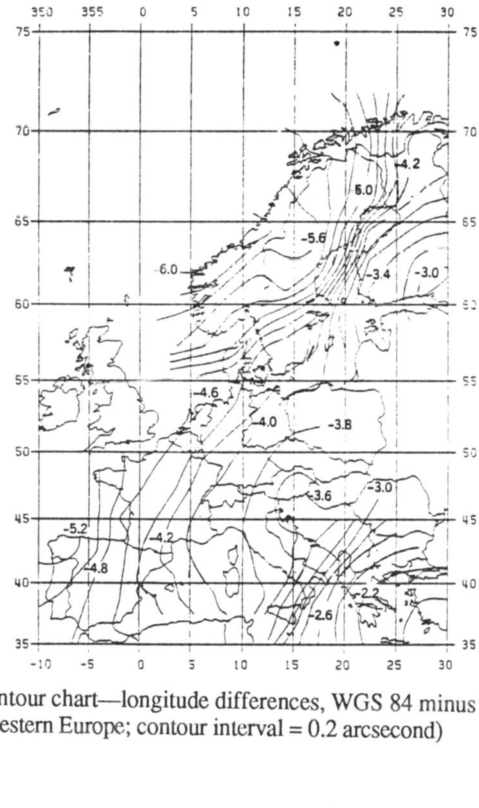

Figure 2b. Contour chart—longitude differences, WGS 84 minus ED 50 (Western Europe; contour interval = 0.2 arcsecond)

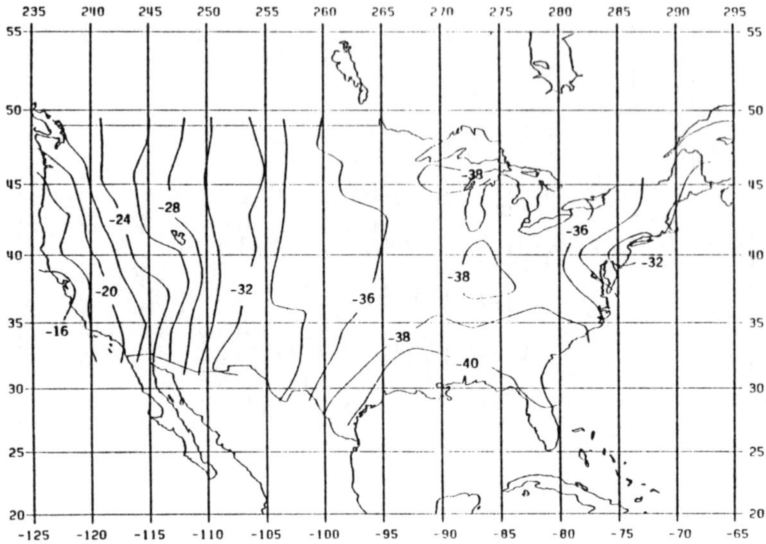

Figure 3a. Contour chart—geodetic height differences, WGS 84 minus NAD 27 (US; contour interval = 2 meters)

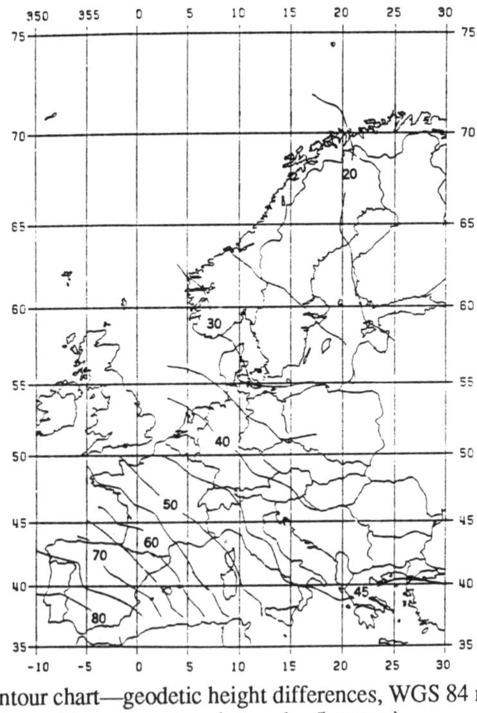

Figure 3b. Contour chart—geodetic height differences, WGS 84 minus ED 50 (Western Europe; contour interval = 5 meters)

Table 1 WGS 84 – Local Datum Parameter Differences

Local Datum	Δa *(m)*	$\Delta f \times 10^4$	ΔX *(m)*	ΔY *(m)*	ΔZ *(m)*
NAD 1927	-69.4	-0.372 646 39	-8	160	176
ED 1950	-251.0	-0.141 927 02	-87	-98	-121
Australian 1966	-23.0	-0.000 812 04	-133	-48	148
South Am. 1969	-23.0	-0.000 812 04	-57	1	-41
Tokyo	739.845	0.100 374 83	-128	481	664
NAD 1983	0	0.000 000 16	0	0	0

5. VERTICAL DATUMS

The height above the ellipsoid, the geodetic height, although rigorously defined, is not the height conventionally used for mapping. Instead, it is common practice to introduce, as a vertical reference datum a surface that is associated in some average way with mean sea level or the mean ocean surface. Heights, now called orthometric heights, are measured with respect to this mean sea level surface. There are different ways in which mean sea level may be defined and determined, and various ways in which heights, given with respect to this surface can be defined. The following gives more specific details on the definition and realization of vertical reference systems.

5.1 THE GEOID

A fundamental surface of gravimetric geodesy, and of high importance to vertical reference systems, is the geoid. The geoid is a specific equipotential surface of the earth's gravity field. In this discussion we will adopt a geoid definition that excludes the direct effects of the Sun and Moon although for some applications (e.g., in oceanography) it is appropriate to consider such effects. The gravity potential on the surface of the geoid is defined to be W_o. By definition, there should be only one geoid, although the estimation of the location of the geoid yields many values. The geoid can be located with respect to a reference ellipsoid through geoid undulations, N. These undulations can be determined from knowledge of the gravity field of the earth. Calculation of N with such data can only be done up to some constant value which is known within one meter. Since the geoid is determined by variations in the gravity field, the geoid is an irregular surface with a maximum positive geoid undulation of 78 meters and a maximum negative undulation of -108 m.

The gravity potential on the geoid is not directly observable. However, it can be computed given knowledge of the Earth's gravity field, and the position of a point on the geoid. One could write:

$$W_o = f(r, \psi, \lambda, C_{nm}, S_{nm}, kM, \omega) \tag{4}$$

Here, r, ψ, λ are the geocentric radius, geocentric latitude, and longitude of a point on the geoid. C_{nm} and S_{nm} are potential coefficients of degree n and order m in a spherical harmonic expansion of the Earth's gravitational field; kM is the gravitational constant times the mass of the Earth, and ω is the rotational velocity of the Earth. (See Chapter 4.) The calculation of W_o is hindered by the lack of knowledge of the physical location of the geoid (so that r, ψ, λ cannot be accurately determined) and lack of knowledge of the potential coefficients. If one could identify points on the geoid, on a global basis, the averaging of many W_o values could lead to an estimate of the potential of the geoid where random errors have been reduced but systematic effects remain.

5.2 THE MEAN SEA LEVEL

Mean sea level is a surface defined by averaging sea level over time and in some cases spatially. Tide gauge stations are the principal source of information on sea level. Such stations continually monitor the rise and fall of sea level. The largest signature will be that of tides. Averaging, at a point, over appropriate time intervals (one year to 18 years) yields an average location of local mean sea level. Mean sea level is not constant as it can be affected by ice cap melting, wind variations and changing ocean current (e.g., El Niño) patterns. The determination of mean sea level in coastal regions may be sensitive to the location of the site. For example, the location near a river discharge to the ocean could give unreliable readings in time of drought or excessive rainfall.

Mean sea level is not an equipotential surface. This is due to the fact that currents exist in the ocean where water will flow from one equipotential level to another. The geoid can now be defined as the equipotential surface that has the same physical location as a global mean sea surface when tidal, atmospheric and current effects are removed. The separation between the mean sea level and the geoid is called sea surface topography (SST). Sea surface topography can be estimated from oceanographic information (such as water density, salinity, pressure, current flow, etc.) in conjunction with assumptions on a level of no motion in the oceans. However, its determination on a global basis is complex due to the need for substantial information that is difficult to collect on a large, ocean-wide scale. The estimation of long-term sea surface topography has been discussed by Lisitzen (1974), Levitus (1982), and others. The estimates of SST made by these authors indicate the separation between the geoid and mean sea level to be on the order of ±1 m. At this time there is no uniform agreement on the deep reference levels in the oceans to be used in SST computations. In addition, SST is especially difficult to compute in the coastal waters where tide gauge measurements are made. Future prospects for SST determination would be enhanced with improved gravity field information from special satellite missions (e.g., using a gradiometer) and through the direct measurements to the ocean surface from satellite altimeters.

To summarize this discussion, consider Fig. 4 which shows a meridian section of the ellipsoid and various surfaces of interest. The ellipsoidal (h) and orthometric (H) heights are shown. The orthometric height is formally measured along the curved vertical between the point P and the reference equipotential surface, the geoid. Fig. 5 portrays the information at a tide gauge station and its connection to a reference benchmark.

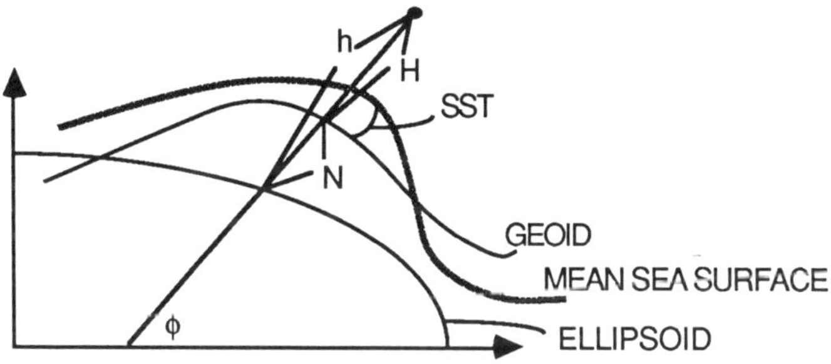

Figure 4. Location of ellipsoid height (h), orthometric height (H), geoid undulation(N), and sea surface topography (SST).

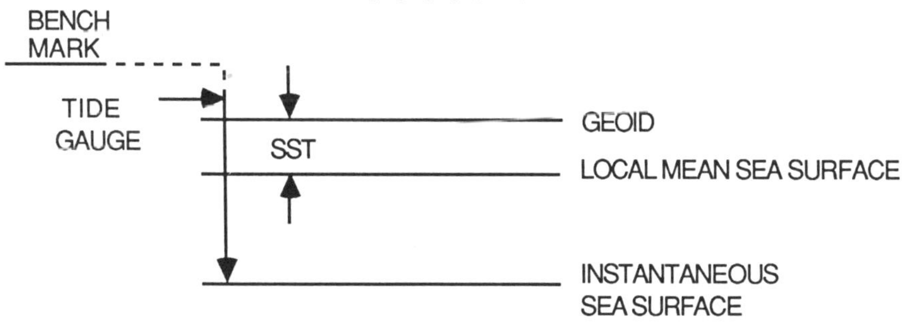

Figure 5. Measurements at a tide gauge site.

5.3 DETERMINATION OF ORTHOMETRIC HEIGHTS

In order to determine orthometric heights, we must determine a reference surface from which these heights are measured. Ideally, this surface should be the same for the whole world; and, therefore, conceptually the geoid is the appropriate surface. Since it is essentially impossible to physically determine the geoid, mean sea level is used. There are several ways in which MSL is introduced into a vertical network. The simplest procedure is one in which mean sea level, at one site, is transferred to a nearby fundamental benchmark. The elevation of this benchmark is found by measuring the small elevation difference between the MSL determination at the tide gauge and the benchmark. This benchmark then becomes the fundamental point of the vertical network. That is, the reference equipotential surface is that surface passing through the benchmark. This surface is traceable to the MSL at the nearby tide gauge site.

Starting from this point, vertical control measurements, consisting of leveled height differences and gravity measurements, are made. With this information, the potential difference or orthometric height, with respect to the reference surface can be determined.

Consider the determination of the elevation of MSL at a site some distance from the fundamental tide gauge. The elevation at this new point would not be expected to be zero because we have previously noted that mean sea level at various locations does not define an equipotential surface.

In contrast to using a single tide gauge station to define the fundamental reference surface, an alternate technique incorporates multiple tide gauge determinations of mean sea level. In this case, a vertical network is adjusted to maintain consistency between the various loops of the network. In addition, constraints are imposed on the adjustment to force the equivalent of a zero elevation at each of the local mean sea levels. This procedure has the advantage that elevations near coast line will be close to zero. However, it has the disadvantage that the datum surface is no longer associated with a single station. In fact, the reference surface is no longer an equipotential surface due to the warping necessary to uphold the constraints of the adjustment.

Another procedure for beginning a vertical datum is to carry out a preliminary adjustment with one station held fixed. At the completion of the adjustment, the heights of the local mean sea levels throughout the network are examined. A mean discrepancy is computed and applied to the station originally held fixed. This procedure forces the average elevation of all local MSL determinations to be zero. It does leave the reference surface unattached to any specific station.

We should finally emphasize that this discussion has ignored the time variations of mean sea level determinations. A noted earlier, mean sea level can change with time so that it is appropriate to associate a vertical datum with a mean sea level at a specified epoch. An alternative is to refer the datum to a defined elevation at a specified datum benchmark. Another complication relates to the motion of the crust, which for this discussion is assumed fixed.

With this discussion in mind, it is clear that there will be many vertical datums in the world. Each datum may be traced to some local mean sea level determination, or to some fixed reference point, or to some implicit surface defined by an adjustment procedure. Each country (or region) may have its own datum

6. SPECIFIC VERTICAL DATUMS

The vertical datum now used in the United States is called the National Geodetic Vertical Datum of 1929 (NGVD 29). The date reflects the time at which the leveling data of the United States and Canada were adjusted. At that time there was 75,159 km of leveling in the U.S. and 31,565 km in Canada. The adjustment was carried out by holding local mean sea level at zero elevation at 21 tide gauge stations in the U.S. and five in Canada. This procedure led to a datum that is warped to local mean sea level. The development of a new vertical network for North America will take into account the fact that local mean sea levels do not fall on the same equipotential surface.

The height system in Australia is called the Australian Height Datum (AHD). It was developed in 1971 through the adjustment of 97,320 km of leveling, holding mean sea level, for the 1966-68 epoch, fixed at zero at 30 tide gauge sites around the coast of Australia. This procedure was analogous to that used in the development of NGVD 29. The reference surface for the AHD is not an equipotential surface, but a surface warped to mean sea level around the continent.

In Europe one finds two types of vertical datums. The first type is that associated with a particular country or region. This type has grown from the historical need for height information. The second type of datum is that associated with the development of the United European Levelling Net (UELN).

In 1973, a new subcommission of the International Association of Geodesy was formed with the task of continuing the work of prior groups involved with the United European Levelling Net (Ehrnsperger *et al.*, 1982). The purpose of the Net was to combine all leveling data from the

European countries into one consistent system. For datum definition purposes, a single point was held fixed at a specified elevation (or geopotential number). This station is No. 4019, Normal Amsterdam Piel (NAP). The datum for the UELN-73 network is the equipotential surface which is a defined potential below the surface that passes through NAP. Since no other constraints have been imposed at tide gauge stations, the UELN-73 datum is free of internal distortions caused by the departure of local mean sea levels from the same equipotential surface.

Arur and Baveja (1984) have discussed the vertical datum for India. The First Level Net of India was adjusted in 1909 holding the elevation zero at mean sea level at nine tide gauge stations. Later preliminary adjustments showed that there were significant (0.3 meters) elevation differences between local mean sea levels between the east and west coast of the country. It was then decided to define the vertical datum origin at a single tide gauge station in Bombay based on a local mean sea level determined from 38 years of observations.

It should be clear from these discussions that most countries have adopted varying procedures for the definition of their vertical datums. Such procedures make it impossible to have a vertical datum that is truly global in nature at this time. Fortunately, since vertical datums are tied to local mean level, the inconsistency of the reference levels should not exceed 2 meters, which is the range of sea surface topography described by Lisitzen or Levitus.

7. FUTURE VERTICAL DATUMS

The above discussion indicates the variety of vertical datums that exist in the world. This leads one to ask two questions:
1) Is it possible to determine the height (or potential) difference between two or more vertical datums?
2) Is it possible to construct a world vertical datum? A general discussion of possible solutions to these questions is found in (Rapp 1983).

The calculation of a potential difference between two datums has been discussed by Colombo (1985), Hajela (1985), and others. In these discussions, several different types of information are brought together. This information includes the geocentric position of fundamental benchmarks as derived from laser tracking of satellites; global gravity field models; detailed gravity surveys within several hundred kilometers of the benchmarks whose geocentric coordinates are known; and potential difference determinations between the geocentrically positioned benchmarks. The simulation studies of Hajela indicated that it would be possible to determine the height difference between Europe and the United States to an accuracy of about ±0.5 m. Since this is about the accuracy we could obtain with existing oceanographic data, it appears that we need to wait for more accurate gravity field models to determine the height difference more accurately.

Of great future interest is the need for a common surface that is ultimately accessible to all countries for vertical reference purposes. Cartwright (1985) has suggested that such a surface may be a surface of no motion in the oceans. Such a surface may exist at locations where the pressure reaches some defined value. One such surface might be the 4000 decibar surface. Using oceanographic measurements, it is possible to calculate the sea surface topography in the open oceans with respect to this reference surface. This information is then brought into the tide gauge stations through satellite altimeter measurements and geostrophic leveling using current measuring devices.

This proposal would enable the local mean sea level heights to be converted to refer to the deep pressure surface. This method could be an important step in defining a world vertical datum. An error analysis of the procedure needs to be done to verify that the accuracy would be substantially better than that which could be accomplished using ellipsoidal heights and geoid undulations.

8. CONCLUDING REMARKS

The following conclusions may be drawn from the above discussion:

1) *Astrogeodetic datums*

Advantage:
Better local fit to the geoid, thus the errors of the development 'method' may not be overpowering; also map requirements for scale may be easier to meet.

Disadvantages:
a) Relationship with respect to the geocenter and the geoid is uncertain, resulting in the need for separating the horizontal from the vertical control and also in inaccurate three-dimensional Cartesian coordinates;
b) Unsuitability for global applications in geodesy, astronomy, etc.;
c) Inconsistency with the gravity reference (normal) field;
d) Since it is defined through parameters valid only at the initial point, distortions arise throughout the network. The practitioners need to use sets of distorted station coordinates as their 'definition' of the datum, while the datum is accurately defined only at the initial point.

2) *Absolute datums*

Advantages:
a) Convenient use for worldwide applications in geodesy, astronomy, etc.,
b) Can be made consistent with the gravity reference field,
c) Can be conveniently used in conjunction with three-dimensional (time-variant) geodetic reference frames ((4) below).
Disadvantages:
a) Local fit to geoid in a given country may be unsatisfactory; thus accurate reductions of observations, i.e., the correct application of the projection method, is a requirement.
b) Same as 1(a) and 1(d) above.

3) *Existing network distortions*

a) To reduce the distortions, the network needs to be readjusted, but only after weak areas have been strengthened with additional observations and all observations have been properly reduced to the reference ellipsoid or to the geoid. A good example is the just completed NAD 83 (Bossler *et al.*, 1982).
b) To account for the distortions, if readjustment is not feasible, the systematic parts (such as those shown in Figs. 1 to 3) may be absorbed by suitable mathematical modeling techniques.

4) *Conventional terrestrial reference frames*

The most effective way of reducing the network distortions would be to change the above-described concept of the geodetic datum to a system similar to that used by astronomers in connection with star positioning. Such a quasi-earth-fixed coordinate system would be defined through the adopted coordinates of 'super control' stations, determined with high precision from satellite, lunar, and Very Long Baseline Interferometric (VLBI) observations. The following criteria seem desirable for such a coordinate system:
(a) The system should be as invariant as possible with respect to changes in the number, distribution,and data acquisition rates of observing stations in different parts of the world;
(b) The system should facilitate the rapid determination of sudden changes in the position or motion of individual stations, as well as sudden changes in polar motion and the rotation of the earth;
(c) The system should be approximately fixed with respect to the mantle in some average sense. These criteria apparently lead to the choice of a coordinate system, based on all of the available observing stations, but in which our previous knowledge of the motion of the stations with respect to each other and to the mantle is used to model out the station motions due to plate movements, earth tides, etc.

The adoption of such a set of coordinates together with the motion model and the parameters therein as the definition of geodetic datum is consistent with geophysical and astronomic needs and also eliminates the necessity of separating, rather artificially, the horizontal control from vertical control. Within such a framework, countries could still use a reference ellipsoid of their choice in accordance with local need without interrupting the unity of the system. Details on such terrestrial reference frames are given in the following chapter.

5) *Vertical datums*

The adoption of the above three-dimensional terrestrial reference frames as replacements for conventional geodetic datums does not resolve the issue of the orthometric heights referenced to vertical datums or to a world vertical datum. Such heights may still be needed continually, mainly for engineering projects.

REFERENCES

Arur, M.G. and Baveja, S.D., 1984, Status of the establishment of vertical datum and the level-net in India, unpubl. manus., Survey of India, Dehradun.

Bomford, G., 1971, *'Geodesy,' 3rd ed.*, Oxford Univ. Press, London.

Bossler, J.D. and 23 others, 1982, AAGS Memo 2, ACSM, Falls Church, Va.

Cartwright, D., 1985, in *'Proc. 3rd Int. Symp. on the North American Vertical Datum,'* National Geodetic Information Center, NOAA, Rockville, Md., 155.

Colombo, 1985, in *'Proc. 3rd Int. Symp. on the North American Vertical Datum,'* National Geodetic Information Center, NOAA, Rockville, Md., 137.

DMA, 1987, Dept. of Defense World Geodetic System 1984, DMA TR 8350.2, Washington, D.C.

Ehrnsperger, W., Kok, J.J. and van Mierlo, J., 1982, Status and provisional results of the 1981 adjustment the United European Levelling Network - UELN-73, in Proc. Int.Symp. on Geodetic Networks and Computations, Deutsche Geodatische Kommission, Reihe B: No. 258/II.

Hajela, D.P., 1985, in *'Proc. 3rd Int. Symp. on the North American Vertical Datum,'* National Geodetic Information Center, NOAA, Rockville, Md.

Heiskanen, W.A. and Moritz, H., 1967, *"Physical geodesy,'* W.H. Freeman and Co., San Francisco and London.

Hirvonen, R.A., 1960, *'Annales Ac. Sci. Fennicae,'* Ser. A, III, No. 56.

Hotine, M., 1969, ESSA Monograph 2, U.S. Govt. Printing Office, Washington, D.C.

Levitus, S., 1982, NOAA Professional Paper 12, NOAA Geophysical Fluid Dynamics Lab., Rockville, Md.

Lisitzin, E., 1974, *'Sea Level Changes,'* Elsevier Science Publ. Co., Amsterdam and N.Y.

Makinen, J., 1987, The Fennoscandian land uplift - a case study, manus.

Molodenskii, M.S., Eremeev, V.F. and Yurkina, M.I., 1960, *"Methods for the study of the external gravitational field and the figure of the earth,'* Israel Program for Scientific Translations, Jerusalem, 1962.

Mueller, I.I., 1969, *'Spherical and practical astronomy as applied to geodesy,'* F. Ungar Publ., N.Y.

Rapp, R.H., 1983, in *'Proc. of Int. Assoc. of Geodesy Symposia, XVIII General Assembly, Hamburg,'* Dept. of Geodetic Science and Surveying, Ohio State Univ., p. 432.

Wilkins, G.A. and Mueller, I., 1986, *EOS, Trans. Am. Geophys. Union*, **67**, 601.

CONVENTIONAL TERRESTRIAL REFERENCE FRAMES

IVAN I. MUELLER
Ohio State University, Columbus, Ohio

1. MOTION OF THE POLE

Polar motion is the motion of the true celestial pole as defined by the theory of precession and nutation, e.g., of the Celestial Ephemeris Pole (CEP, see Chapter 8), with respect to the pole (third axis) of a conventionally selected earth "fixed" (terrestrial) reference frame (CTS). The latter is usually selected to be near the average position of the CEP taken over a certain time interval and is called the Conventional Terrestrial Pole (CTP). In geodesy both geodetic and astronomic coordinates (latitude and longitude) are referred to the CTS. The terrestrial Cartesian coordinate system thus has its origin at the center of the earth: the first axis is oriented towards the Greenwich Mean Astronomic Meridian of zero longitude, the third axis towards the CTP, and the second axis is perpendicular to both and forms a right-handed coordinate system.

As mentioned in the Introduction to this book and in the previous chapter, the frame of the CTS is in some "prescribed way" attached to observatories located on the surface of the earth. The connection between the CTS and the frame of a Conventional Inertial System (CIS) by tradition (to be preserved) is through the rotations expressed by (see Chapter 11, eq. (7)),

$$[\mathrm{C}\vec{\mathrm{T}}\mathrm{S}] = \mathbf{SNP}\ [\mathrm{C}\vec{\mathrm{I}}\mathrm{S}] \qquad (1)$$

where P is the matrix of rotation for precession, N for nutation, and S for earth rotation. Polar motion is thus defined as the angular separation of the third axis of the CTS, the Conventional Terrestrial Pole (CTP), and the axis of the earth for which the nutation (N) is computed (e.g., instantaneous rotation axis or Celestial Ephemeris Pole (see Chapter 8)).

J. Kovalevsky et al. (eds.), Reference Frames, 163–169.

Geodynamic requirements for CTS may be discussed in terms of global or regional problems. The former are required for monitoring the earth's rotation, while the latter are mainly associated with crustal motion studies in which one is predominantly interested in strain or strain rate, quantities which are directly related to stress and rheology. Thus for these studies, global reference systems are not particularly important although it is desirable to relate regional studies to a global frame.

For the rotation studies one is interested in the variations of the earth's rotation rate and in the motions of the rotation axis both with respect to space (CIS) and to the crust (CTS). The problem therefore is threefold:

(i)- To establish a geometric description of the crust, either through the coordinates of a number of points fixed to the crust, or through polyhedron(s) connecting these points whose side lengths and angles are directly estimable from observations using the new space techniques (laser ranging or VLBI). The latter is preferred because of its geometric clarity.

(ii)- To establish the time-dependent behavior of the polyhedron due to, for example, crustal motion, surface loading or tides.

(iii)- To relate the polyhedron to both the CIS and the CTS. For the global tectonic problems only the first two points are relevant although these may also be resolved through point (iii).

In the absence of deformation, the definition of the CTS is arbitrary. Its only requirement is that it rotates with the rigid earth, but common sense suggests that the third axis should be close to the mean position of the rotation axis and the first axis be near the origin of longitudes.

In the presence of deformations, particularly long-periodic or secular ones, the definition is more problematic, because of the inability to separate rotational (and translational) crustal motions of the crust from those of the CTS.

One geophysical requirement of the reference system is that other geophysical measurements can be related to it. One example is the gravity field. The reference frame generally used when giving values of the spherical-harmonic coefficients is tied to the mean axes of figure of the earth. This frame should be simply related with sufficient accuracy to the CTS as well as to the CIS in which, for example, satellite orbits are calculated (cf. Chapter 4). Another example is height measurements with respect to the geoid (cf. Chapter 6).

The vertical motions may require some special attention, because absolute motions with respect to the center of mass have an immediate geophysical interest and are realizable. Again, if the center of mass has significant motions with respect to the crust, such a motion will be absorbed in the future CTS, if defined as suggested above. At present there is no compelling evidence that the center of mass is displaced significantly, at least at the decade time scale.

Apart from the geometrical considerations, the configuration of observatories should be such that
(i)- there are stations on most of the major tectonic plates in sufficient number to provide the necessary statistical strength,
(ii)- the stations lie on relatively stable parts of the plate so as to reduce the possibility that tectonic shifts in some stations will not overly influence, at least initially, the parameters defining the CTS frame.

Finally one should realize that the problem of the geometric origin of the CTS is linked to that of a geocentric ephemeris frame. The center of mass of the earth is directly accessible to dynamical methods and is the natural origin of a geocentric satellite-based dynamical system. But, as such, it is model dependent. And, unless the terrestrial reference frame is also constructed from the same satellites (as is the case in various earth models such as GEM, SAO, GRIM), there may be inconsistencies between the assumed origin of a kinematically obtained terrestrial system and the center of mass. A time-dependent error in the position of the center of mass, considered as the origin of a terrestrial frame, may introduce spurious apparent shifts in the position of stations that may then be interpreted as erroneous plate motions. To avoid this problem the parameters defining the CTS frame should include translational terms as well.

2. CTS THROUGH THE MID 1980'S

Until 1984 the internationally accepted Woolard series of nutation was used to compute the position of the instantaneous rotation axis of the rigid earth, and the CTP was the Conventional International Origin (CIO), defined by the adopted astronomic latitudes of the five International Latitude Service (ILS) stations (cf. Chapter 9).

From 1984 onward the IAU 1980 (Wahr, 1981) series of nutation for the nonrigid earth gives the space position of the Celestial Ephemeris Pole (CEP). The CTP officially remained the same as before. Thus, conceptually, polar motion was to be determined from latitude observations only at the ILS stations. As described in Chapter 9, this had been done for over 80 years, and the results are the best available *long-term* polar motions, properly, but not very acurately, determined. The first axis of the CTS, the Greenwich Mean Astronomical Meridian, was defined by the adopted astronomic longitudes of time observatories participating in the work of the Bureau International de l'Heure (BIH).

For reasons explained elsewhere, the use of the CIO was no longer a reality. The common denominator being the series of nutation, observationally the CTP was defined by the coordinates of the pole as published by the International Polar Motion Service (IPMS) or by the BIH. Thus it was legitimate to speak of IPMS and BIH CTP's. The situation has become even more complicated because Doppler and laser satellite tracking, VLBI observations, and lunar laser ranging also can determine earth rotation parameters (ERP's), some of which were incorporated in the BIH computations. Further confusion arose due to the fact that the BIH had two systems: the BIH 1968 and the BIH 1979, the latter due to the incorporation of certain annual and semiannual variations of polar motion determined from the comparisons of astronomical (optical) results with those from Doppler and lunar laser observations.

Though naturally every effort has been made to keep the IPMS and BIH poles of the CTS as close as possible to the CIO, the situation could not be considered satisfactory from the point of view of the geodynamic acuracy requirement of a few parts in 10^{-8}.

3. PRESENT CONCEPT OF THE CTS

There seemed to be general agreement that the new CTS frame conceptually be defined similarly to the CIO-BIH system (Bender and Goad, 1979; Guinot, 1979; Kovalevsky, 1979; Mueller, 1975, 1981; Kovalevsky and Mueller, 1981), i.e., it should be attached to observatories located on the surface of the earth. The main difference in concept is that these can no longer be assumed motionless with respect to each other. Also they must be equipped with advanced geodetic instrumentation like VLBI or lasers, which are no longer referenced to the local plumblines as the astrometric instruments are. Thus the transformation formula may have the form

$$[\,\overrightarrow{OBS}\,]_j = \overrightarrow{L}'_j + [\,\overrightarrow{CTS}\,]_j + \overrightarrow{v}_j \tag{2}$$

where L'_j is the vector of the 'j' observatory's movement on the deformable earth with respect to the CTS. Other terms are explained below.

The $[\,OBS\,]_j$ is related to the observatory coordinates (X_j^o), determined in the terrestrial frame inherent in the observational technique 'o', through the well-known transformations involving three translation components (δ^o), three (usually very small) rotations (β^o) and a differential scale factor (c):

$$[\,\overrightarrow{OBS}\,]_j = \overrightarrow{X}_j^o + \overrightarrow{\delta}^o + \mathbf{R}_1(\beta_1^o)\,\mathbf{R}_2(\beta_2^o)\,\mathbf{R}_3(\beta_3^o)\overrightarrow{X}_j^o + c\overrightarrow{X}_j^o \qquad (3)$$

Naturally in the case of techniques which observe directions only (e.g., astrometry), the terms containing translation and scale will be omitted. Equations (2) and (3) together with (4) (and possibly others, cf. (Mueller, 1988)) may form the observation equations to be used when realizing the new type of CTS. Equations (4) derived in (Zhu and Mueller, 1983) relate an ERP series determined by the technique 'o' within its own frame of reference, to the series referenced to the CTS:

$$\begin{aligned} x_p - \beta_2^o + \alpha_1^o \sin\theta + \alpha_2^o \cos\theta &= x_p^o + v_{x_p} \\ y_p - \beta_1^o - \alpha_1^o \cos\theta + \alpha_2^o \sin\theta &= y_p^o + v_{y_p} \qquad (4) \\ \omega_d \mathrm{UT1} + \beta_3^o - \alpha_3^o &= \omega_d \mathrm{UT1}^o + v_{\mathrm{UT1}} \end{aligned}$$

where x_p^o, y_p^o and UT1^o are the observed ERP's, ω_d the conversion factor, and θ the sidereal time. The small rotations α^o are between the CIS (of the service) and that associated with the technique o.

The unknowns in the above system of equations to be solved for, in a least squares solution minimizing the square sum of the residuals v_j, are [CTS]$_j$ and L_j' for the observatories; δ^o, β^o and c^o for the terrestrial frames of the techniques; α^o for their inertial frames; and finally, the ERP (x_p, y_p and UT1) for the service. If, however, in eq. (4) the ERP's (x_p^o, y_p^o, UT1^o) are mean values averaged over intervals longer than a day, α_1^o and α_2^o cannot be determined, because the sin θ and cos θ terms average to zero in one sidereal day.

As mentioned, the parameters pertaining to the observatories ([CTS]$_j$ and L_j') define the CTS. The others give the relationship of the CTS to the technique 'o' terrestrial frame (δ^o, β^o, c^o); to the CIS (x_p, y_p, UT1); and the latter's relationship to the technique 'o' inertial frame (α^o).

The rotations in eq. (3) can either be determined from the Cartesian coordinates (e.g., Moritz, 1979)) or, for possibly better sensitivity, since the rotation is least sensitive to variations in height, only from those of the horizontal coordinates (geodetic latitude and longitude) (e.g., (Bender and Goad, 1979)). It is, however, unlikely that the rotations will be determined from astronomical coordinates, i.e., from the direction of the vertical, for the reason of inadequate observational accuracy. Note that when using this method, the deformations (and the residuals) by definition cannot have common rotational (or translational) components.

As far as the origin of the CTS is concerned, it could be centered at the center of mass of the earth, and its motion with respect to the stations can be monitored either through observations to satellites or the moon, or, probably more sensitively, from continuous global gravity observations at properly selected observatories (Mather *et al.*, 1977). For the former method, the condition

$$\sum_{D} w_D \, \vec{\delta}^{\,o}_{D} = 0$$

could be imposed on the above adjustment. The summation would be extended to all the above dynamic techniques D with given relative weights w_D. A similar condition could also be imposed on the orientation and/or the scale extended to techniques defining the best orientation and/or scale (probably VLBI) (Mueller, 1988).

The above method of determining ERP or some variation there of needs to be initialized in a way to provide continuity. This could be done through the IPMS or BIH poles, and the BIH zero meridian, at the selected initial epoch (or averaged over a well-defined time interval, say 1 to 1.2 years), uncertainties in their definition mentioned in Chapter 9 being mercifully ignored.

It is probably not useless to point out that if such a system is established, the most important information for the users will be the ERP and the transformation parameters, but for the scientist new knowledge about the behavior of the earth will come from the analysis of the residuals after the adjustment.

The IAU and IUGG recently made practical recommendations on the establishment of such a (or very similar) Conventional Terrestrial System, including the necessary plans for supporting observatories and services by establishing the International Earth Rotation Service, effective 1 January 1988 (Wilkins and Mueller, 1986). The goal of the service is the determination of the total transformation between the CTS and CIS. The service will publish not only ERP determined from the repeated comparisons (the past situation), but also the models and parameters discussed above, i.e., the parameters defining the whole system.

The repeatedly determined coordinates (or coordinate differences) of the observatories, suitably corrected for those variations which are due to well-established (especially periodic) deformations, will serve as the basis of the CTS, and from the practical point of view, the definition of the CTS frame will be the form proposed by Guinot (1981):

> The pole of the conventional terrestrial frame (CTP) is the origin of the polar motion derived by a future international service. The first axis of the frame (CTO) is the point on the equator of the CTP used by such a service for deriving UT1. In these derivation the assumption is made that the progressive changes of the reference coordinates of the observatories contributing to the determination of the earth rotation (position of the instantaneous rotation axis and UT1) do not represent statistically a rotation (and a translation).

The practical work initiated by the BIH in the mid-1980's along the above guidelines, and which is continued by the new IERS, is described in Chapter 14. For more details on the establishment and maintenance of the CTS, see (Moritz and Mueller, 1987; Mueller, 1988).

REFERENCES

Bender, P. and Goad, C., 1979, in *'The Use of Artificial Satellites for Geodesy and Geodynamics, Vol. II,'* G. Veis and E. Livieratos (eds.), National Technical Univ., Athens, 145.

Guinot, B., 1979, in *'Time and the Earth's Rotation,'* D.D. McCarthy and J.D.H. Pilkington (eds.), Reidel Publ., Dordrecht, 7.

Guinot, B., 1981, in *'Reference Coordinate Systems for Earth Dynamics,'* E. M. Gaposchkin and B. Kolaczek (eds.), Reidel Publ., Dordrecht, 125.

Kovalevsky, J. and Mueller, I., 1981, in *'Reference Coordinate Systems for Earth Dynamics,'* E.M. Gaposchkin and B. Kolaczek (eds.), Reidel Publ., Dordrecht, 375.

Kovalevsky, J., 1979, in *'Time and the Earth's Rotation,'* D.D. McCarthy and J.D.H. Pilkington (eds.), Reidel Publ., Dordrecht, 151.

Mather, R. *et al.*, 1977, Uniserv G 26, Univ. of New So. Wales, Australia.

Moritz, H., 1979, Ohio State Univ. Dept. of Geodetic Science and Surveying Rep. 294, Columbus.

Moritz, H. and Mueller, I., 1987, *'Earth Rotation: Theory and Observation,'* Ungar Publ., New York.

Mueller, I., 1975, *Geophys. Surveys*, **2**, 243.

Mueller, I., 1981, in *'Reference Coordinate Systems for Earth Dynamics,'* E.M. Gaposchkin and B. Kolaczek (eds.), Reidel Publ., Dordrecht, 1.

Mueller, I., 1988, in *'Theory of Satellite Geodesy and Gravity Field Determination,'* R. Rummel and F. Sansò (eds.), Springer Verlag, Heidelberg.

Wahr, J., 1981, *Geophys. J. R. Astr. Soc.*, **64**, 705.

Wilkins, G. and Mueller, I., 1986, *EOS, Trans. Am. Geophys. Union*, **67**, 601.

Zhu, S.Y. and Mueller, I., 1983, *Bull. Geodes.*, **57**, 29.

PART 3

ROTATION OF THE EARTH AND THE TERRESTRIAL AND CELESTIAL FRAMES

THEORETICAL ASPECTS OF THE EARTH ROTATION

HIROSHI KINOSHITA
Tokyo Astronomical Observatory, Mitaka, Tokyo, Japan
TETSUO SASAO
International Latitude Observatory of Mizusawa, Mizusawa, Iwate, Japan

The motion of the Earth is composed of an orbital motion of the centre of mass around the Sun and a rotational motion about figure axes passing through this centre of mass. Both the orbital motion and the rotational motion determine the fundamental reference frames. Both motions are permanently perturbed by gravitational attractions and torques of other members of the solar system, and the orbital motion disturbs the rotational motion and also the rotational motion disturbs the orbital motion, which is very small but is now detectable by high accuracy observations.

The rotational motion has five components: a uniform rotation about the figure axis, a forced secular motion (precession), Euler motion (Chandler motion), periodic motions (nutation), and irregular motions. Among them, if the disturbing bodies do not exist and the Earth is an axially symetric rigid body, there are only the uniform rotation and Euler motion. The external torques due to the disturbing bodies cause the precession and forced periodic motion (nutation). The non-rigidity of the Earth changes the amplitude of the nutation and the phase of the nutation due to the dissipative mechanisms, but does not change the precessional motion. Then, the geophysical phenomena on the surface and within the Earth produce irregular and unpredictable motions.

The precession was discovered by the Greek astronomer Hipparchus from the proper motion of stars. Copernicus was the first to ascribe precession to the Earth's axis motion and gave the kinematical explanation. Newton gave the dynamical explanation of the precession and calculated the rate of precession due to the Moon and the Sun (Principia). D'Alembert gave the first analytical theory of precession and nutation, which had been discovered shortly before by Bradley from observation. Euler first showed that the axis of rotation does not

J. Kovalevsky et al. (eds.), Reference Frames, 173–211.

necessarily coincide with the axis of figure, which is observed as a periodic variation in the astronomical latitudes. This variation (polar motion) was discovered about one century after Euler's prediction and confirmed by Chandler.

In Section 1, we give the fundamental equations of motion for a rigid Earth and discuss briefly Woolard's theory. In sections 2 to 4 we discuss how precession and nutations for a rigid Earth are obtained in a theory based on canonical perturbation method using Andoyer's variables in some detail, because non-rigid nutation series are based on the rigid nutation series. The relation between the fundamental reference frames and the adopted reference frame where the rotational motion is described is also discussed here. In Section 5, we give an outline of non-rigid theories of rotation of the Earth. At the end of Sections 4 and 5, in order to get a more precise theory of the rotation of the Earth than the present theory, we summarize what we should do in the near future.

1. BASIC EQUATIONS AND WOOLARD'S THEORY OF A RIGID EARTH

The basic equations of the rotational motion of a rigid body are

$$\frac{dL}{dt} = N, \tag{1}$$

where L is the angular momentum vector and N is the external torque due to the disturbing bodies. When we express the equations in the body fixed coordinates, equation (1) takes the following form:

$$\frac{dL}{dt} + \vec{\omega} \times L = N, \tag{2}$$

where $\vec{\omega}$ is the angular velocity vector of the body. When the body fixed coordinates coincide with the principal axes of the body, we have the following relations between the angular momentum vector and the angular velocity vector:

$$L_x = A\omega_x,\ L_y = B\omega_y,\ L_z = C\omega_z \tag{3}$$

where A, B, C are the principal moments of inertia. From now on, we assume without loss of generality $A \leq B < C$. The components of the angular velocity vector referred to the body fixed frame are expressed in terms of Eulerian angles, ψ, θ, φ (see figure 1):

$$\omega_x = \dot{\psi} \sin\theta \sin\varphi + \dot{\theta} \cos\varphi,$$

$$\omega_y = \dot{\psi} \sin\theta \cos\varphi - \dot{\theta} \sin\varphi,$$

$$\omega_z = \dot{\psi} \cos\theta + \dot{\varphi}. \tag{4}$$

Substituting (3) into (2), we get

$$A\frac{d\omega_x}{dt} + (C\text{-}B)\omega_y\omega_z = N_x,$$

$$B\frac{d\omega_y}{dt} + (A\text{-}C)\omega_z\omega_x = N_y,$$

$$C\frac{d\omega_z}{dt} + (B\text{-}A)\omega_x\omega_y = N_z, \tag{5}$$

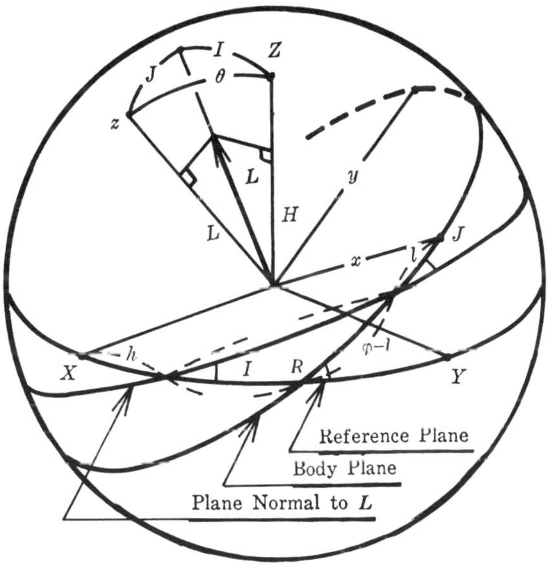

Figure 1. The coordinate system

X-Y-Z;	inertial frame.
x-y-z;	cartesian coordinates fixed to the rigid body.
ψ,θ,φ;	Eulerian angles.
L;	angular momentum vector.
$L;z$	component of the angular momentum vector.
$H;Z$	component of the angular momentum vector.

which are the Eulerian dynamical equations of a rigid body. The external torque N is expressed in the following form:

$$N_x = \frac{\sin\varphi}{\sin\theta}\left(\cos\theta\frac{\partial U}{\partial\varphi} - \frac{\partial U}{\partial\psi}\right) - \cos\varphi\frac{\partial U}{\partial\theta},$$

$$N_y = \frac{\cos\varphi}{\sin\theta}\left(\cos\theta\frac{\partial U}{\partial\varphi} - \frac{\partial U}{\partial\psi}\right) + \sin\varphi\frac{\partial U}{\partial\theta},$$

$$N_z = -\frac{\partial U}{\partial\varphi}, \tag{6}$$

where U is the disturbing potential due to the external bodies (see Section 3). The disturbing potential U is a complicated function which depends on the positions of the disturbing bodies and the orientation of the body. In order to know the orientation of the body with repsect to the reference frame, we have to solve (5) together with equation (4).

Oppolzer (1880) got the solutions of (5) in a systematic and comprehensive way with use of Le Verrier and Hansen's (1857) theories of the Sun and the Moon's motions. Woolard (1953) treated equations (5) by a similar method as Oppolzer and improved the precision of the rotational motion by adopting Newcomb and Brown's theories as the Sun and Moon's motion instead of Le Verrier (1858) and Hansen's theories. Woolard first solved Poisson equations (30) in Section 3, which are derived from equations (5) and (6) with (4), and then gave small corrections to the solutions of Poisson's equations.

Euler Motion

When the Earth is axially symetric ($A=B$) and there is no disturbing body, equation (5) can be easily solved:

$$\omega_x=f\sin l,\quad \omega_y=f\cos l,\quad \omega_z=constant,\quad l=(1\text{-}C/A)\omega_z t+l_0. \tag{7}$$

The rotational axis referred to the body frame moves counter-clock wise. Its orbit is circular and its period is about 10 months. This motion is called Eulerian motion. The observed period, however, is about 14 months, since the actual Earth is not rigid, and is called Chandler period (see Section 5). The torque free motion in case of $A\neq B$ is discussed in Section 2.

The Departure Point

In case of $A=B$, φ does not appear in U and, from the third equation of (5), we have

$$\omega_z = \textit{constant}. \tag{8}$$

The departure point X^* in the instantaneous equator is defined by

$$\overset{\frown}{X^*R} = \int \cos\,\theta d\psi, \tag{9}$$

where the point R is the ascending node of the osculating equator (see figure 1). In other words, as the osculating equator moves under the influence of the disturbing bodies, the locus of X^* is always perpendicular to the trace of the osculating plane on the celestial sphere. The speed of x axis measured from the departure point along the instantaneous equator is

$$\frac{d}{dt}\overset{\frown}{X^*x} = \frac{d}{dt}\,(\overset{\frown}{X^*R}+\overset{\frown}{Rx}) = \dot{\psi}\,\cos\,\theta+\dot{\varphi} = \omega_z \tag{10}$$

In the definition of Universal Time that has been used for many years and has been officially adopted by IAU (see Aoki *et al.*, 1982), the concept of the uniform rotation measured from the departure point is implicitly included (Aoki and Kinoshita, 1983).

2. DESCRIPTION OF A ROTATIONAL MOTION WITH USE OF ANDOYER VARIABLES

The rotational motion of the rigid Earth under the disturbing torque due to the Moon and Sun is a conservative system, and can be systematically treated by a perturbation theory based on canonical transformations, which have been successfully used for a system of particles in the various fields of Celestial Mechanics.

In describing a rotational motion of a rigid body, Andoyer variables (Andoyer 1923, Kinoshita 1972) are suitable canonical variables for the treatment of a rotational motion by a canonical perturbation method, even though any canonical variables are mathematically equivalent in the canonical perturbation theory. Using Andoyer variables as canonical sets, it is easy to apply a perturbation theory based on canonical transformation, to separate secular perturbations (precessional motion) from periodic perturbations (nutations), and to treat separately the motions of the figure, rotation, and angular momentum axes, as seen in later discussions.

We introduce the intermediate plane normal to the angular momentum vector of the body. The node and the inclination of this plane with respect to the inertial plane are denoted by h and I, and those of the body fixed plane (the equatorial plane) with respect to the previous plane by g and J. Finally, l is the longitude of the x axis measured from the nodal line of these two planes (see figure 1). Canonical variables conjugate to g,l,h are

G = the angular momentum of the body,
L = z component of the angular momentum vector = $G \cos J$,
H = Z component of the angular momentum vector = $G \cos I$.

The components of the angular momentum vector referred to the body fixed coordinate are written as

$$L_x = G \sin J \sin l, \quad L_y = G \sin J \cos l, \quad L_z = L = C\omega_z. \quad (11)$$

The relationships between Andoyer variables and Eulerian angles, are easily obtained from the spherical triangle PQR (see figure 1). The order of magnitude of the angle J between the angular momentum vector and the figure axis (z axis) is about 10^{-6}, and therefore, neglecting the second order of J, we have

$$\psi = h + \frac{J}{\sin I} \sin g + o(J^2),$$

$$\theta = I + J \cos g + o(J^2),$$

$$\varphi = l + g - J \cot I \sin g + o(J^2). \quad (12)$$

The direction of the angular velocity vector ω is expressed by the longitude of the node and the inclination of the plane perpendicular to the rotational axis, h_r and I_r:

$$h_r = h + (1 - \frac{C}{2A} - \frac{C}{2B}) \frac{J}{\sin I} (\sin g - e \sin (2l+g)) + o(J^2),$$

$$I_r = I + (1 - \frac{C}{2A} - \frac{C}{2B}) J (\cos g - e \cos (2l+g)) + o(J^2). \quad (13)$$

where

$$e = \tfrac{1}{2} (1/B-1/A)/(1/C-(1/A+1/B)/2M). \quad (14)$$

The parameter e is a measure of the triaxiality of the Earth and its value is about 3×10^{-3}. These relations (12) and (13) are valid in perturbed motion as well as in a moving reference frame (non inertial).

In the case of $A=B$, three axes - the figure axis, the angular momentum vector, and the angular velocity vector - are in a same plane, which is derived from (3). This is also valid whether or not the motion is disturbed. The angle j between the rotational axis and angular momentum axis is

$$j = (C/A\text{-}1)\, J+o(J^2), \tag{15}$$

which is about 3 cm on the Earth surface.

The Hamiltonian for torque-free motion, which is kinetic energy of the rotation, is

$$\mathrm{F}_0 = \tfrac{1}{2}\,(A\omega_x^2+B\omega_y^2+C\omega_z^2))$$

$$= \tfrac{1}{2}\,(\frac{\sin^2 l}{A} + \frac{\cos^2 l}{B})\,(G^2\text{-}L^2) + \tfrac{1}{2}\,\frac{L^2}{C}, \tag{16}$$

which is much simpler than the Hamiltonian in terms of Eulerian angles:

$$F_0 = \frac{1}{2A}\,(\, p_\theta \cos\varphi + \frac{1}{\sin\theta}\,(\, p_\psi\text{-}p_\varphi \cos\theta)\sin\varphi)^2$$

$$+ \frac{1}{2B}\,(\text{-}p_\theta \sin\varphi + \frac{1}{\sin\theta}\,(\, p_\psi\text{-}p_\varphi \cos\theta)\cos\varphi)^2 + \frac{1}{2C}\,p_\varphi^2, \tag{17}$$

where p_ψ, p_θ, p_φ, are conjugate momenta of ψ, θ, φ.

Since the angular variables g and h are cyclic, the number of degrees of freedom is only one. Therefore the general solution is easily obtained by quadrature with use of elliptic functions (Whittaker, 1964, Kinoshita, 1972). Furthermore, H does not appear in the Hamiltonian, which means the torque-free motion is degenerate and the angular momenturm vector is a constant vector, because G, H and h are constant. From Kinoshita (1972), we have the torque free motion to the first order of J and e:

$$\omega_x = \frac{C}{A}(1-\tfrac{1}{2}e)\,J\sin\tilde{l}, \quad \omega_y = \frac{C}{B}(1+\tfrac{1}{2}e)\,J\cos\tilde{l}, \tag{18}$$

$$\psi = h + \frac{J}{\sin I}\,(\cos\tilde{g} + \tfrac{1}{2}e\,\sin(\tilde{g}+2\tilde{l})),$$

$$\theta = I + J(\cos\tilde{g} + \tfrac{1}{2}e\,\cos(\tilde{g}+2\tilde{l})),$$

$$\varphi = \tilde{l} + \tilde{g} - \frac{J}{\cot I}\,(\sin\tilde{g} + \tfrac{1}{2}e\,\sin(\tilde{g}+2\tilde{l})), \tag{19}$$

where $\tilde{l}$ and $\tilde{g}$ are angular variables corresponding to l and g:

$$\tilde{l} = (1 - \frac{C}{2A} - \frac{C}{2B})\omega t + l_0,$$

$$\tilde{g} = \tfrac{1}{2}(\frac{C}{A} + \frac{C}{B})\omega t + g_0. \tag{20}$$

Equations (18) to (20) show that the Eulerian motion in case of $A \neq B$ is elliptic, and is nearly diurnal when the Earth is observed from the inertial frame. The angular velocity component about C axis (z axis) is

$$\omega_z = \omega(1 - \tfrac{1}{2}e\, J^2 \cos 2\tilde{l}). \tag{21}$$

The rotational rate is periodic due to the triaxiality, and its period is twice of the Eulerian period, but the amplitude of this periodic term is of order 10^{-14}, which is not detectable.

The uniform rotation about the principle axis is a stationary solution of the Hamiltonian (16) and the rotational energy of the stationary solutions is

$$F_{0C} = G^2/2C < F_{0B} = G^2/2B < F_{0A} = G^2/2A, \tag{22}$$

where the suffix indicates the corresponding principal axis. The stationary solution about B axis corresponds to a saddle point and its rotation is unstable. The stationary solution about A axis is stable but seculary unstable, which means this rotation becomes unstable if the dissipative mechanism exists. On the other hand, the stationary solution about C axis, which is the largest principal moment of inertia, is seculary stable. When the dissipative mechanism exists, the Eulerian motion dumps out. The actual Earth as a dynamical system is a dissipative system and the Euler motion (Chandler motion) does exist, which means that some kind of excitation mechanism works continuously (see Chapter 10).

3. DISTURBING POTENTIAL AND THE HAMILTONIAN REFERRED TO A MOVING PLANE

The disturbing potential due to a disturbing body takes the following forms with use of spherical harmonics:

$$U = U_1 + U_2,$$

$$U_1 = \frac{\kappa^2}{r^3} M' \left\{ \frac{2C-A-B}{2} P_2 (\sin \delta) + \frac{A-B}{4} P_2^2 (\sin \delta) \cos 2\alpha \right\},$$

$$U_2 = \sum_{n=3}^{\infty} \frac{\kappa^2 M' M_\oplus}{r^{n+1}} \{J_n P_n (\sin \delta) -$$

$$- \sum_{m=1}^{n} P_n^m (\sin \delta) (C_{nm} \cos m\alpha + S_{nm} \sin m\alpha)\}, \quad (23)$$

where the disturbing body is assumed to be a point mass; and α and δ are the geocentric longitude and latitude of the disturbing body M' referred to the principal axes of the Earth; and J_n, C_{nm} and S_{nm} are the coefficients of the geopotential, which depends on the density distribution of the Earth; κ^2 is the gravitational constant. The order of magnitude of the disturbing function U_1 is

$$U_1/F_0 \simeq \frac{M'}{M'+M_\oplus} \left(\frac{n'}{\omega}\right)^2 \frac{2C-A-B}{2C}$$

$$\simeq 6 \times 10^{-8} \; (Moon), \; 3 \times 10^{-8} \; (Sun), \quad (24)$$

where n' is the mean motion of the disturbing body. The effect due to the Sun is about one half of that due to the Moon. The order of magnitude of U_2 is

$$U_2/U_1 \simeq \frac{J_3 a_\oplus}{J_2 a'} \simeq 4 \times 10^{-5} \; (Moon), \; 10^{-7} \; (Sun). \quad (25)$$

In order to get a theory with a high precision, we have to take into account U_2 due to the Moon as the second order perturbation, however, we can neglect U_2 duc to the Sun because of its smallness.

The disturbing potential (23) is a complicated function of the orientation of the Earth and the positions of the disturbing bodies. We have to develop this disturbing function in terms of Andoyer variables and orbital elements of the disturbing bodies, which is simple in principle but laborious part in constructing a rotational theory of the Earth (Woolard 1953, Kinoshita 1977).

The motion of the Sun is analytically expressed as a sum of the secular part and the periodic part due to planetary perturbations (Le Verrier 1858, Newcomb 1895a, and Bretagnon 1982). The mean orbital plane defined by the secular part is called the ecliptic of date, to which the rotational motion is referred. The motion of the ecliptic is expressed in polynomials of time

$$\sin \pi_1 \sin \Pi_1 = pt + p't^2 + p''t^3 + ...,$$
$$\sin \pi_1 \cos \Pi_1 = qt + q't^2 + q''t^3 + ..., \tag{26}$$

where π_1 is the inclination and Π_1 is the longitude of the ecliptic of date referred to the fixed ecliptic and the fixed equinox at some epoch, which is the inertial reference system. The motion of the Moon, which is strongly perturbed by the Sun, is usually referred to the moving ecliptic of date.

If the disturbing function is developed in the inertial system, mixed secular terms appear which originate from the slowly moving ecliptic (26), and the development of the disturbing function is complicated, and the equations of motion are also complicated (see Woolard's equations (44) and (45), 1953). Another demerit of adopting the inertial system is that final solutions referred to the fixed ecliptic have to be transformed to those referred to the moving ecliptic, since what we need in the reduction of observations are nutations referred to the moving ecliptic.

If we adopt the ecliptic as a reference plane, we have to add an additional term E to the Hamiltonian:

$$E = G \sin I \,(\frac{d\Pi_1}{dt} \sin \pi_1 \cos (h\text{-}\Pi_1) - \frac{d\pi_1}{dt} \sin (h\text{-}\Pi_1)m)$$

$$+ \mathrm{H}\,(1-\cos\pi_1)\,\frac{d\Pi_1}{dt}\,. \tag{27}$$

Even if the Hamiltonian seems to become more complicated than the Hamiltonian referred to the inertial system, the additional term contributes only to secular motion (precession) and does not affect periodic motion (nutation) directly. In summary, using the non-inertial system as a reference system, the development of the disturbing function is greatly simplified, and it is easy to separate secular perturbations from periodic perturbations, and the transformation from the fixed system to the ecliptic is not necessary.

The Hamiltonian K for the rotational motion of the rigid Earth referred to the non-inertial system from equations (16), (23) and (26), is

$$K = F_0 + U + E, \tag{28}$$

where the longitude is measured from the fixed mean equinox of epoch along the fixed ecliptic and then along the ecliptic of date. The equations of motion with use of Andoyer variables are

$$\frac{d}{dt}(L,G,H) = -\frac{\partial K}{\partial(l,g,h)}, \frac{d}{dt}(l,g,h) = \frac{\partial K}{\partial(L,G,H)} . \tag{29}$$

From (29), we have the equations of motion for I and h:

$$\frac{d}{dt}I = \frac{1}{c\omega \sin I}\frac{\partial}{\partial h}U(J=0) + o(J),$$

$$\frac{d}{dt}h = -\frac{1}{c\omega \sin I}\frac{\partial}{\partial I}U(J=0) + o(J). \tag{30}$$

The second term of the righthand side of (30) compared with the first term is of order J, which is about 10^{-6} for the Earth. Equations of motion without the second term are called Poisson equations. Therefore, Poisson equations well represent the equations of the angular momentum axis for the Earth but not the equations of motion of the figure axis. The statement that Poisson equations give a very close approximation to the motion of the axis of figure is misleading. This statement is only valid for a very fast spinning body, where Oppolzer terms are negligible. When J is not small like Comet Halley, Poisson equations do not represent the motion of the angular momentum axis.

Strict Definition of the Ecliptic

Strictly speaking we have two kinds of definition of the ecliptic. The first one by Le Verrier is a mean orbital plane determined from the secular parts of the longitude of the ascendng node and the inclination of the Sun with respect to an inertial reference plane. If we adopt Le Verrier's definition, the observed declinations of the Sun near the solstices have a constant perturbation and the observed declination of the Sun is not zero near the equinoxes. The second one by Newcomb is a mean orbital plane such that the latitude with respect to this ecliptic does not have cos u or sin u terms (u is the argument of latitude) (see Standish 1981, Kinoshita and Aoki 1983, Chapter 3 for more detailed discussion). IAU has used Newcomb's definition so far. The relation between these two ecliptics is

$$\begin{aligned} \varepsilon_N &= \varepsilon_L + 0.''00329, \\ \Omega_N &= \Omega_L + 0.''09351 \quad \text{at } J2000.0, \end{aligned} \tag{31}$$

where ε and Ω are the obliquity and the longitude of node of the ecliptic, respectively, and the suffixes indicate Newcomb and Le Verrier.

Development of the Disturbing Function

The dominant part U_1 of the disturbing function is expressed in the following form:

$$\left(\frac{a}{r}\right)^3 P_2(\sin\delta) = \tfrac{3}{2}(3\cos^2 J-1)\sum_\nu B_\nu \cos\Theta_\nu -$$

$$-\tfrac{3}{2}\sin 2J \sum_{\varepsilon=\pm1}\sum_\nu C_\nu(\varepsilon)\cos(g-\varepsilon\Theta_\nu)$$

$$+\tfrac{3}{4}\sin^2 J \sum_{\varepsilon=\pm1}\sum_\nu D_\nu(\varepsilon)\cos(2g-\varepsilon\Theta_\nu) \tag{32}$$

and

$$\left(\frac{a}{r}\right)^3 P_2^2(\sin\delta)\cos 2\alpha = -\tfrac{9}{2}\sin^2 J\sum_\nu B_\nu\cos(2l-\varepsilon\Theta_\nu)$$

$$-3\sum_{\rho=\pm1}\sin J\,(1+\rho\cos J)\sum_{\varepsilon=\pm1}\sum_\nu C_\nu(\varepsilon)\cos(g+2\rho l-\varepsilon\Theta_\nu)$$

$$-\tfrac{3}{4}\sum_{\rho,\varepsilon=\pm1}(1+\rho\cos J)^2\sum_\nu D_\nu(\varepsilon)\cos(2g+2\rho l-\varepsilon\Theta_\nu), \tag{33}$$

in which

$$B_\nu = -\tfrac{1}{6}(3\cos^2 I-1)A_\nu^{(0)} - \tfrac{1}{2}\sin 2I A_\nu^{(1)} - \tfrac{1}{4}\sin^2 A_\nu^{(2)},$$

$$C_\nu(\varepsilon) = -\tfrac{1}{4}\sin 2I A_\nu^{(0)} + \tfrac{1}{2}(1+\varepsilon\cos I)(-1+\varepsilon\cos I)A_\nu^{(1)} +$$

$$+\tfrac{1}{4}\varepsilon\sin I\,(1+\varepsilon\cos I)A_\nu^{(2)},$$

$$D\nu(\varepsilon) = -\tfrac{1}{2}\sin^2 I A_\nu^{(0)} + \varepsilon\sin I\,(1+\varepsilon\cos I)A_\nu^{(1)} -$$

$$-\tfrac{1}{4}(1+\varepsilon\cos I)^2 A_\nu^{(2)}. \tag{34}$$

Here $A_\nu^{(i)}$ and Θ_ν are functions of positions of the disturbing bodies. Θ_ν are linear with respect to time, since they are linear combinations of the angle variables of the disturbing bodies. The coefficients $A_\nu^{(i)}$ are not constant, since the eccentricity of the Sun changes secularly due to the planetary perturbations, which, however, can be treated as constant in the periodic perturbations.

Kinoshita (1977), whose theory of the rigid Earth is implicitly used in 1980 IAU theory of nutation, adopted Newcomb's theory (1895a) for the Sun and Brown's Improved Lunar Theory (Eckert *et al.* 1966) in the calculation of $A_\nu^{(i)}$. In order to improve 1980 IAU theory of nutation and obtain much higher precision of nutation series, we have to use more elaborate and accurate theories, that is, VSOP82 (Bretagnon 1982) for the Sun and planets and ELP2000 (Chapront-Touze and Chapront 1983) for the Moon, which have been available.

4. NUTATION AND PRECESSION

In order to solve equations (29), Kinoshita (1977) adopted a canonical perturbation method (Hori 1966), which eliminates short periodic terms using Lie transformations and an averaging method. In the present problem, l, g, and the angular variables appearing in the motions of the disturbing bodies, the Moon and the Sun, are assumed to be short periodic: the period of g and l are about 1 day and 1 year, respectively; the longitude of the Moon's node has the longest period, 18.6 years, which is also considered to be short periodic in this treatment. Here we show only the first order solution (see Kinoshita, (1977) for more complete discussion).

4.1. NUTATION

First we discuss the periodic perturbation due to the symmetric part of the disturbing function U_1 (equation 23). The motion of the angular momentum axis is

$$\Delta_S h = -k \sum_{\nu \neq 0} \frac{E_\nu}{N_\nu} \sin \Theta_\nu + o(J),$$

$$\Delta_S I = -\frac{k}{\sin I} \sum_{\nu \neq 0} i_5 \frac{B_\nu}{N_\nu} \cos \Theta_\nu + o(J), \tag{35}$$

where

$$E_\nu = \frac{1}{\sin I} \frac{\partial B_\nu}{\partial I}, \quad N_\nu = \frac{d}{dt} \Theta_\nu,$$

$$k = 3 \frac{\kappa^2 M'}{a'^3 \omega} \tilde{H}, \quad \tilde{H} = \frac{2C-A-B}{2C}.$$

Here the suffix S denotes the symmetric part, a' is the semi-major axis of the disturbing body, and $\tilde{H}$ is the dynamical ellipticity of the Earth. Common factors k for the Moon and the Sun in (35) are expressed as

$$k_{☽} = 3\tilde{H} \frac{M_{☽}}{M_{☽}+M_{\oplus}} \frac{1}{F_2^3} \frac{n_{☽}^2}{\omega},$$

$$k_{\odot} = 3\tilde{H} \frac{M_{\odot}}{M_{\odot}+M_{☽}+M_{\oplus}} \frac{n_{\odot}^2}{\omega}, \tag{36}$$

where F_2 is the factor for the mean distance of the Moon in Brown's theory.

Equations (35) can be considered as solutions of Poisson equations (30) and these terms are called Poisson terms: $\Delta_S h$ and $\Delta_S I$ are nutations in longitude and obliquity of the angular momentum axis, respectively. Since N_ν appears in the denominator, the term with a longer period has a larger amplitude than the term with a shorter period. The largest term in (35) has an argument of $\Omega_{☽}$ (18.6 years) and its amplitude is about $17''$ for h and $9''$ for I. Therefore the second term of (35) whose amplitude depends on the amplitude of Euler motion is of order $1'' \times 10^{-5}$, which has not been detectable so far.

The periodic change of the orientation of the Earth can be obtained from equations (12):

$$\Delta_S\psi = \Delta_S h + \frac{k}{\sin I} \sum_{\nu} \sum_{\varepsilon=\pm 1} \frac{\varepsilon C_\nu(\varepsilon)}{n_g - \varepsilon N_\nu} \sin \Theta_\nu + o(J),$$

$$\Delta_S\theta = \Delta_S I + k \sum_{\nu} \sum_{\varepsilon=\pm 1} \frac{C_\nu(\varepsilon)}{n_g - \varepsilon N_\nu} \cos \Theta_\nu + o(J),$$

$$\Delta_S\varphi = -\cos I \Delta_S\psi + o(J). \tag{37}$$

The second terms in (37) are called Oppolzer terms. $\Delta_S\psi$ and $\Delta_S\theta$ are nutations of the figure axis and $\Delta_S\varphi$ is the equation of equinoxes. The ratio of Oppolzer terms to Poisson terms is of order N_ν/ω. Oppolzer terms become larger as the body rotates more slowly. The largest Oppolzer term for the Earth has an argument $2L_{☽}$ and its amplitude is about $0''.017$ for ψ and $0''.006$ for θ. Oppolzer terms for a slowly rotating body like Mercury or Venus are comparable with Poisson terms. On the other hand, Poisson terms for a fast rotating body like the Earth or Mars are dominant. Oppolzer terms referred to the body fixed frame have a nearly diurnal period, and are called forced diurnal polar motion or forced diurnal nutation. They cause the spurious phenomenon of dynamical variation of latitude and longitude, if we adopt the pole of the rotational axis as a celestial ephemeris pole (Atkinson 1973, 1975, Kinoshita *et al.* 1979, Seidelmann 1982, and Chapters 7 and 11 of this book).

Periodic perturbations of the orientation of the Earth due to the asymmetric part of the disturbing function (23) have the following form:

$$\Delta_A\psi = \frac{1}{\sin I}\sum_{\nu}\sum_{\varepsilon=\pm 1} F_\nu(\varepsilon)\sin(2g+2l-\varepsilon\Theta_\nu) + o(J),$$

$$\Delta_A\theta = \sum_{\nu}\sum_{\varepsilon=\pm 1} F_\nu(\varepsilon)\cos(2g+2l-\varepsilon\Theta_\nu) + o(J),$$

$$\Delta_A\varphi = -\cos I\,\Delta_A\psi + o(J), \tag{38}$$

where

$$F_\nu(\varepsilon) = \frac{B-A}{2C-A-B}kC_\nu\left(\frac{1}{n_g+2n_l-\varepsilon N_\nu} - \frac{1}{2n_g+2n_l-\varepsilon N_\nu}\right).$$

These terms are nearly semi-diurnal and the principal terms are

$$\Delta_A\psi = 0''.000037\sin 2(g+l) - 0''.000029\sin(2g+2l-L_☽) - 0''.000012\sin(2g+2l-L_\odot),$$

$$\Delta_A\theta = 0''.000015\cos 2(g+l) - 0''.000012\cos(2g+2l-L_☽) - 0''.000005\cos(2g+2l-L_\odot), \tag{39}$$

in which 1/300 is used as the approximate value of $(B-A)/(2C-A-B)$. These terms are too small to be detected. Woolard (1953) gave nutations due to the triaxiality, which are not for the axis of figure but for the axis of angular momentum.

4.2. PRECESSION

Precession is a secular perturbation from the view point of perturbation theory. Secular perturbations of the orientation of the Earth (Lieske 1967, Kinoshita 1975, Lieske *et al.* 1977) are

$$\psi^* = -(f-d\cos I_0)t - (f'-d'\cos I_0 + \tfrac{1}{2}pq)t^2 + o(t^3),$$

$$\theta^* = I_0 - qt - (q'-\tfrac{1}{2}pf+\tfrac{1}{2}pd\cos I_0)t^2 + o(t^3),$$

$$\varphi^* = \omega t + \{(f\cos I_0-d)t+(f'\cos I_0-d')+t^2\} + o(t^3), \tag{40}$$

where $ft+f't^2$ and $dt+d't^2$ are luni-solar precession and planetary precession, respectively, and p, q, q' are the coefficients of the expression of the ecliptic of date (26): $-\psi^*$ is the general precession in longitude, and the second term of φ^* is the general precession in right ascension. The luni-solar precession is a direct effect of the Moon and the Sun, and the planetary precession is an indirect effect of planets which disturb the motion of the ecliptic. The linear term of t of θ^* originates from the choice of the moving ecliptic as a reference plane and this term does not appear if a fixed ecliptic is chosen as a reference plane. The quadratic term is a purely dynamical effect of coupling between the rotational motion and the orbital motion of the Earth. Precession is not affected by the non-rigidity of the Earth (see eq. (84) for $n=0$ or eq. (97) for $\omega+\sigma=0$). In a longer time span, the right hand side of (26) is expressed as a sum of long periodic terms whose periods range from 10^5 years to 2×10^6 years.

Finally we have the orientation of the Earth as a sum of nutation and precession from equations (37), (38), and (40):

$$\psi = \psi^* + \Delta_S\psi + \Delta_A\psi,$$

$$\theta = \theta^* + \Delta_S\theta + \Delta_A\theta,$$

$$\varphi = \varphi^* + \Delta_S\varphi + \Delta_A\varphi, \tag{41}$$

4.3. NUTATIONAL COEFFICIENTS

In order to get nutational coefficients, we have to know the numerical values of common factors $k_☽$ and $k_\odot$ in (35), which are functions of the dynamical ellipticity and the ratios of the masses of the Moon, the Earth, and the Sun. Because of the large uncertainty of $\mu=M_☽/M_\oplus$ at the time, Woolard adopted 9.″21 as the coefficient of $\cos\,\Omega_☽$, which was obtained by Newcomb (1895b). Using 9.″21, Woolard first obtained $k_☽$ and then derived $k_\odot$ using 5037.″08 per tropical century at 1900 of the lunisolar precession in longitude. The value 9.″21 determined from observations includes an effect due to the non-rigidity of the Earth.

The lunisolar precession in longitude is of the form (Kinoshita 1975, 1977):

$$f_{2000} = 3\tilde{H}\Bigg\{\frac{\mu}{1+\mu}\,\frac{1}{F_2^3}\,\frac{n_☽^2}{\omega}\,\Big[(M_0-M_2/2)\cos\varepsilon+M_1\frac{\cos 2\varepsilon}{\sin\varepsilon}$$

$$+M_3\frac{\mu}{1+\mu}\,\frac{n_☽^2}{\omega\Omega_☽}\tilde{H}(6\cos^2\varepsilon-1)\Big] + \frac{M_\odot}{M_\odot+M_☽+M_\oplus}\,\frac{n_\odot^2}{\omega}S_0\cos\varepsilon\Bigg\}_{2000}-p_g, \tag{42}$$

with

$$M_0 = 496303.3 \times 10^{-6},$$
$$M_1 = -2.07,$$
$$M_2 = 0.1,$$
$$M_3 = 3020.2,$$
$$S_0 = 500209.1,$$

which are values at J2000.0 and depend only the orbital moiton of the Moon and the Sun. The terms having M_1, M_2 and M_3 as factors are not included in Newcomb's (1906) precessional theory; M_1 and M_2 come from the long-periodic terms in the Moon's motion, and M_3 arises from the second-order secular perturbation. p_g is geodesic precession, which is general relativistic effect and is included in the numerical value of lunisolar precession observationally determined (see Chapter 11). As seen from (42), if we know μ, we can determine the dynamical ellipticity, and then $k_☽$ and $k_\odot$. The ratio μ is well determined from recent data obtained by lunar and planetary spacecraft. To obtain the dynamical ellipticity, we adopt the following values of IAU (1976) System of Astronomical Constants:

$$\mu = 0.01230002,$$
$$\varepsilon_{2000} = 23^\circ\ 26'\ 21''.448,$$
$$f_{2000} = 5038.''7784/\textit{Julian century},$$
$$p_g = 1.''92/\textit{Julian century}.$$

Using these values, we get from (36) and (42)

$$\tilde{H} = 0.0032739935 = 1/305.43738,$$
$$k_☽ = 7567.''8292/\textit{Julian century},$$
$$k_\odot = 3475.''4416/\textit{Julian century},$$
$$N_{2000} = 9.''22777,$$

whcre N_{2000} is the largest nutation in obliquity with argument $\Omega_☽$. The dynamical ellipticity can be obtained, in principle, from the density distribution in the Earth. However, because we do not have accurate enough density distribution in the Earth to calculate principal moments of inertia, the dynamical ellipticity $\tilde{H}$ can be determined only from the luni-solar precession in longitude, which is obtained from the analysis of proper motions of nearby stars (Fricke 1971).

The inaccuracies of f_{2000} and μ have effects on $\tilde{H}$, $k_☽$, $k_\odot$ and N_{2000} by the following amounts:

$$\Delta\tilde{H} = 6.5 \times 10^{-7}\ \Delta f - 1.8 \times 10^{-1}\ \Delta\mu,$$

$$\begin{aligned} \Delta k_{☽} &= 1.5\ \Delta f + 2.0 \times 10^{-5}\ \Delta\mu, \\ \Delta k_{\odot} &= 0.7\ \Delta f - 1.9 \times 10^{5}\ \Delta\mu, \\ \Delta N_{2000} &= 1.8 \times 10^{-3}\ \Delta f - 2.4 \times 10^{2}\ \Delta\mu. \end{aligned} \tag{43}$$

The order of the accuracy we now have is about 0."15 per century for f_{2000} and 10^{-8} for μ. The accuracy of f_{2000} has more effect on the determination of these values than that of μ does. As seen from (43), the present accuracy of f_{2000} and μ are sufficient to determine nutational coefficients for a rigid Earth within 0."0001 other than the term with argument $\Omega_{☽}$. In order to improve the present 1980 Theory of nutation by a factor of 10, we have to know the lunisolar precession with an accuracy of 0."01 per century, which will be available in the near future by VLBI.

In the near future, we will have an observational accuracy of submilliarcsecond by VLBI. In order for a theory to be compatible with this accuracy, we have to improve the present rigid theory of rotation by one order of magnitude by taking into account of following factors:

(i)–use of more precise theories of the Moon and the Sun (for example ELP2000 and VSOP82) than those by Brown and Newcomb,

(ii)–take full account of second order effects such as disturbing potentials due to J_3 and J_4, interactions among nutations, reactions from the Moon disturbed by Earth rotation,

(iii)–direct torques from planets.

5. EFFECTS OF NON-RIGIDITY

Deviations of actual nutations and polar motions from those predicted by the rigid-body theory are interesting from a geophysical point of view, because they might tell us something valuable about physical properties of the Earth's interior. This is a major motivation for theoretical efforts to clarify effects of various nonrigidity factors in the Earth upon the nutations and polar motions.

The outer core of the Earth is the largest fluid layer of the Earth accounting for one third of total mass and 10 percent of the moment of inertia. The mobility of the core fluid thus necessarily has profound effects upon the Earth's rotation in general. Moreover, a resonance occurring in the core motion in response to the torque due to the Sun and Moon makes core effects upon some of the nutation components detectable even by the earliest international Earth rotation observations with optical instruments. Elastic deformation of the Earth changes the moment of inertia tensor and affects variations of the Earth's rotation including the nutations and polar motions.

Dynamical effects of an ideal fluid core contained in a spheroidal rigid mantle were examined in the classical works by Hopkins (1839), Kelvin (1876), Hough (1895), Sludsky (1896) and Poincaré (1910). Theoretical modelling of the nutations and polar motions based on achievements of modern seismology on radial distributions of elastic parameters within the mantle was first undertaken by Jeffreys and Vicente (1957a, b) with the aid of two simplified core models. Later, Molodensky (1961) developed a theory applicable to any stratified core. Among subsequent studies we could mention, as representative, Po-Yu Shen and Mansinha (1976), Smith (1974, 1977) and Wahr (1981a, b, c). In the following, we would like to give a brief review on the basic problems and guiding ideas in theoretical studies on the nutations and polar motions of the nonrigid Earth.

5.1 TISSERAND MEAN SYSTEM

When we consider rotational motion of a deformable body, we must somehow specify a 'body-fixed' system, whose rigid rotation could be regarded as rotation of the body itself. A widely accepted practice is to use the so-called Tisserand mean system. A Tisserand mean system for a deformable body is a rectangular reference system.

(i)–relative to which any motion of the material in the body does not contribute to the net angular momentum (Munk and MacDonald, 1960), and

(ii)–of which the axes coincide with principle axes of the body when the deformation is removed or averaged.

Hereafter, we will consider rotation of a Tisserand mean system of a body as rotation of the body and any motion of the body material relative to the system as deformation.

5.2. GENERAL FORMULATION OF THE PROBLEM

Motion of a self-gravitating, hydrostatically prestressed and dissipation-free elastic body with an isotropic stress-strain relation can be described in a rotating reference system, with bases (i_1, i_2, i_3), angular velocity vector $\vec{\omega}$ and an origin at the centre of mass, in terms of the elastic equation of motion:

$$\rho(\frac{d^2\boldsymbol{u}}{dt^2} + \frac{d\vec{\omega}}{dt} \times \boldsymbol{r} + 2\vec{\omega} \times \frac{d\boldsymbol{u}}{dt} + \vec{\omega} \times (\vec{\omega}\times\boldsymbol{r})) = \rho\nabla U+\nabla\boldsymbol{\cdot}\boldsymbol{T}, \tag{44}$$

and continuity:

$$\rho - \rho_0 + \nabla\bullet(\rho \boldsymbol{u}) = 0 \tag{45}$$

Poisson equation:

$$\nabla^2 U = -4\pi G\rho, \tag{46}$$

and the stress-strain relation:

$$T_{ij} = -(P_0+\boldsymbol{u}\bullet\nabla P_0)\delta_{ij}+\lambda\nabla\bullet\boldsymbol{u}\delta_{ij}+\mu(\frac{\partial u_i}{\partial x_j} + \frac{\partial u_j}{\partial x_i}), \tag{47}$$

where u is an infinitesimal displacement vector, U is a gravitational potential including both self-gravitation of the body and external tide-generating potential, T is a stress tensor including hydrostatic pressure, G is the gravitational constant, ρ_0 and P_0 are mass density and pressure at hydrostatic equilibrium, and λ and μ are Lamé parameters. In application to the motion of the Earth, we can directly use the above equations for the elastic mantle and inner core. For the fluid core, we can use either hydrodynamical equations or the above equations with rigidity $\mu=0$. As the rotating reference system, we choose a system rotating with an angular velocity vector close to the one of the actual Earth. Such a reference system could be a Tisserand mean system of the elastic mantle (Molodensky, 1961; Po-Yu Shen and Mansinha, 1976) or a system uniformly rotating around an axis fixed in space that is sufficiently close to a mean direction of the rotation axis of the Earth (Jeffreys and Vicente, 1957a, b; Smith, 1974, 1977; Wahr, 1981a, b, c). In either case, we assume that, in the basic state with no external torques, the Earth uniformly rotates around a fixed axis in space with a constant angular velocity $\vec{\omega}=\omega i_3$, and conditions of the hydrostatic equilibrium:

$$- \nabla P_0 + \rho_0\nabla\ (U_0+\tfrac{1}{2}\omega^2 l^2) = 0, \tag{48}$$

and

$$\nabla^2 U_0 = - 4\pi G\rho_0$$

are satisfied, were ω is the mean rotatation rate of the Earth, $l^2=x^2+y^2$, and U_0 is the potential of self-gravitation in the basic state. We further assume that the density, gravitational plus centrifugal potential and Lamé parameters in the basic state have axially symmetric distribution with small ellipticity $\varepsilon(r)$:

$$\rho_0=\rho_0(r_0),\quad \Phi_0\equiv u_0+\tfrac{1}{2}\omega^2l^2=\Phi_0(r_0),\quad \lambda=\lambda(r_0),\quad \mu=\mu(r_0), \tag{49}$$

with

$$r_0=r[1+\tfrac{2}{3}\varepsilon(r)\ P_2(\cos\,\theta)],$$

where θ is the colatitude. We regard any deviation of the Earth from the basic state as an infinitesimal perturbation.

We can readily linearize equations (44) to (47), using equations (48) for the basic state. If we were able to solve the linearized equations with suitable boundary conditions, we could rigorously calculate any nutation, polar motion and associated deformation of the nonrigid Earth. Unfortunately, effects of rotation, as presented by the third term in the left hand side of equation (44) (the Coriolis term), and ellipticity, as expressed by equation (49), cause a substantial difficulty in this problem.

Let us expand the displacement vector $\boldsymbol{u}$ and scalar variables s, like perturbed density or potential, as

$$\boldsymbol{u} = \sum_{l=0}^{\infty}\ \sum_{m=0}^{l} \vec{\sigma}_l^m + \vec{\tau}_l^m,$$

$$s = \sum_{l=0}^{\infty}\ \sum_{m=0}^{l} S_l^m(r)Y_l^m, \tag{50}$$

in terms of spheroidal and toroidal vectors of degree l and order m:

$$\vec{\sigma}_l^m = \tilde{U}_l^m(r)\ \frac{\boldsymbol{r}}{r}\ Y_l^m + \tilde{V}_l^m(r)\ r\nabla\tilde{Y}_l^m,$$

$$\vec{\tau}_l^m = -\tilde{W}_l^m(r)\ \boldsymbol{r}\times\nabla Y_l^m\ , \tag{51}$$

and surface spherical harmonics:

$$\tilde{Y}_l^m(\theta,\varphi) = P_l^m(\cos\,\theta)\ exp(-im\varphi), \tag{52}$$

where $\tilde{U}_l^m$, $\tilde{V}_l^m$, $\tilde{W}_l^m$ and $\tilde{S}_l^m$ are complex scalar functions of radius r, P_l^m is the associated Legendre function and φ is the east longitude. In the expansion with the form of equations (50) to (52), we imply real parts for $\boldsymbol{u}$ and s.

Substitution of equations (50) into the linearized version of equations (44) to (47) yields a system of equations for the scalar functions. Resultant equations are fairly complicated and we will not reproduce them here (see, for a complete expression of the system of equations, Smith, 1974). It is well-known that, in the similar system of scalar equations for a simpler model of a spherically symmetric and non-rotating Earth, the spheroidal and toroidal modes are decoupled, and a displacement field $\boldsymbol{u}$ due to an external potential of degree l and order m is completely described by a single spheroidal vector field $\boldsymbol{u}+\vec{\sigma}_l^m$, In our more general model, however, the Coriolis force and ellipticity cause coupling of the spheroidal and toroidal modes with the same order and, as a result, the displacement $\boldsymbol{u}$ due to l, m potential can only be described by an infinite chain:

$$\boldsymbol{u} = \vec{\sigma}_m^m + \vec{\tau}_{m+1}^{\,m} + \dots + \vec{\tau}_{l-1}^{\,m} + \vec{\sigma}_l^{\,m} + \vec{\tau}_{l+1}^{\,m} \dots \text{ when l–m is even,}$$

$$\boldsymbol{u} = \vec{\tau}_m^m + \vec{\sigma}_{m+1}^{\,m} + \dots + \vec{\tau}_{l-1}^{\,m} + \vec{\sigma}_l^{\,m} + \vec{\tau}_{l+1}^{\,m} \dots \text{ when l–m is odd,} \quad (53)$$

(Smith, 1974). Therefore, it is impossible to solve the system of scalar equations in a straightforward way. On the other hand, since the Coriolis force and ellipticity of the Earth obviously play essential roles in the nutations and polar motions, we are not allowed, generally speaking, to neglect them. Thus we meet a serious difficulty.

Depending on how the difficulty is handled, existing theories are divided into two groups: one includes Jeffreys and Vicente (1957a, b), Molodensky (1961) and similar theories, and another, Smith (1974, 1977) and Wahr (1981a, b, c).

5.3. MOLODENSKY'S APPROACH

As an example of the theories of the first group, we will examine Molodensky's (1961) theory as reformulated and simplified by Sasao *et al.* (1980) and Moritz (1982, 1985).

5.3.1. Separation of Rotation and Deformation

Molodensky's approach is conceptually similar to and could be regarded as an extension of the well-known theory for solving problems related with the nutations and polar motions of a wholly elastic, axially symmetric and rotating Earth (Munk and MacDonald, 1960). In fact, a basic idea for avoiding the above mentioned difficulty, 'separation of rotation and deformation', is the same in these theories. It would therefore be worthwhile for a moment to follow a general line of thought of the approach in the case of this simpler model.

Let us choose, as a reference system, the Tisserand mean system of the elastic Earth rotating with $\vec{\omega}$. The basic idea is to use the general equation (2) of the angular momentum balance instead of directly using equation (44) (note, however, that we indirectly use the equation of motion because equation (2) can be derived from equation (44)) by taking a vector product with $\boldsymbol{r}$ and taking an integral over the volume of the Earth. If we ignore the excitation functions due to the atmosphere and oceans, the only difference from the rigid Earth case is the appearance of a small perturbation c_{ij} due to the deformation in the moment of inertia tensor C_{ij}:

$$C_{ij} = \begin{bmatrix} A & 0 & 0 \\ 0 & A & 0 \\ 0 & 0 & C \end{bmatrix} + c_{ij} \tag{54}$$

where A and C are the equatorial and polar moments of inertia in the basic state. The total angular momentum vector L in the first order approximation with respect to small variables $|\vec{\omega}-\omega \boldsymbol{i}_3|/\omega$ and $|c_{ij}|/C$ is

$$\boldsymbol{L} = A\vec{\omega} + (C-A)\ \omega \boldsymbol{i}_3 + c_{31}\omega \boldsymbol{i}_1 + c_{32}\omega \boldsymbol{i}_2. \tag{55}$$

Since nutations and polar motions are completely decoupled with variations of axial spin rate in the first order theory for an axially symmetric Earth, we assume here that $\omega_3=\omega=$constant and $c_{33}=0$. Substituting equation (55) into equation (2) and introducing complex notations:

$$\tilde{\omega} = \omega_1 + i\omega_2, \quad \tilde{c} = c_{31} + ic_{32}, \quad \tilde{N} = N_1 + iN_2, \tag{56}$$

we obtain a complex equation:

$$[A\frac{d}{dt} - i(C-A)\omega]\tilde{\omega} + (\frac{d}{dt} + i\omega)\omega\tilde{c} = \tilde{N}. \tag{57}$$

It is remarkable that effects of deformation appear in this equation only through the product of inertia $\tilde{c}$. In the present problem, there are two sources causing the deformation: one is a $\tilde{Y}_2^1$-mode external tide-generating potential U_e which also causes the torque $\tilde{N}$:

$$U_e = \omega^2(\varphi_1 xz + \varphi_2 yz), \tag{58}$$

where φ_1 and φ_2 are dimensionless coefficients which can be obtained from equation (23), and another is a perturbation in the centrifugal potential due to the variable rotation of our reference system associated with the nutation and polar motion, which is often referred to as 'pole tide' U_p and also represented by the $\tilde{Y}_2^1$-mode:

$$U_p = -\omega(\omega_1 xz + \omega_2 yz). \tag{59}$$

As reminded from success of earlier theoretical calculations of the Earth tides and seismic free oscillations, the elastic deformation under the potentials can well be calculated with a reasonable accuracy by means of the familiar equations ('Earth tide equations') for the quasi-static $\vec{\sigma}_2^1$ spheriodal deformation of a spherically symmetric and non-rotating Earth, which is free from the mode-to-mode coupling. Equations (44) to (47) can easily be reduced to the simpler equations if we neglect the first three terms in the left hand side of equation (44) and put $\varepsilon(r)=0$ in equation (49). In this particular case, we even do not need to integrate the equations, because the well-known MacCullagh's theorem allows us to relate the products of inertia with the source potentials through the body tide Love number k_b. Then we obtain an expression for $\tilde{c}$:

$$\tilde{c} = -A\kappa(\tilde{\varphi} - \tilde{m}), \tag{60}$$

with

$$\tilde{\varphi} = \varphi_1 + i\varphi_2, \quad \tilde{m} = \tilde{\omega}/\omega, \tag{61}$$

where $\tilde{\varphi}$ is a normalized complex coefficient of the tide-generating potential, and $\tilde{m}$ is a 'wobble' defined as a terrestrial motion of the instantaneous rotation axis. A dimensionless coefficient κ is related with the Love number k_b by the equation:

$$\kappa \; k_b a^5 \omega^2/(3GA), \tag{62}$$

where a is the mean radius of the Earth.

Now, requiring $\tilde{\varphi}=0$, $\tilde{N}=-i(C-A)\omega^2\tilde{\varphi}=0$ and $\tilde{\omega}\propto exp(i\sigma t)$, and substituting equation (60) into equation (57), we can easily get an eigenfrequency of the well-known rotational eigenmode of the Earth, the Chandler free polar motion, which equals in the first order approximation with respect to small values of $\sim 10^{-3}$,

$$\sigma_1 = (e-\kappa)\omega, \tag{63}$$

where e is a ratio

$$e = \frac{C-A}{A}\;. \tag{64}$$

We adopt, throughout this section, a sign convention for σ that positive

σ corresponds to prograde motion.

Note here that rotation and deformation are effectively separated in this method: effects of Coriolis force and ellipticity are fully taken into account when nutation and polar motion are calculated by the exact angular momentum balance equation (57), whereas elastic deformation is calculated by the approximate Earth tide equations for the spherical non-rotating Earth. Thus the difficulty is avoided as far as we are satisfied with accuracy of the approximate scheme.

5.3.2. Elastic Mantle and Fluid Core

In order to apply the approach shown above to the case of more general Earth model containing a fluid core, we must first formulate an angular momentum balance equation for the fluid core. For this purpose, we adopt, as our reference system, the Tisserand mean system of the elastic mantle rotating with an angular velocity $\vec{\omega}$. Then linearized hydrodynamical equations of motion and continuity, which correspond to the elastic equations (44) and (45), are

$$\rho_0\left(\frac{\partial \boldsymbol{v}_f}{\partial t} + \frac{d\vec{\omega}}{dt} \times \boldsymbol{r} + 2\omega \boldsymbol{i}_3 \times \boldsymbol{v}_f\right) = -\nabla P_1 + \rho_1 \nabla U_0 + \rho_0 \nabla U_1, \tag{65}$$

and

$$\frac{\partial \rho_1}{\partial t} + \nabla \cdot (\rho_0 \boldsymbol{v}_f) = 0, \tag{66}$$

where $\boldsymbol{v}_f$ is velocity of core flow relative to the mantle frame, ρ_1, P_1, U_1 are perturbed density, pressure and potential, respectively. On the core flow $\boldsymbol{v}_f$, Molodensky (1961) introduced an assumption, which plays a fundamental role in the present approach. He assumed that $\boldsymbol{v}_f$ is composed of a dominant rigid-rotation term $\vec{\omega}_f \times \boldsymbol{r}$ and a small remainder $\boldsymbol{v}$ due to non-sphericity and deformation of the equipotential surfaces and compressibility of the core fluid:

$$\boldsymbol{v}_f = \vec{\omega}_f \times \boldsymbol{r} + \boldsymbol{v}, \tag{67}$$

where $\vec{\omega}_f = \overset{f}{\omega}_1 \boldsymbol{i}_1 + \overset{f}{\omega}_2 \boldsymbol{i}_2$ (we assume, for our axially symmetric model, $\overset{f}{\omega}_3 = 0$) is chosen to be an incremental 'angular velocity vector' of the fluid core in Tisserand's sense, so that an incremental angular momentum due to the core flow $\boldsymbol{v}_f$ is

$$\boldsymbol{h}_f = \int_{V_f} \rho \boldsymbol{r} \times \boldsymbol{v}_f dV = A_f \vec{\omega}_f. \tag{68}$$

A_f here stands for the equatorial moment of inertia of the fluid core in the basic state. Substituting equation (67) into the equation of motion in (65), taking a vector product of the resultant equation with r and integrating it over the volume of the fluid core with the aid of the equation of continuity, we obtain an equation for an angular momentum $\boldsymbol{L}_f$ of the fluid core:

$$\frac{d\boldsymbol{L}_f}{dt} - \vec{\omega}_f \times \boldsymbol{L}_f = 0, \tag{69}$$

which describes the angular momentum balance of the fluid core (Sasao *et al.*, 1980). In the derivation of equation (69), we adopted an approximation where products of the small term $\boldsymbol{v}$ with small values of the order of ellipticity (~1/400) are neglected. Comparing equation (69) with a general equation of the angular momentum conservation:

$$\frac{d\boldsymbol{L}_f}{dt} + \vec{\omega} \times \boldsymbol{L}_f = \boldsymbol{K} + \boldsymbol{N}_f, \tag{70}$$

where $\boldsymbol{N}_f$ is an external tide-generating torque acting on the fluid core, we obtain an expression for a torque $\boldsymbol{K}$ due to pressure force at the core-mantle boundary (inertial coupling torque):

$$\boldsymbol{K} = (\vec{\omega}+\vec{\omega}_f) \times \boldsymbol{L}_f - \boldsymbol{N}_f. \tag{71}$$

This expression of the inertial coupling torque is exactly the same as that derived from the classical theory for the homogeneous and incompressible core (Rochester, 1976). Moritz (1982, 1985) derived an equation equivalent to equation (69) in a very general and symmetric way based on an elegant variational formalism.

Let us denote now the moment of inertia tensor of the fluid core in a form similar to equation (54):

$$C^f_{ij} = \begin{bmatrix} A_f & 0 & 0 \\ 0 & A_f & 0 \\ 0 & 0 & C_f \end{bmatrix} + c^f_{ij} \tag{72}$$

where C_f is the polar moment of inertia of the fluid core in the basic state and c^f_{ij} is the perturbed moment of inertia tensor. Then, taking into account equations (54), (68) and (72), we obtain first order expressions for angular momenta for the whole Earth $\boldsymbol{L}$ and the fluid core $\boldsymbol{L}_f$:

$$\boldsymbol{L} = A\vec{\omega} + (C-A)\omega\boldsymbol{i}_3 + A_f\vec{\omega}_f + c_{31}\omega\boldsymbol{i}_1 + c_{32}\omega\boldsymbol{i}_2,$$

$$\boldsymbol{L}_f = A_f(\vec{\omega}+\vec{\omega}_f) + (C_f-A_f)\omega\boldsymbol{i}_3 + \overset{f}{c}_{31}\omega\boldsymbol{i}_1 + \overset{f}{c}_{32}\omega\boldsymbol{i}_2, \tag{73}$$

where we again assumed $\overset{f}{c}_{33}=0$ in view of the axial symmetry. Substituting equations (73) into equations (2) and (69) and introducing complex notations:

$$\tilde{\omega}_f = \overset{f}{\omega}_1 + i\overset{f}{\omega}_2, \quad \tilde{c} = \overset{f}{c}_{31} + i\overset{f}{c}_{32}, \tag{74}$$

we obtain complex equations of the angular momentum balance for the whole Earth:

$$[A\frac{d}{dt} - i(C-A)\omega]\tilde{\omega} + (\frac{d}{dt} + i\omega)(A_f\tilde{\omega}_f+\omega\tilde{c}) = \tilde{N}, \tag{75}$$

and for the fluid core:

$$A_f\frac{d\tilde{\omega}}{dt} + (A_f\frac{d}{dt} + iC_f\omega)\tilde{\omega}_f + \omega\frac{d\tilde{c}_f}{dt} = 0. \tag{76}$$

Here we neglect the effects of rotation of the solid inner core in view of the smallness of its moment of inertia. Similarly to the case of the wholly elastic Earth, effects of deformation appear in these equations only through the products of inertia $\tilde{c}$ and $\tilde{c}_f$. Now we have three sources of elastic deformation, i.e. the external tide-generating potential proportional to $\tilde{\varphi}$ (equation (55)), the pole tide potential proportional to $\tilde{m}$ (equation (59)) and dynamical pressure at the core-mantle boundary due to the incremental rotation of the fluid core which is proportional to $\tilde{m}_f=\tilde{\omega}_f/\omega$. All of them are characterized by $\tilde{Y}_2^1$-mode spherical harmonics. We then obtain from the familiar Earth tide equations, the reduced version of equations (44) to (47), and boundary conditions for quasi-static $\vec{\sigma}_2^1$-deformation of an spherical non-rotating Earth,

$$\tilde{c} = -A[\kappa(\tilde{\varphi}-\tilde{m}) - \xi\tilde{m}_f], \tag{77}$$

and

$$\tilde{c}_f = -A_f[\gamma(\tilde{\varphi}-\tilde{m}) - \beta\tilde{m}_f], \tag{78}$$

where κ, ξ, γ and β are dimensionless parameters obtainable from numerical integration of the Earth tide equations for an appropriate Earth model. κ, in particular, can be calculated by equation (62). ξ and γ are related with each other by a reciprocity relation:

$$A\xi = A_f \gamma. \tag{79}$$

For a particular model by Wang (1972), we have

$$\begin{aligned} A &= 8.013 \times 10^{44}\ g\ \mathrm{cm}^2, & A_f &= 9.152 \times 10^{43}\ g\ \mathrm{cm}^2, \\ e &= 3.245 \times 10^{-3}, & e_f &= 2.525 \times 10^{-3}, \\ \kappa &= 1.045 \times 10^{-3}, & \xi &= 2.252 \times 10^{-4}, \\ \gamma &= 1.971 \times 10^{-3}, & \beta &= 6.270 \times 10^{-4}, \end{aligned} \tag{80}$$

where $e_f=(C_f-A_f)/A_f$.

Equations (75) to (78) now form a closed system and can be used for calculations of nutations and polar motions of the Earth. Here again rotation and deformation are conveniently separated. In terms of the expansion of equation (53), this approach corresponds to a truncation of the infinite chain with $\boldsymbol{u}=\vec{\tau}_1^1+\vec{\sigma}_2^1$ in the fluid core ($\vec{\omega}_f \times \boldsymbol{r}$ rotation and $\vec{\sigma}_2^1$ deformation) and with $\boldsymbol{u}=\vec{\sigma}_2^1$ in the mantle.

Po-Yu Shen and Mansinha (1976) used a more general truncation $\boldsymbol{u}=\vec{\tau}_1^1+\vec{\sigma}_2^1+\vec{\tau}_3^1$ for the core, but they still used the spherical and non-rotating model for the mantle ($\boldsymbol{u}=\vec{\sigma}_2^1$). Their results are generally agreeing with Molodensky's (1961) despite the different degrees of approximation. Therefore, their results can be regarded as evidences for the general appropriateness of Molodensky's (1961) approach.

5.3.3. Free Core Nutation and Fluid Core Resonance

From equations (75) to (78) with $\tilde{\varphi}=0$ and $\tilde{N}=0$, we obtain two rotational eigenmodes of the Earth. One is the Chandler polar motion with an approximate eigenfrequency for the oceanless Earth:

$$\sigma_1 = \frac{A}{A_m}(e-\kappa)\omega, \tag{81}$$

where $A_m=A-A_f$ is the moment of inertia of the mantle. Note that the fluid core reduces the Chandler period by 10 percent compared with the value for the wholly elastic Earth model. Since this reduction is almost compensated by the increase due to the oceans, the actual Chandler period is close to the elastic Earth value. Another eigenmode is called 'free core nutation' (FCN) and has an approximate eigenfrequency:

$$\sigma_2 = -\omega + n_0, \tag{82}$$

with

$$n_0 = -\frac{A}{A_m}(e_f-\beta)\omega,$$

where e_f is

$$e_f = \frac{C_f-A_f}{A_f}. \tag{83}$$

We adopted here a first order approximation with respect to the values of the order of e and e_f or smaller. Equation (82) shows that the FCN frequency is close to $-\omega$. Hence the mode is often referred to as 'nearly diurnal free wobble'. However, the frequency is with respect to the rotating reference frame. Viewed from the inertial space, the FCN is a slow retrograde motion with spatial frequency n_0, kinematically very similar to the forced luni-solar nutations. For a physical picture of the eigenmode, one can conceive a fluid core rotating around an axis inclined to the symmetry axis of the spheroidal core-mantle boundary. Since the dynamical pressure due to the oblique rotation asymmetrically pushes the boundary surface, the resulting force gives rise to a non-zero net torque, which is nothing but the inertial (or pressure) coupling torque. Retrograde nutations of both mantle and core under the coupling torque is the FCN.

In the same first order approximation with respect to the values of the order of e and e_f or smaller, the equations (75) and (78) yield amplitude of the wobble $\tilde{m}$ normalized by the rigid Earth value $\tilde{m}_R$, which are associated with a circular motion component of a nutation induced by a torque $\tilde{N}=\tilde{N}\exp[i(-\omega+n)t]$,

$$\frac{\tilde{m}}{\tilde{m}_R} = 1 - \frac{\kappa}{e}\frac{n}{\omega} + \frac{A_f}{A}\frac{n}{\omega-n}\frac{\tilde{m}_f}{\tilde{m}_R}, \tag{84}$$

and

$$\frac{\tilde{m}_f}{\tilde{m}_R} = \frac{A}{A_m}\frac{\omega-n}{n-n_0}\left(1-\frac{\gamma}{e}+\frac{\gamma-\kappa}{e}\frac{n}{\omega}\right), \tag{85}$$

$$\tilde{m}_R = \frac{i\tilde{N}}{A\omega(\omega-n)}. \tag{86}$$

We see a resonance at $n=n_0$ in equation (85) and, therefore, also in equation (84). This resonance in the nutational response of the Earth to the external tide-generating force is called 'fluid core resonance'. The resonance is the major reason why the fluid core has the largest influence upon the nutations among the non-rigidity factors.

As reminded from the well-known Poinsot representation, we have following simple kinematical relations between wobbles $\tilde{m}$, $\tilde{m}_R$ and spatial motions of the figure axes $\tilde{\zeta}$, $\tilde{\zeta}_R$ for each circular motion component

$$\tilde{m} = -(n/\omega)\tilde{\zeta} \quad \text{and} \quad \tilde{m}_R = -(n/\omega)\tilde{\zeta}_R,$$

and hence

$$\tilde{m}/\tilde{m}_R = \tilde{\zeta}/\tilde{\zeta}_R. \tag{87}$$

(Sasao *et al.*, 1977). $\tilde{\zeta}$ here stands for the spatial motion of the axis of figure defined for mean shape of the mantle (the third axis of our Tisserand mean system). Thus, using the ratio $\tilde{\zeta}/\tilde{\zeta}_R$ calculated by equations (84), (85) and (87), together with the nutation coefficients for the axis of figure of the rigid Earth (Kinoshita, 1977; Kinoshita *et al.*, 1979 and Section 4), we can easily obtain nutation coefficients for the axis of figure defined for the mean shape of the mantle (see, for more details, Kinoshita *et al.*, 1979).

The fluid core resonance occurs also in the diurnal tides due to external potential of $\tilde{Y}_2^1$-mode, because the deformation induced by the dynamical pressure of the core fluid is resonant at the FCN frequency. Equations (77) and (78) with equation (85) are examples of the resonance in the tidal deformations. Tidal Love numbers for the diurnal tides, in particular, are also resonant at the FCN frequency.

5.4. THEORY OF WAHR

Smith (1974, 1976, 1977) and Wahr (1981a, b, c) worked out a theoretical scheme which is capable of calculating normal modes and nutational and tidal responses of an elliptical rotating Earth with a great accuracy suitable to recent precision geodetic measurements. According to them, let us now adopt a new reference system which is uniformly rotating around an axis fixed in space. It is assumed that the axis and rotating rate of the system are sufficiently close to the mean direction of the rotation axis and the mean rotation rate, respectively, of the Earth, so that any (rotational and/or deformational) displacement of a particle in the Earth can be regarded as infinitesimal. In this reference frame, $\vec{\omega}$ in equation (44) is a constant vector $\vec{\omega}=\omega \boldsymbol{i}_3$ and the second term in its left hand side $(d\vec{\omega}/dt) \times \boldsymbol{r}$ is identically zero.

5.4.1. Normal Modes of an Elliptical Rotating Earth

Smith (1974) derived a general expression for the infinite set of coupled equations and boundary conditions for the scalar functions in equations (50) and (51) as applied to a rotating slightly elliptical Earth. In order to solve the equations for a particular problem on nutation, polar motion and related deformation, Smith (1977) introduced a truncated representation of equation (53):

$$\boldsymbol{u} = \vec{\tau}_1^1 + \vec{\sigma}_2^1 + \vec{\tau}_3^1 \,. \tag{88}$$

In this truncated scheme, we get a finite set of equations. Numerical solution of the equations under suitable boundary conditions with no external forcing yields elastic-gravitational normal modes of the elliptical rotating Earth. They are

(i)-infinite number of seismic free oscillation modes with eigenperiods less than one hour,

(ii)-rotational eigenmodes, which include
  tilt-over-mode, i.e. a uniform rotation around a fixed axis inclined to the $\boldsymbol{i}_3$-axis of our reference frame, which does not accompany any deformation,
  Chandler polar motion,
  FCN,
  free polar motion of the inner core,

(iii)-infinite number of internal core modes, which are inertial-gravitational oscillations almost confined in the fluid core.l

For DG579 model with squared Brunt-Väisälä frequency $N^2=8.1\times10^{-9}s^{-2}$ for a stably stratified fluid core, we obtain for the Chandler and FCN frequencies:

$$\begin{aligned} \sigma_1 &= (1/403.6)\omega, \\ \sigma_2 &= -(1/0.9978)\ \omega, \end{aligned} \tag{89}$$

(Smith, 1977). Accuracies of the calculations are checked by comparing numerical solutions with known analytical solutions for a number of simple models. It is then confirmed that this method gives correct solutions for the Chandler and FCN modes. The only discrepancy is observed between numerical and analytical solutions for the internal core modes and free polar motion of the inner core, for which probably the truncation in equation (88) is too simplified. Note that Smith's (1974, 1977) and subsequently Wahr's (1981a, b, c) theories do not use any artificial separation of rotation and deformation, and even do not use explicitly the general angular momentum balance equation (2). The only assumption adopted in their theories is the truncation of solution given in equation (88). As a result, the theories consistently take into account effects of rotation and ellipticity in every material layer and at every boudary of the Earth.

On the basis of the normal mode theory, Smith and Dahlen (1981) gave the most thorough theoretical expression of the Chandler period taking into account effects of the equilibrium oceans, and showed that the slight anelasticity of the mantle is likely to account for both period and Q-value of the Chandler motion.

5.4.2. Normal Mode Expansion of Forced Responses of an Elliptical Rotating Earth

When a complete set of normal modes of an elliptical rotating Earth is given, we can calculate any rotational and deformational response of the Earth to an external force:

$$\boldsymbol{f} = \rho_0 \nabla U_e \propto exp\ (i\sigma t), \tag{90}$$

by means of a normal mode expansion. Wahr (1981a) showed that in the frequency domain the response $\boldsymbol{u}(\boldsymbol{r},\sigma)$ can be expressed in a form:

$$\boldsymbol{u}(r,\sigma) = \sum_n a_n(\sigma)\ \boldsymbol{u}_n(\boldsymbol{r}), \tag{91}$$

with

$$a_n = \frac{1}{2} \frac{(\boldsymbol{f}, \boldsymbol{u}_n)}{(\sigma - \sigma_n)[(\boldsymbol{u}_n, i\vec{\omega}\times\boldsymbol{u}_n) - \sigma_n(\boldsymbol{u}_n, \boldsymbol{u}_n)]},$$

where $\boldsymbol{u}_n(\boldsymbol{r})$ is n-th displacement eigenvector and $(\boldsymbol{u},\boldsymbol{v})$ denotes the inner product:

$$(\boldsymbol{u},\boldsymbol{v}) = \int_{V_E} \rho \boldsymbol{u} \cdot \boldsymbol{v}^* dV$$

with * denoting complex conjugation and V_E the volume of the Earth. We assume here that the force $\boldsymbol{f}$ does not torque the Earth around the $\boldsymbol{i}_3$ axis. Note that Wahr (1981a, b, ac) adopts a sign convention for σ which is opposite to ours. In order to perform the normal mode expansion, Wahr first recalculated the normal modes for the 1066A model of Gilbert and Dziewonski (1975) and three other Earth models, slightly extending Smith's (1977) algorism. Among the normal modes, we ignore the uncertain internal core modes and the inner core mode, because they are most likely to have no significant effect upon the mantle motion.

The general infinitesimal displacement vector $\boldsymbol{u}(\boldsymbol{r},\sigma)$ can be separated into rotational and deformational terms in the following fashion:

$$\begin{aligned}\boldsymbol{u}(\boldsymbol{r},\sigma) &= \tilde{\eta}(r,\sigma)\boldsymbol{T}_1^1 + \boldsymbol{b}(r,\sigma) \\ &= \tilde{\eta}(r,\sigma)(\boldsymbol{i}_1 - i\boldsymbol{i}_2) \times \boldsymbol{r} + \boldsymbol{b}(r,\sigma),\end{aligned} \tag{92}$$

where $\tilde{\eta}(r,\sigma)=\eta_1(r,\sigma)+i\eta_2(r,\sigma)$ is a radially dependent nutation angle (η_1 around $\boldsymbol{i}_1$-axis and η_2 around $\boldsymbol{i}_2$-axis) and $\boldsymbol{b}(\boldsymbol{r},\sigma)$ stands for a body-tide deformational displacement. Here again the separation of rotation and deformation follows the Tisserand mean concept, namely we choose $\tilde{\eta}$ in such a way that the complex angular velocity vector of a spheroidal shell of effective radius r (r_0 in equation (49)) in Tisserand's sense will be $i\sigma\tilde{\eta}$. If we denote a unit vector of the axis of figure for the Tisserand mean system of the shell as $\boldsymbol{i}_3+\vec{\zeta}(r,\sigma)$, then the 'spatial motion of the figure axis of the shell' $\tilde{\zeta}(r,\sigma)=\xi_1(r,\sigma)+i\zeta_2(r,\sigma)$ equals

$$\zeta(r,\sigma) = -i\tilde{\eta}(r,\sigma). \tag{93}$$

Similarly to (92), the eigenvector $\mathbf{u}$ is also presented as

$$\boldsymbol{u}_n(\boldsymbol{r}) = \tilde{\eta}_n(r)(\boldsymbol{i}_1 - i\boldsymbol{i}_2) \times \boldsymbol{r} + \boldsymbol{b}_n(\boldsymbol{r}), \tag{94}$$

where the first and second terms correspond to nutational and deformational displacements in the normal mode. Then we get from equations (91) and (94) an expansion for the forced nutations:

$$\tilde{\eta}(r,\sigma) = \sum_n a_n(\sigma)\, \tilde{\eta}_n(r). \tag{95}$$

Direct use of equation (95) for calculating the nutations would require a summation of the infinite number of the seismic free oscillation modes. In order to avoid the infinite summation, Wahr (1981b, c) first directly integrated the truncated equations at the frequency of the O_1 tide $\sigma_{01}=-0.927\omega$ to get the response $\tilde{\eta}(r,\sigma_{01})$ and then expressed the response $\tilde{\eta}(r,\sigma)$, at any other frequency σ, in the following form:

$$\tilde{\eta}(r,\sigma) = \tilde{\eta}(r,\sigma_{01}) + \sum_n [a_n(\sigma) - a_n(\sigma_{01})]\, \tilde{\eta}_n(r). \tag{96}$$

Since the seismic free oscillation contributions to equation (95) are very nearly frequency-independent at the diurnal frequency band, they almost cancel out by the subtraction in the second term at the right hand side of equation (96). Then fewer modes are needed in equation (96) than in equation (95). According to Wahr (1981c), nine normal modes (six seismic free oscillations and three rotational eigenmodes) are sufficient to guarantee an accuracy better than 1/300. Now we can calculate $\tilde{\eta}(a,\sigma)$ for the 1066A model and take a ratio $\tilde{\eta}(a,\sigma)/\tilde{\eta}_R(a,\sigma)$ which equals $\tilde{\zeta}(a,\sigma)/\tilde{\zeta}_R(a,\sigma)$ according to equation (93), where, as before, $r=a$ is the mean radius of the Earth and the suffix R implies the rigid Earth value. In an approximation with an accuracy better than 1/300, we have (Wahr, 1981b)

$$\frac{\tilde{\zeta}(a,\sigma)}{\tilde{\zeta}_R(a,\sigma)} = 1 + \left\{0.416 + (\sigma-\sigma_{01})\left[\frac{0.810}{\sigma-\sigma_1} + \frac{0.665}{\sigma-\sigma_2} - \frac{1.06}{\sigma-e\omega}\right]\right\} \frac{(\sigma+\omega)(\sigma-e\omega)}{\omega^2}, \tag{97}$$

where e (equation (64)) is

$$e = 0.00328,$$

the Chandler frequency for the oceanless Earth is

$$\sigma_1 = 0.002478\omega = (1/403.6)\omega$$

and the FCN frequency is

$$\sigma_2 = -1.0021714\omega = -\omega+n_0, \tag{98}$$

with

$$n_0 = -(1/460.53)\omega,$$

in the 1066A model. Multiplying the normalized amplitudes obtained above with the rigid Earth values (Kinoshita, 1977; Kinoshita *et al.*, 1979), we get the nutation coefficients for the spatial motion of the axis of figure of the Tisserand mean system of the surface layer of the Earth ($r=a$). The body tide displacement $\boldsymbol{b}(\boldsymbol{r},\sigma)$ is also calculated in a similar way (Wahr, 1981c).

Now we can compare equation (97) with equations (84) and (85), noting that $\sigma=-\omega+n$. Both equations show the same characteristic resonance in the nutation amplitudes due to the FCN. Both show a similar tendency $\tilde{\zeta}\to\tilde{\zeta}_R$ in a limiting case of $\sigma\to-\omega(n\to 0)$. This is a demonstration of the well-known fact that the non-rigidity almost does not affect the precession. Despite the big difference in fundamental structures of the theories, the numerical results derived from the two equations are in fairly good agreement (typically in 1% level, or better than 1 millisecond of arc (mas) in the nutation coefficients). It must be kept in mind however that Molodensky's (1961) and similar theories are dependent on an ad hoc assumption on the core flow and use equations for the deformation of a spherical non-rotating Earth, which are out of date in modern theoretical calculations of the tides and free oscillations. Wahr's (1981a, b, c) approach is definitely more advanced and more accurate than Molodensky's (1961) and similar approaches. Wahr's theoretical nutations and tides are the only standards available at present which can be directly compared with the recent high precision Earth rotation and tide measurements. Equations (84) and (85), on the other hand, could still be useful in physical interpretations of some observed results, because they provide analytical expressions for the nutational and tidal responses in terms of the Earth model parameters.

5.4.3. IAU 1980 Theory of Nutation

Wahr's nutation and tide series are officially adopted as the IAU and IUGG constants to be used in the reduction of geodetic and astrometric observations. The nutation series, which are given in a table of coefficients in longitude and obliquity, are called 'IAU 1980 Theory of Nutation'. As mentioned above, the coefficients are calculated for the axis of figure for the Tisserand mean system of the surface layer of the Earth (Wahr's B-axis), which is proven to correspond to the actual nutation observable in existing geometrical observation techniques (Wahr, 1981b; see also Kinoshita *et al.*, 1979). The internal precision of the theoretical nutation coefficients is estimated to be 0.1 mas.

5.5. RECENT OBSERVATIONS AND THEIR IMPLICATIONS

It has long been confirmed that the theoretical nutations obtained from the earlier studies by Jeffreys and Vicente (1957a), Molodensky (1961) and others were in general agreement with the optical astrometry observations (as a summary, see Kinoshita *et al.*, 1979). Wako (1970), in particular, successfully explained Kimura's annual z-term in terms of the effect of non-rigidity as appeared in the semi-annual nutation component, thus essentially clarifying this 70-year old mystery in the international latitude observations. Yokoyama (1976) later showed that the non-rigidity effect in the semi-annual nutation also affects results of the time observations and gives rise to so-called τ-term similar to the z-term in the latitude. The nutation observations gave an independent, though not very precise, confirmation of the seismologically constructed Earth models.

Herring *et al.* (1986) recently analyzed results of VLBI (Very Long Baseline Interferometry) observations from July 1980 to December 1984 carried out by the POLARIS-IRIS network and estimated corrections to Wahr's (1981b) theoretical nutation coefficients. The largest deviation of $-1.80\pm0.18-i(0.42\pm0.18)$ mas was found in the retrograde annual nutation known as the term closest to the FCN. Gwinn *et al.* (1986) interpreted the result as an indication of a departure of the actual FCN from the theoretical model. They estimated period and damping time of the FCN to be 435.0±2 sidereal days and 7000±3500 sidereal days. Note that the estimated eigenperiod differs from the theoretical value by as much as 25 days. Also, the finite damping time is not, of course, expected in the dissipation-free theory. This must be explained in terms of some dissipative processes occurring in the Earth.

Prior to the discovery by Herring *et al.* (1986), Goodkind (1983), Sato (1984) and Zürn *et al.* (1986) had estimated the same FCN parameters from the Earth tide measurements. The results could be summarized as follows:

	Gwinn *et al.* (1986)	Goodkind (1983)	Sato (1984)	Zürn *et al.* (1986)
Technique	VLBI	Cryogenic gravimeter	Strainmeter	Cryogenic gravimeter
Data span	53 months	18 months	48 months	33 months
Components in analysis	annual nutation	$P_1, K_1, \psi_1, \varphi_1$	p_1, K_1	$P_1, K_1, \psi_1, \varphi_1$
FCN period in s.d.	435.0±2	477±?	not estimated	431.2±3.1
FCN damping time in s.d.	7000±3500	270 (160 ~ 450)	700±300	990±100

The FCN period by Zürn *et al.* is in a striking agreement with that of Gwinn *et al.* However, the damping time of Zürn *et al.*, which agrees fairly well with Sato's, is an order of magnitude smaller than the corresponding value of Gwinn *et al.* Goodkind's results show somewhat anamolous disagreement with others' and might be affected by poorly modeled ocean tidal loadings or other sources of disturbance.

Although the observations are still to some extent contradictory, one thing seems clear: the recent precision VLBI and tide measurements are now achieving enough accuracy to be sensitive to the physical properties of the Earth's deep interior which have not been foreseen in the best available seismologically constructed Earth models. The observed deviation from the theory might be due to the departure of the core-mantle boundary shape from the hydrostatic equilibrium figure (Gwinn *et al.*, 1986), the dissipative core-mantle coupling (Toomre, 1974; Rochester, 1976; Sasao *et al.*, 1977, 1980), the ocean-induced nutations (Wahr and Sasao, 1981; Molodensky, 1981), the mantle anelasticity (Molodensky, 1981; Wahr and Bergen, 1986) or some combination of them. To date, none of the theories could adequately explain the reported damping time of the FCN, even in order of magnitude. Apparently, more efforts are required in theory, as well as in observation, with regard to this point. In particular, it seems important to develop a theory of the core-mantle dissipative coupling which properly takes into account the effects of the bumps at the core-mantle boundary (see, for example, Hide, 1986). Also, a refinement of the rigid-Earth theory to achieve the internal precision better than 0.1 mas is urgently needed in view of the rapid improvement of the accuracy in the VLBI and other high-precision techniques.

It will be very interesting to detect the FCN itself. From a theoretical point of view, the detection seems plausible (Sasao and Wahr, 1981; see related discussions in Herring *et al.*, 1986). The free mode, if detected, will give us further information on the physical processes at the core-mantle boundary.

REFERENCES

Andoyer, H., 1923, *Cours de Mécanique Céleste*, vol. 1, Gauthier-Villars, Paris, p. 54.

Aoki, S., Guinot, B., Kaplan, G.H., Kinoshita, H., McCarthy, D.D. and Seidelmann, P.K., 1982, *Astron. Astrophys.*, 108, 359.

Aoki, S. and Kinoshita, H., 1983, *Celes. Mech.*, 29, 335.

Atkinson, R.d'E., 1973, *Astron. J.*, 78, 147.

Atkinson, R.d'E., 1975, *Monthly Notices Roy. Astron. Soc.*, 71, 381.

Bretagnon, P., 1982, *Astron. Astrophys.*, 114, 278.

Chapront-Touze, M. and Chapront, J., 1983, *Astron. Astrophys.*, 124, 50.

Eckert, W.J., Walker, M.J. and Ecker, D., 1966, *Astron. J.*, 71, 314.

Fricke, W., 1971, *Astron. Astrophys.*, 13, 298.

Gilbert, F., and Dziewonski, A.M., 1975, *Phil. Trans. R. Soc.*, 278A, 187.

Goodkind, J.M., 1983, *Proc. 9th Int. Symp. Earth Tides*, J. Kuo (ed.),

Schweizerbart, Stuttgart, 569.
Guinot, B., 1979, *Proc. 82th IAU Symp. Time and Earth's Rotation*, D.D. McCarthy and J.D.H. Pilkington (eds.), p. 7.
Gwinn, C.R., Herring, T.A. and Shapiro, I.L.I., 1986, *J. Geophys. Res.*, 91, 4755.
Hansen, P.A., 1857, *Tables de la Lune construites d'après le principe newtonien de la gravitation universelle*, 4to, London.
Herring, T.A., Gwinn, C.R., and Shapiro, I.I., 1986, *J. Geophys. Res.*, 91, 4745.
Hide, R., 1986, *Q.J.R. Astr. Soc.*, 27, 3.
Hopkins, W., 1839, *Phil. Trans. R. Soc.*, 129, 381.
Hori, G., 1966, *Publ. Astron. Soc. Japan*, 18, 287.
Hough, S.S., 1895, *Phil. Trans. R. Soc.*, A186, 469.
Jeffreys, H., and Vicente, R.O., 1957a, *Mon. Not. R. Astr. Soc.*, 117, 142.
Jeffreys, H., and Vicente, R.O., 1957b, *Mon. Not. R. Astr. Soc.*, 117, 162.
Kelvin, Lord, 1876, *Brit. Assoc. Report.*
Kinoshita, H., 1972, *Publ. Astron. Soc. Japan*, 24, 423.
Kinoshita, H., 1975, *Smithsonian Astrophys. Obs. Special Report*, no. 364.
Kinoshita, H., 1977, *Celes. Mech.*, 15, 277.
Kinoshita, H., Nakajima, K., Kubo, Y., Nakagawa, I., Sasao, T., and Yokoyama, K., 1979, *Publ. Int. Latit. Obs. Mitzusawa*, 12, 2.
Kinoshita, H. and Aoki, S., 1983, *Celes. Mech.*, 31, 329.
Le Verrier, U.J., 1858, *Ann. Observatoire*, 3, 13.
Lieske, J.H., 1967, *NASA Tech. Report* 31-1044.
Lieske, J.H., Lederle, T., Fricke, W., and Morando, B., 1977, *Astron. Astrophys.*, 58, 1.
Molodensky, M.S., 1961, *Comm. Obs. R. Belg.*, 188, 25.
Molodensky, S.M., 1981, *Izvestiya, Earth Physics*, 17, 395.
Moritz, H., 1982, *Bull. Geod.*, 56, 364.
Moritz, H., 1985, *Obs. R. Belg.*, 10, 99.
Munk, W.H. and MacDonald, G.J.F., 1960, *The Rotation of the Earth*, Cambridge University Press, Cambridge, London.
Newcomb, S., 1895a, *Astron. Papers Amer. Ephemeris*, 15, pt. 1.
Newcomb, S., 1895b, *Supplement to the American Ephemeris for* 1897.
Newcomb, S., 1906, *A Compendium of Spherical Astronomy*, New York, MacMillan, chap. 9.
Oppolzer, T., 1880, *Bahnbestimmung der Kometen und Planeten*, 2nd ed., vol. 1, chap. 5.
Poincaré, H., 1910, *Bull. Astr.*, 27, 321.
Po-Yu Shen and Mansinha, L., 1976, *Geophys. J. R. Astr. Soc.*, 46, 467.
Rochester, M.G., 1976, *Geophys. J. R. Astr. Soc.*, 46, 109.
Sasao, T., Okamoto, I., and Sakai, S., 1977, *Publ. Astron. Soc. Japan*, 29, 83.
Sasao, T., Okubo, S., and Saito, M., 1980, *Proc. IAU Symp. 78, Nutation and the Earth's Rotation*, E.P. Fedorov, M.L. Smith and P.L. Bender (eds.),

Reidel, Dordrecht, p. 165.
Sasao, T., and Wahr, J.M., 1981, *Geophys. J. R. Astr. Soc.*, 64, 729.
Sato, T., 1984, *Proc. Symp. Relativistic Framework and New Techniques in Astrometry and Geodesy*, Kashima, Japan, 547.
Seidelmann, P.K., 1982, *Celes. Mech.*, 27, 79.
Sludsky, F., 1896, *Bull. Soc. Natur. Moscou*, 9, 285.
Smith, M.L., 1974, *Geophys. J. R. Astr. Soc.*, 37, 491.
Smith, M.L., 1976, *J. Geophys. Res.*, 81, 3055.
Smith, M.L., 1977, *Geophys. J. R. Astr. Soc.*, 50, 103.
Smith, M.L., and Dahlen, F.A., 1981, *Geophys. J. R. Astr. Soc.*, 64, 223.
Standish, E.M., 1981, *Astron. Astrophys.*, 101, L18.
Toomre, A., 1974, *Geophys. J. R. Astr. Soc.*, 38, 335.
Wahr, J.M., 1981a, *Geophys. J. R. Astr. Soc.*, 64, 651.
Wahr, J.M., 1981b, *Geophys. J. R. Astr. Soc.*, 64, 705.
Wahr, J.M., 1981c, *Geophys. J. R. Astr. Soc.*, 64, 677.
Wahr, J.M., and Bergen, Z., 1986, *Geophys. J. R. Astr. Soc.*, 87, 633.
Wahr, J.M., and Sasao, T., 1981, *Geophys. J. R. Astr. Soc.*, 64, 747.
Wako, Y., 1970, *Publ. Astr. Soc. Japan*, 22, 525.
Wang, C.Y., 1972, *J. Geophys. Res.*, 77, 4318.
Whittaker, E.T., 1964, *A Treatise on the Analytical Dynamics of Particles and Rigid Bodies*, Cambridge University Press, p. 144.
Woolard, E.W., 1953, *Astron. Papers Amer. Ephemeris*, 15, pt. 1.
Yokoyama, K., 1976, *Astron. Astrophys.*, 47, 333.
Zürn, W., Rydelek, P.A., and Richter, B., 1986, *Proc. 10th Int. Symp. Earth Tides* (ed. R. Vierira), Conserjo Superior de Investigaciones Cientificas, Madrid.

EARTH ROTATION MONITORING

B. KOLACZEK
Space Research Centre, Polish Academy of Sciences
Warsaw, Poland

1. INTRODUCTION

The full description of the Earth rotation in space is given by:
- the motion of the Earth rotation axis with respect to the Earth shown by polar motion;
- the motion of the Earth rotation axis in space, defined by lunisolar precession and astronomical nutation (see Chapter 8);
- the angular position around the Earth rotation axis that is the universal time UT1.

The polar motion and UT1 were and are unpredictable due to the complexity of the physical phenomena involved and require continuous monitoring. The need of continuous monitoring of the polar motion was recognized at the end of the 19-th century, when in 1890-1891 F. Kustner detected observationally polar latitude variations of the order of 0.″3 with opposite phases in Berlin and Honolulu. At the same time S.C. Chandler, analyzing available latitude variations of different stations found that the pole motion contains two main periodical components, free and forced nutation with periods of about 1.2 years and annual one respectively.

Since 1899, when the International Association of Geodesy - IAG established the International Latitude Service - ILS, there were a few reorganisations of the international services monitoring the pole motion and the universal time, called the Earth rotation parameters - ERP in the second half of this century (see table 1). It was caused by the increase in numbers of astrometric instruments (zenith telescopes, PZTs, transit instruments, Danjon astrolabes, circumzenit instruments), the necessity of increasing the accuracy of determinations and the development of new observational techniques such as Doppler, laser and radio-interferometric ones. Reviews of international services and their ERP

J. Kovalevsky et al. (eds.), Reference Frames, 213–240.

TABLE 1. The international services of the Earth rotation

Service	Period of the activity	Main task
International Latitude Service - ILS	1899.0-1982.0 Established by the IAG in 1899	Determinations of instanteneous pole coordinates x, y
Bureau International de l'Heure - BIH	Since 1912	Organized for establishing and publishing time of radiosignal emissions
a) Rapid Latitude Service - RLS of the BIH	1955-1967 Established by the IAU (IAU, 1955)	Quick determinations and predictions of the instanteneous pole rotation for providing time corrections
b) BIH International Service of the ERP	1967.0-1988.0 (BIH, 1967-1987)	Simultaneous determination of x, y, UT1 from all available observational data
International Polar Motion Service - IPMS	1962-1988 Established by the IAU and the IUGG (IAU, 1961; IUGG, 1960; IPMS, 1962-1984)	Determinations of x, y from all available latitude data. Since 1977 the IPMS determined also x, y, UT1 from all available φ and UT0 data (IPMS, 1974)
Monitoring of the Earth Rotation and Intercomparison of the Techniques-MERIT Project	09-11.1980; 10.1983-11.1984 Established by the IAU (IAU, 1979; IUGG, 1979)	Determinations of the ERP by the use of all available observational techniques)
International Earth Rotation Service - IERS	1988 onwards. Established by the IAU (IAU, 1985, IUGG, 1987)	Determinations of the ERP, astronomical nutation and precession by the VLBI and laser ranging techniques

Stations	Publications
5 ILS stations on the parallel of φ = 39°08′(Table 4) equipped with TZ	Results of the ILS 1903-1978 IPMS Annual Report (IPMS, 1962-1984) and IPMS monthly notes (IPMSa 1975-1987)
All available astrometric instruments	Bulletin Horaire (BIH, 1955-1967)
Changeable number of astrometric instruments and stations. In 1967 and in 1983 there were 51 stations and 67 instruments (TZ, PZT, A, TI), as well as 61 stations and 88 instruments respectively. BIH global solution contains also data of all other observational techniques	BIH Annual Report (BIH, 1967-1987) BIH Circulars A-F (BIH a, 1968-1987)
Similarly as in the case of the BIH. In 1962 and in 1983 there were 54 stations and 75 instruments (TZ, PZT, A, T1) as well as about 60 stations and 80 instruments respectively	IPMS Annual Report (IPMS, 1962-1984) IPMS Monthly Notes (IPMSa, 1975-1987)
About 120 stations, 5 observational techniques participated in the MERIT. Campaignes (Wilkins, Mueller, 1986).	Project MERIT LETTERS (Wilkins, 1980-1985) Project MERIT CIRCULAR (BIH, 1984-1986) The MERIT FINAL REPORT (Feissel, 1985c; Mueller, 1985; Wilkins *et al.*, 1986, 1987)
About 30 satellite lasers, 3 lunar lasers and several radiotelescopes	IERS publications

monitoring are given by B. Guinot (1978) and H. Moritz, I.I. Mueller (1987).

At present, the ERP are currently monitored by laser and radio-interferometric (VLBI) techniques. The astronomical nutation is also monitored together with the ERP by VLBI (IRIS, 1984-1987; Schutz, 1983-1987). The accuracy of the ERP determinations by laser observations of LAGEOS and of the Moon and VLBI is one magnitude better than the accuracy obtained by astrometric and Doppler techniques. It achieves the order of 1-2 mas in pole coordinates and a few tenths of a millisecond of time in UT1 (Feissel, 1985). Thus, the weights of the ERP determined by these techniques in the global BIH solution considering all available series of the ERP have increased strongly in the last years (Fig. 1). The new International Earth Rotation Service, established in 1988, is based on laser and radio-interferometric observational techniques only.

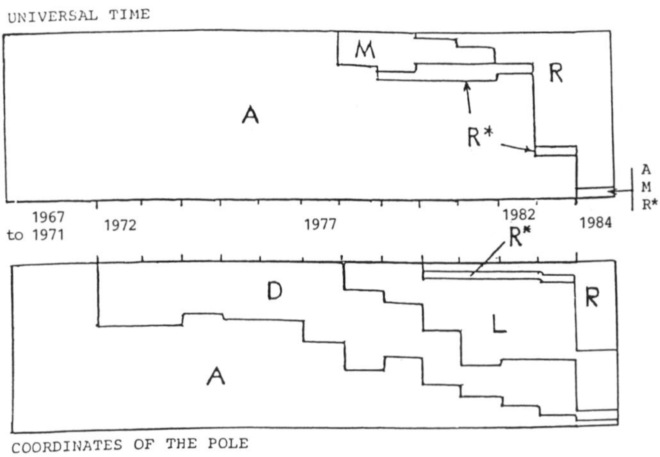

Figure 1. BIH combined solution for the Earth's rotation; relative weights of the techniques (Feissel, 1985a).

A Optical astrometry
L Satellite laser ranging
R VLBI
D Doppler tracking of satellites
M Lunar laser ranging
R* Connected Radio Interferometry

2. MONITORING OF THE ERP BY THE INTERNATIONAL SERVICES ILS, IPMS, BIH

2.1 INTRODUCTION

Determinations of pole coordinates x, y and universal time UT1 from latitude and UT0 (or longitude) data obtained from astrometric observations carried on at stations are based on the following two equations:

$$\Delta\varphi_i = \varphi_i - \varphi_{0,i} = x \cos L_{0,i} + y \sin L_{0,i} + z \qquad (1)$$

$$\mathrm{UT0} = \Delta L = L_i - L_{0,i} = (-x \sin L_{0,i} + y \cos L_{0,i}) \tan \varphi_{0,i} + \tau \qquad (2)$$

where: x, y are pole coordinates in the plane Cartesian coordinate system (Fig. 2)
L_i, φ_i are instantaneous latitude and longitude of a station
$L_{0,i}$, $\varphi_{0,i}$ are adopted conventional latitude and longitude of a station
z is the Kimura empirical term
τ is an additional empirical term.

Formulae 1 and 2 can be obtained from the spherical triangle shown in the Fig. 2. The emperical term z was introduced in 1922 by Kimura to equation (1). This term is named after him and contains all the nonpolar effects common in the observed latitudes of the stations. Similar empirical term τ was introduced later to the equation (2).

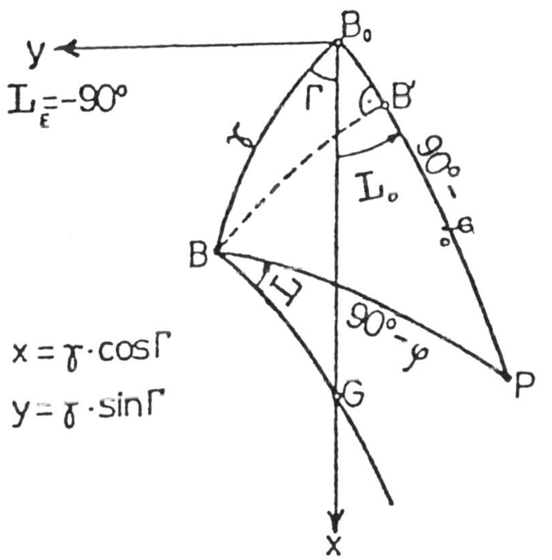

Figure 2. Mean and instantaneous geographical coordinates L_0, φ_0 and L, φ, respectively.

TABLE 2. Conventional poles used in pole determinations by international Services

International Service	Period	Conventional terrrestrial pole
ILS	1899-1949	Mean poles of consecutive six year periods
	1949-1968	The new system 1900-1905*
	1968-1982	Conventional International Origin - CIO
BIH	1955-1958	Cechini pole**
	1958.65-1959.15	Transformation from Cechini pole to the mean pole of the date
	1959.2-1968.0	Mean pole of the date
	1968.0-1979.0	The 1968 BIH System pole coinciding with the CIO in 1967 and defined on the basis of latitude and time data of 34 stations
	1979.0-1984.0	The 1979 BIH System pole
	1984.0-	The BIH Terrestrial System
IPMS	1962-1988	The IPMS System pole coinciding with the CIO at the beginning of the seventies defined on the basis of the latitude data of 54 instruments
	1974-1988	The IPMS System pole defined on the basis of the latitude and time data of 88 instruments in agreement with CIO and UT1(BIH) in 1967-1975

* Mean position of the true celestial pole in 1900-1905 (IUGG, 1960) named CIO in 1968.

** Mean terrestrial pole of 1949-1958 (Cechini, 1973).

2.2. REFERENCE COORDINATE SYSTEM APPLIED IN THE ERP MONITORING

Determinations of the ERP with a high accuracy require a precise definition of the applied coordinate system. Traditionally in the Earth rotation investigations the Cartesian plane coordinate system is used. Its origin is located at the designated conventional terrestrial pole, for instance the mean pole of the epoch. The x-axis is positive south, towards the zero meridian, for instance the zero mean Greenwich meridian, and y towards the West (Fig. 2).

Different conventional origins of this coordinate system have been used in practice (Table 2), (BIH, 1968).

Conventional International Origin - CIO was defined by the IAU and IUGG in 1967 as the conventional terrestrial pole by adopting the reference latitude of the 5 ILS stations equal to their mean values in the period 1900-1905 (Table 3). Because of local crustal motions of the ILS stations as well as the local seasonal effects (referaction, instruments, underground water) the CIO was not realized accurately in practice neither by the ILS, nor by the BIH and the IPMS based on the nets of several tens of stations.

TABLE 3. The ILS stations

Station	Longitude	Latitude
Carloforte, Italy	8°18′44″E	39°08′ 09.″941
Geithersburg, Maryland	77 11 57 W	13.202
Kitab, USSR	66 52 51 E	01.850
Mizusawa, Japan	141 07 51 E	03.602
Ukiah, California	123 12 35 W	12.836

In 1968 the BIH difined the 1968 BIH system, its conventional pole and the zero meridian, by adopting reference latitudes, longitudes and weights of 68 instruments of 47 stations participating in ERP determinations in 1967 (BIH, 1968). The conventional origin of this system coincides with the CIO at the beginning of 1968. It was ensured by adding the constants to the initial latitudes and longitudes which were obtained from the comparison of the BIH ERP data with ILS ones for the years 1964-1966. These differences of the ILS and the BIH pole coordinates in this period are the following:

$$\begin{aligned} x_{CIO} - x_M &= x_{ILS} - x_M = +0.''007 \\ y_{CIO} - y_M &= y_{ILS} - y_M = +0.''233 \end{aligned} \tag{3}$$

In order to refer the previous ERP data determined by the BIH to the 1968 BIH System, it is necessary to introduce corrections due to the application of the different reference systems in the past which are published in the BIH Annual Report for 1968.

In 1979 the BIH defined the new 1979 BIH System by introducing the following corrections to the 1968 BIH System (BIH, 1979).

$$x_{1979,BIH} - x_{1968,BIH} = 0.''024 \sin [2\pi(t\text{-}0.158)] +$$
$$0''.007 \sin [4\pi(t\text{-}0.289)]$$
$$y_{1979,BIH} - y_{1968,BIH} = 0 \quad (4)$$
$$\mathrm{UT1}_{1979,BIH} - \mathrm{UT1}_{1968,BIH} = 0.0007 \sin [2\pi(t\text{-}0.477)] +$$
$$0.0007 \sin [4\pi(t\text{-}0.397)]$$

where t is expressed in Besselian years and UT1 in seconds.

The necessity of introducing these corrections was caused by the existence of the seasonal variations of the station coordinates defining the 1968 BIH System which were not eliminated fully by the adjustment. They were found by the comparison of the BIH astrometric ERP data with the data obtained by the Doppler observations of artificial satellites and lunar laser ranging.

The IPMS defined its own system of coordinates several times. In 1962 the IPMS defined a preliminary system of latitudes and longitudes of 54 instruments of 44 stations using the residual latitudes obtained in respect to the CIO in several years period (IPMS, 1972). The IPMS pole coordinates from the period 1962-1972 were transformed to this IPMS system of coordinates (IPMS, 1972).

In 1975, the IPMS began to compute pole coordinates and UT1-UTC from latitude and UT0 data, additionally to the series of pole coordinates determined from latitude data only. The IPMS longitudes of 88 instruments of 60 stations were determined from UT0 data of the period 1967-1975 ensuring the continuity of UT1 determined by the IPMS with UT1 determined by the BIH (IPMS, 1974). It ensured the coincidence of the zero meridians of these two systems. Thus, this new IPMS system is defined by the latitudes of stations referred to the CIO, longitudes referred to the BIH zero meridian and its own system of instrument weights published in the IPMS Annual Report (IPMS, 1974).

The high accuracy of the ERP determined by the laser and VLBI techniques, which is one order of magnitude higher than the accuracy of the astrometric ERP, causes that recently the BIH global solution of the ERP depends in more than 90% on laser and VLBI techniques and only in a few percent on the astrometric observations. In this situation it was necessary to change the reference system of the BIH global solution of the ERP. In 1983 the BIH defined the new BIH Terrestrial System, the BTS, 1984. In 1986 the BTS-1985 was defined by the geocentric coordinates of 35 stations carrying on observations during several years by at least two new observational techniques. It contains 28 satellite laser stations, 3 lunar laser stations, 14 VLBI stations. Doppler observations were carried out at all of them. The BTS-1986 contains 51 stations (Boucher, 1987). The geocentric coordinates of the stations were determined by the least square method together with the ERP of this period and 7 parameters of transformations of the coordinates sytem of each technique to the BTS (Boucher, Feissel 1984; Boucher *et al.*, 1985, 1986). The transformation parameters defining the BTS are given in Table 4. The geocentric coordinates of stations are published by the BIH in the Annual Report (BIH, 1986). An agreement of orientations of the systems of laser and VLBI techniques is on the level of 0.″01. The coordinate system of the Doppler technique shows the largest discrepancy.

TABLE 4. Transformation parameters: translation (*T*), rotation (*A*) and scale (*D*) between the individual terrestrial coordinate systems and the BTS, 1985 (Boucher, *et al.*, 1986).

Network	Solution	T1	T2	T3	D	R1	R2	R3
		CM	CM	CM	10^{-8}	0.″001	0.″001	0.″001
VLBI-NGS	83 R 02	163.1	-98.9	38.7	-3.9	-6.0	9.5	-3.6
VLBI-GSFC	84 R 01	142.2	-101.3	36.7	-3.6	-5.7	8.2	14.2
VLBI-JPL	83 R 05	16.0	-47.3	-4.0	-5.0	-6.2	-7.7	-5.0
LLR-JPL	86 M 02	.0	.0	.0	-3.5	4.6	1.3	-9.9
SLR-CSR	84 L 01	.0	.0	.0	.0	-4.2	-2.7	0.8
DOPPLER-IMA	77 D 01	6.1	-36.3	-437.2	60.4	20.0	1.6	-806.0

2.3. SERIES OF THE ERP DETERMINED BY THE INTERNATIONAL SERVICES ILS, IPMS, BIH FROM CLASSICAL ASTROMETRIC OBSERVATIONS

The international services ILS, IPMS, BIH were determining the ERP from latitudes and universal times - UT0 obtained from astrometric observations of cooperating stations. Reviews of monitoring the ERP by ILS, IPMS, BIH are given by B. Guinot (1978) and H. Moritz and I. I. Mueller (1987).

Determinations of latitudes by the use of zenith telescopes and photographic zenith tubes are based on micrometric measurements of differences of zenith distances of chosen star pairs or of zenith distances of zenith stars respectively in station meridian.

Determinations of UT0 by the use of transit instruments or photographic zenith tubes are based on visual, photoelectric or photographic registrations of star transits through station meridians.

Determinations of latitudes and UT0 of stations by the use of

Danjon astrolabe and the method of equal zenith distances are based on registrations of star transits through the same almucantarat with the zenith distance equal for instance 60 degrees.

These methods of latitude and UT0 determinations are described in detail by I.I. Mueller (1969) and H. Moritz and I.I. Mueller (1987).

The main sources of errors of astrometric methods of latitude and UT0 determinations are errors of star coordinates of astrometric refraction and some instrumental errors.

Each international service determining the ERP from latitudes and UT0 of cooperating stations creates its own system of the ERP by the adoption of

- a reference coordinate system (see Section 2.2),
- a system of weights according to the quality of observations,
- a way of removing some systematic errors,
- an averaging time,
- smoothing techniques of observational data and evaluated results.

2.3.1. International Latitude Service

In the period 1899-1982 the ILS were determining the pole coordinates x, y from latitude data of 5 ILS stations located on the common parallel of $\varphi = 39^\circ 08'$ (Table 3) using least square solutions of equations 1 (ILS, 1903 - 1978, IPMS, 1962 - 1982). The ILS pole coordinates from the period 1899.0 - 1979.0 were revised and homogeneous systems of pole coordinates were computed by Yumi and Yokoyama (1980).

Variations of the revised x_{ILS} and y_{ILS} are shown in Fig. 3 (Dickman, 1981) from which variations of amplitudes and phase of Chandler nutation in the past were computed by Dickman (1981). Variations of mean values of pole positions computed every six years presented in Fig. 4 show the secular motion of the pole of the order of $0.''003$ per year in the direction of $L = 80^\circ\ W$ and the Markowitz wobble with the period of about 40 years (Markowitz 1968).

Small number of stations of the ILS, large secular variations of latitudes of Mizusawa and Ukiah located near the tectonic plate boundaries as well as large refraction anomalies disturb strongly the ILS polar motion.

2.3.2. International Polar Motion Service

The IPMS was established by the IUGG and the IAU and it began its activity in 1962 (IUGG, 1960; IAU, 1961; IPMS, 1962). The main task of the IPMS was determinations of pole coordinates from latitude data of all latitude stations as well as continuation of the task of determining pole coordinates from latitude data of the 5 ILS stations.

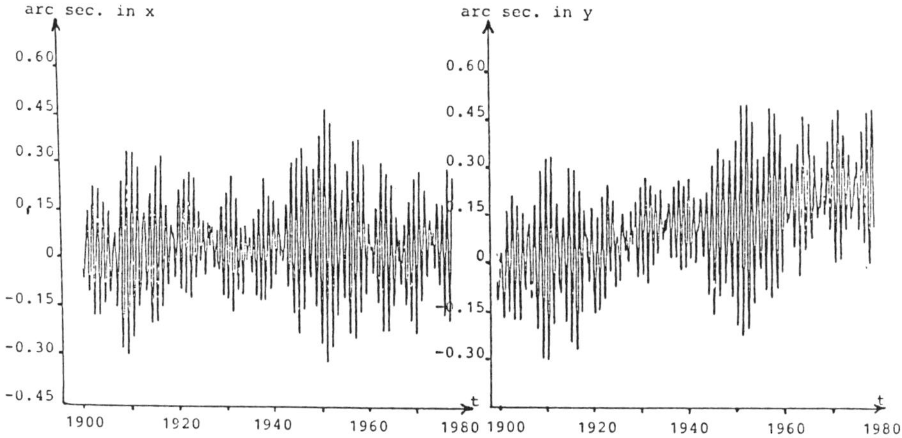

Figure 3. Variations of the ILS pole coordinates 1900-1978 (Dickman, 1981)

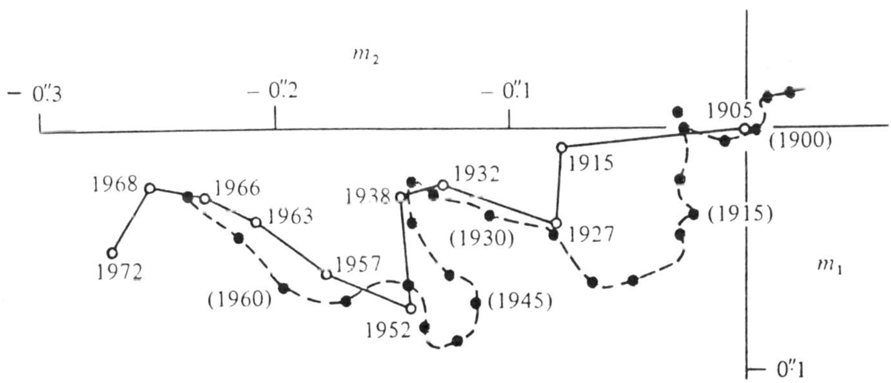

Figure 4. The secular motion of the mean pole computed by Stoyko and Markowitz (Lambeck, 1980).

The IPMS has been producing two series of the ERP:

- the IPMS, L series of pole coordinates determined from all available latitude data of cooperating stations beginning in 1962 (IPMS, 1962-1984; IPMSa, 1975-1987);

- the IPMS, $L+T$ series of the ERP from all available latitude and UT0 data of cooperating stations beginning in 1967 (IPMS, 1974-1984; IPMS a, 1975-1987).

The IPMS pole coordinates are computed at 0.05 year intervals as the least square solutions of equations (1) or equations (1) and (2) respectively.

The new IPMS solutions of the ERP for the period 1962-1986 were given by Manabe and others (Manabe et al, 1987). The IPMS ERP were estimated simultaneously with the coordinates of optical astrometric stations. Longitudes and latitudes of stations were determined every 2 months. An agreement of this new set of the IPMS ERP with results obtained by the space techniques increased much in comparison with the previous IPMS ERP.

2.3.3. Bureau International de l'Heure

The BIH began to monitor the universal time as the parameter of the Earth rotation after establishing the Atomic time scale, presently designated by International Atomic Time TAI or its sister time scale UTC (see Chapter 16). Since 1955 the BIH has computed routinely the pole coordinates and the rate of the Earth rotation, UT1. First the Rapid Service RLS was established by the IAU at the BIH in 1955 in order to compute the pole positions from the available latitude data and to predict the pole motion in order to provide time corrections on a nearly current basis. The results were published by the BIH and in Bullelin Horaire (BIHb, 1955-1967).

Beginning in 1965 the BIH started to compute the ERP simultaneously from all available latitude and UT0 data by the least square method using the following equations (BIH, 1967-1968):

$$\varphi_i - \varphi_{0,i} + R_i = x_{BIH} \cos L_{0,i} + y_{BIH} \sin L_{0,i} + z_{BIH}$$

$$(\mathrm{UT0} - \mathrm{UTC})_i + S_i = (-x_{BIH} \sin L_{0,i} + y_{BIH} \cos L_{0,i}) \tan \varphi_{0,i} + \\ + (\mathrm{UT1} - \mathrm{UTC})_{BIH} \qquad (5)$$

where:

$L_{0,i}$, $\varphi_{0,i}$ are longitude and latitude of i station adopted by the BIH, which together with the adopted weight of i station define the 1968 BIH system

R_i , S_i are station the corrections introduced by the BIH in order to preserve the system.

The corrections R_i and S_i are represented by the formula

$$R_i = a_i + b_i \sin 2\pi t + c_i \cos 2\pi t + d_i \sin 4\pi t + e_i \cos 4\pi t$$
$$S_i = a_i' + b_i' \sin 2\pi t + c_i' \cos 2\pi t + d_i' \sin 4\pi t + e_i' \cos 4\pi t \qquad (6)$$

where t is the date expressed in years and they are determined from previous years observations, recently from previous 4 years (Feissel, 1980). The coefficients a_i-e_i, a_i'-e_i' and the raw and smoothed values of x, y and UT1 are published currently in the BIH Annual Reports.

Z.X. Lie revised the BIH astronomic set of the ERP for 1962-1982.0. The main unknows UT1 - UTC and the coordinates of the pole were adjusted at 5-day intervals, together with auxiliary group unknows G and $\dot{G}$ used to improve the accuracy and the classical z and τ term of equations 1 and 2 (Lie Z.X. *et al.*, 1985, BIH 1984).

3. MONITORING OF THE ERP BY NEW OBSERVATIONAL TECHNIQUES

3.1. THE INDEPENDENT SERIES OF THE ERP

New observational techniques such as Doppler and laser tracking artificial satellites - SLR, lunar laser ranging - LLR and very long base interferometry - VLBI have been introduced gradually into determinations of the ERP in the seventies, creating new independent series of the ERP. They have been applied also by the BIH to the BIH global solution (Table 5).

New observational techniques and their applications to regular determinations of the ERP have been developped continuously, especially in the last few years of realization of the MERIT Project (Monitoring of the Earth Rotation and Intercomparison of Techniques Project) and afterwards (Wilkins *et al.*, 1986). At present the most accurate ERP are determined by the VLBI as well as by the LAGEOS and the Moon laser ranging. The typical standard errors of individual techniques in the BIH global solution in the main MERIT campaign are given in the Table 6 (Feissel, 1985b).

TABLE 5. Independent series of the ERP in 1962-1985 (BIH, 1985)

Technique, Stations	Object	Organi-zation	ERP	Period	Remarks
Global solutions of all techniques	Different	BIH	x, y, ΔT1	1976-1987	several solutions
ASTROMETRY					
5 stations	Stars	ILS	x, y	1899-1982	
About 50-100 instruments.		IPMS	x, y	1962-1987	
		IPMS	x, y, ΔT1	1967-1987	
About 50-100 instruments.		BIH	x, y, ΔT1	1967-1987	several solutions
DOPPLER					
About 20 stations	NNSS satellites	DMA	x, y	1977-1985	several solutions
		DPMS	x, y	1972-1977	several solutions
		GRGS	x, y	1977-1980	
		NSCW	x, y		
		BIH	x, y	1975-1984	
SLR					
15-40 stations	LAGEOS	CSR	x, y l.o.d.R.	1976-1987	several solutions
		GSFC	x, y	1976-1982	several solutions
		SAO	x, y	1980-1983	
LLR					
Mc Donald	MOON	CERGA	ΔT1	1971-1985	several solutions
in 1970-1985		JPL	ΔT0R	1970-1985	several solutions
3 stations		MIT	ΔT0	1970-1979	
from 1983		SHA	ΔT1	1983-1985	
VLBI					
12 stations	26 Radiosources	NGS-IRIS	x, y, ΔT1	1980-1987	several solutions
3 stations	20 Radiosources	JPL-DSN	ΔT0	1971-1985	several solutions
15 stations	44 Radiosources	GSFC	x, y, ΔT1	1976-1985	several solutions
CERI					
35 km radio-interferometer		USNO	ΔT0	1981-1985	
where	ΔT1 = UT1–UTC,	ΔT0 = UT0–UTC			

Abbreviations in Table 5:

CERGA	Centre d'Etudes et de Recherches Géodynamiques et Astronomiques
CSR	Center for Space Research
DMA	Defense Mapping Agency
DPMS	Dahlgren Polar Monitoring Service
GRGS	Groupe de Recherches de Géodesie Spatiale
GSFC	Goddard Space Flight Center
JPL	Jet Propulsion Laboratory
MIT	Massachusetts Institute of Technology
NGS	National Geodetic Survey
NNSS	Navy Navigation Satellite System
NSWC	Naval Surface Weapons Center
SAO	Smithsonian Astrophysical Observatory
SHA	Shanghai Observatory
USNO	United States Naval Observatory

TABLE 6. Contribution of the different series to the combined solution (Feissel, 1985b)

series of ERP	average time	normalized standard error		
		x 0.001″	y 0.001″	UT1 0.000″
ERP(NGS) 83 R 01	1d	1.8	1.3	0.7
ERP(JPL) 83 R 01	3h			
Baseline: AC			14.6(1)	8.6(2)
Baseline: SC			12.6(1)	10.4(2)
ERP(USNO) 81 R 01	3d			33.4(2)
ERP(BIH) 84 A 02	5d	13.4	10.4	6.6
ERP(JPL) 84 M 01	1-8h			
Station: MCD				9.5(2)
Station: CER				5.3(2)
ERP(CSR) 84 L 02	5d	1.8	1.8	
ERP(DMA) 77 D 01	2d	14.3	11.6	
ERP(DMA) 79 D 01	2d	17.2	14.0	
ERP(DMA) 82 D 01	2d	9.7	8.5	

(1) latitude variations, (2) UT0-UTC, MCD-McDonald, CER-CERGA
Other abbreviations are explained after Table 5.

However, each organization and each analysing centre, applies its own methods of observation evaluations, their smoothing, adjustments, introducing systematic corrections etc., creating different independent ERP series. Additionally, some analysing centres revise their own ERP series. Thus there are many independent series of the ERP (Table 5). Since 1972 the BIH has been introducing, step by step, the ERP of independent series of new techniques: Doppler (1972), lunar laser ranging - LLR and LAGEOS laser ranging - LALAR (1978), VLBI (1980) into the new global BIH solution (BIH 1972-1985; Feissel 1980). The independent series of the ERP of new techniques are included in the global BIH solution with the weights adopted by the BIH according to their accuracy. The participation of new techniques in the global BIH solution has increased strongly in the last years (Section 1 and Fig. 1). The raw and smoothed values of the ERP of the BIH global solution are published in the BIH Annual Report at 5-day intervals (BIH, 1972-1985). The comprehensive review of these independent series and the methods of their evaluation is given by H. Moritz and I.I. Mueller (1987).

The list of all independent series of the ERP as well as the list of stations of different nets and their geocentric coordinates are published in the BIH Annual Report for 1985. The individual series of the ERP are published "in extenso" in the BIH Annual Reports (BIH, 1978-1986). Usually the ERP are given at 5 day interval. The Doppler ERP are determined in 2 days interval. In the intensive VLBI campaigns, UT1-UTC were determined daily.

It is necessary to add that more analysing centers than those mentioned in the Table 5 were determining the ERP during the main MERIT campaign. These series are listed in the BIH Annual Report for 1985 also (BIH, 1985, Feissel, 1985c). The BIH Circular D (BIHa, 1967-1987), the IRIS Bulletin (IRIS, 1984-1987) and the CSR Circular (Schutz, 1983-1987) publish the ERP monthly.

3.2. METHODS OF EVALUATION OF THE INDEPENDENT SERIES OF THE ERP

An outline of basic ideas of methods of evaluation of the independent series of the ERP is given here for a better understanding of the considered independent series and their accuracy. The comprehensive review of evaluation methods of observations of techniques applied in the ERP determinations, Doppler, laser, radio-interferometric ones are given by H. Moritz and I.I. Mueller (1987).

3.2.1. Applied CTS and CIS

The ERP are determined always through comparison of the orientation of the conventional terrestrial system of coordinates - CTS with that of the conventional inertial system of coordinates - CIS. In practice of the ERP determinations the CTS is always defined by the adopted terrestrial coordinates of stations, for instance defined by the BTS-86 (see Section 2.2.). The CIS is defined differently in different techniques (Table 7). Definitions and practical realizations of the CTS

TABLE 7. The adopted CIS in the ERP determinations

Technique		The adopted CIS	Accuracy
Astrometry		Stars coordinates of fundamental catalogues like FK4 or FK5	Accuracy of the FK4 catalogue is of the order of several hundreds of mas
Satellite tracking	Doppler laser	Satellite ephemeris reference orbits	Typical LAGEOS orbit fits for 40 minutes, 5, 15, 30 days are 3, 10, 15, 20 cm respectively (Moritz and Mueller, 1987)
Lunar laser ranging		The Moon ephemeris	See Section 4
VLBI		Radiostar coordinates of the adopted catalogue	Several thousandths of mas (BIH, 1985)

and the CIS are discussed widely in the Chapters 11-14 and 1-4 of this book respectively.

The determination of coordinates of terrestrial points or celestial objects creating the CTS or the CIS respectively implies the adoption of some standard theories, models of geophysical and astronomical phenomena and fundamental constants, for instance theories of precession and astronomical nutation, tides, the IAU fundamental constants etc.

The MERIT Standards (Melbourne, 1983) consisting of such standard theories, fundamental constants and models of geophysical and astronomical phenomena were elaborated for the MERIT Project in order to get comparable results of large number observations of different techniques as the level of 10^{-8} (see Chapter 18 and Annex).

3.2.2. Satellite tracking techniques

At present, in satellite tracking techniques, the principal observables are either ranges (laser) or range differences (integrated Doppler counts and interferometry). Range difference changes can be derived as the differences between two ranges. Thus, further discussion can be limited to range observations only. The observables ought to be freed from all systematic errors caused by refraction, instrumental errors, time correction etc. Provided that the theory of satellite motion, its mathematical and physical model is correct and that observations (range or range differences) have been freed of systematic errors, the differences between computed C_1 (1, ..., n) and observed O_1 (1, ..., n) values of observables will be due to erroneous geocentric coordinates of the station S_k and the erroneous parameters - P including the constant orbital elements $E_{0,i}$ at the epoch T_0 and the force parameters.

Assuming that O_1–C_1, dS_k, dP_j are differentially small we can write according to the Taylor's theorem

$$O_1 - C_1 = \sum_j \sum_i \frac{\delta C_1}{\delta E_i} \frac{\delta E_i}{\delta P_j} dP_j + \sum_k \frac{\delta C_1}{S_k} dS_k \tag{7}$$

where:

E_1 is a variable orbital element at the instant of observation

dP_j, dS_k are the unknown corrections to the assumed force parameters and station coordinates respectively.

In the case of Earth satellite observations, the CIS is defined through the satellite reference orbit and the CTS is defined by the coordinates of the tracking stations. The ERP can be determined in two ways. In one way, the reference orbits are kept fixed and the effect of ERP manifests itself in the time variations of the station coordinates and the ERP are determined from a model reflecting such changes. In this case the second term on the right hand side of equation (7) are amended by adding terms of the following form:

$$\sum_k \sum_m \frac{\delta C_1}{\delta S_k} \frac{\delta S_k}{de_m} de_m$$

where: e_m is a general symbol for the ERP: x, y, UT1.

In the second way of the ERP determinations through the SLR, the station coordinates in the applied CTS are fixed, and the ERP are viewed as the sources of perturbations of satellite motion with respect to the reference orbit. Here, care must be taken that no other reference frame related effects, as for instance the Earth gravitational field, should mask these perturbations. In this case consideration of the ERP is included in the first term of equation (7). The final formule were given by Lambeck (1971).

In order to obtain a satisfactory solution through the least square adjustment of equation (7) the number of observations m must greatly exceed the number of unknown parameters n, plus the stations coordinate corrections p. The accomplishment of such computations is an enormous task considering that recent solutions included data of tracking of several satellites in two to four week periods from 50-100 stations.

There are thousands of orbital constants, coordinates of stations, gravitational coefficients. Such general solution needs access to large computers and it is expensive. Thus in practice the general solutions are performed for longer span of time, two-four weeks. The partial solutions are applied for shorter span of time (two-five days) in which only positions of new stations, six initial orbital constants $E_{0,i}$, components of the ERP and usually a dray (scalling) parameters are included. Other ones are taken from general solutions.

3.2.3. Lunar laser ranging

Similarly as in the SLR, ranges are observables in the LLR. Four lunar retroreflectors installed on the Moon by three Apollo missions and the Luna mission are observed. The main difference between the SLR and the LLR is caused by the fact that the vector **SR**, station - lunar retroreflector, length of which is measured by the LLR, is the sum of three vectors:

$$\mathbf{SR} = \mathbf{ST} + \mathbf{TL} + \mathbf{LR}$$

where:

- **ST** is the geocentric vector of a station,
- **TL** is the geocentric vector of the Moon's center of mass defined by the Lunar Ephemeris,
- **LR** is the selenocentric vector of a reflector defined by the parameters of the Moon's physical libration.

In the case of the SLR the vector **SR** is the sum of two vectors, the geocentric vector of a station and geocentric vector of a satellite.

Thus determinations of the ERP by the use of LLR can be treated similarly as in the case of SLR. In this case the basic equation (7) has to be amended by adding a third term of the following form

$$\sum_k \sum_n \frac{\delta C_1}{\delta R_k} \frac{\delta R_k}{\delta l_n} dl_n$$

where:

- l_n is a general symbol for the Moon's physical libration parameters;
- R_k are lunar radioreflector coordinates.

At the beginning of lunar laser ranging techniques, parameters and the theory of lunar free libration have been improved enabling the application of these techniques into Earth rotation determinations.

Determinations of three ERP, x, y and UT1-UTC and filtering out the orbital and libration parameters of the Moon require at least 3 terrestrial stations and three retroreflectors.

Mc Donald Observatory, as a single LLR station, has carried on lunar laser ranging of three lunar retroreflectors since 1969, delivering long series of UT1-UTC. Since 1983 three lunar laser stations have been observing regularly.

3.2.4. Very long baseline interferometry

In the VLBI techniques time delay - τ, which is the difference in arrival time of natural radiosource electromagnetic wave recorded at two radiotelescopes of the interferometer, is a direct measure. The second observable is the delay rate.

It is necessary to remember that the measured time delay is composed of three components:
- geometry of stations and radiosources,
- instrumental, mainly clock effects,
- effect of the propagation medium.

The time delay can be written as follows:

$$\tau = \frac{d}{c} = \frac{B}{c} \cos \psi = \frac{\mathbf{B}}{c} \mathbf{e} \tag{8}$$

where vectors **B** and **e** are the base and radiosource unit vectors, d is the length of the base d between two radiotelescopes, c is velocity of the light and ψ is the angle between the base direction of the radiointerferometer and the direction to the radiosource.

Differentating equation (8) we get the delay rate

$$\dot{\tau} = \frac{\dot{\mathbf{B}}}{c} \mathbf{e} \tag{9}$$

Vector **B** rotates with the constant angular velocity of the Earth-Ω, thus variations of τ represent a diurnal sinusoid (Robertson, 1975). Taking into account the differences in clock epochs of two base stations - Δc_0 and their rates Δc_1 we can write:

$$\tau = K_1 \sin (\Omega t + \Phi) + K_2 + \Delta c_0 + \Delta c_1 t \tag{10}$$

where:

$$K_1 = \frac{1}{c} (\Delta U^2 + \Delta V^2) \cos \delta$$

$$K_2 = \frac{1}{c}\,(\Delta W \sin\,\delta)$$

$$\Phi = \alpha - \alpha_0$$

ΔU, ΔV, ΔW are differences of the geocentric coordinates between the radiotelescopes;

δ, α are the right ascension and declination of the radiosource;

α_0 is the right ascension of the baseline of the instrument at epoch $t = 0$.

In the case of time delay, equation (10) has four independent parameters (K_1, Φ, K_2, Δc_0, Δc_1) and seven unknowns (ΔU, ΔV, ΔW, α, δ, Δc_0, Δc_1). Thus, the case is undetermined. Observations of a minimum of three sources are necessary in order to determine all these unknowns. In practice this minimum number of sources and observations is greatly exceeded.

Differentiating equation (10) we get the formula for the delay rates

$$\dot{\tau} = \Omega\, K_1 \cos\,(\Omega + \phi) + \Delta c_1 \tag{11}$$

There are four unknowns (ΔU, ΔV, δ, Δc_1) and only three independent parameters (K_1, Φ, Δc_1). Thus, one unknown has to be determined earlier, separately.

The observation equations of the VLBI techniques applied to the ERP determinations for two observables, τ and $\dot{\tau}$, can be written in the form

$$d(d_{ijk}) = \sum_j A_j\, dP_j$$

$$d(\dot{d}_{ijk}) = \sum_j A'_j\, dP_j \tag{12}$$

where:

d_{ijk}, $\dot{d}_{ijk}$ are the geometric delay and the geometric delay rate, respectively;

A_j, A'_j are the required partial derivatives of the geometric delay and delay rate with respect to the considered parameters (ΔU_i, ΔV_i, ΔW_i, x_j, y_j, $(\text{UT1-UTC})_j$, α_K, δ_K, Δc_0, Δc_1), respectively (Bock 1980, Moritz and Mueller, 1987).

At least two baselines of different configurations are nesessary for ERP determinations. Generally speaking, a baseline is insensitive to ERP components perpendicular to its orientation. Observations from one baseline can be used to estimate only two of three Earth orientation parameters.

4. INTERCOMPARISON OF THE ERP SERIES

The different procedures used by the international services and individual organizations to derive their respective ERP produce systematic differences.

There are systematic ERP differences of various kinds: constant biases, long term drifts, long periodic variations, seasonal variations, shorter periodic variations. Comparisons of the ERP determined by different techniques are shown in figures 5-8. The centers of polhodes of different individual series in the MERIT campaign show discrepancies of the z axis of the CTS applied in these series (Mueller *et al.*, 1985) (Fig. 9).

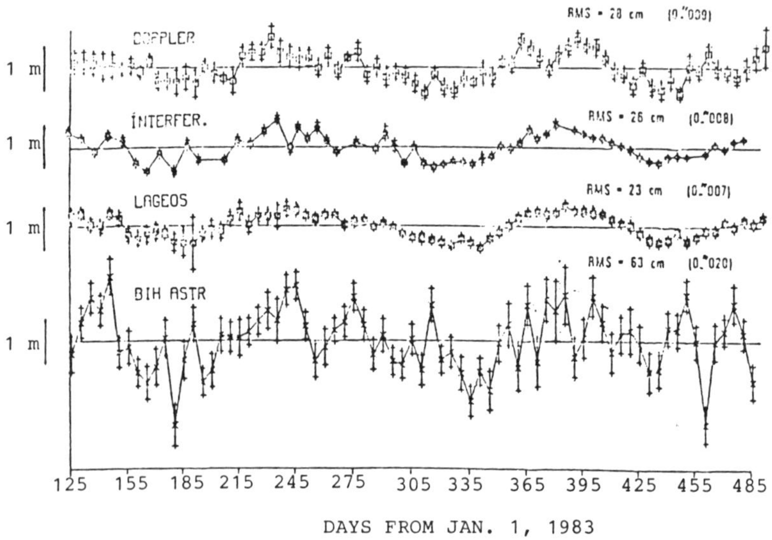

Figure 5. *X*-component of combined Doppler, interferometer, LAGEOS laser ranging, and BIH-Astrometric pole position evaluation (Colquitt *et al.*, 1984)

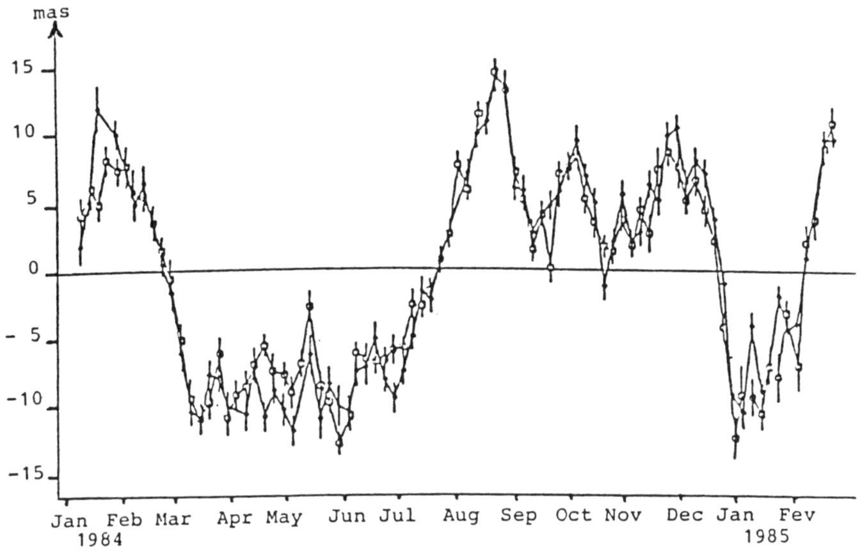

Figure 6. Residuals of x determined by NGS-VLBI (IRIS) and CSR-LALAR in respect to the best fitted polhode (Robertson, 1985

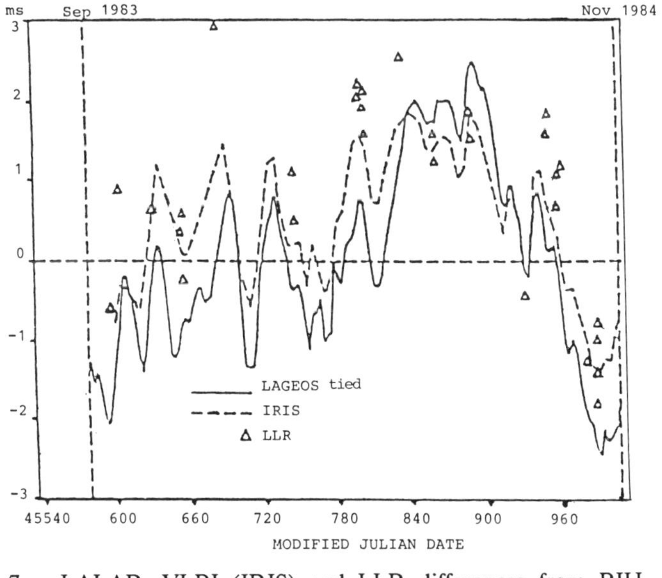

Figure 7. LALAR, VLBI (IRIS) and LLR differences from BIH Circular D during the MERIT Campaign, (Tapley *al.*, 1985).

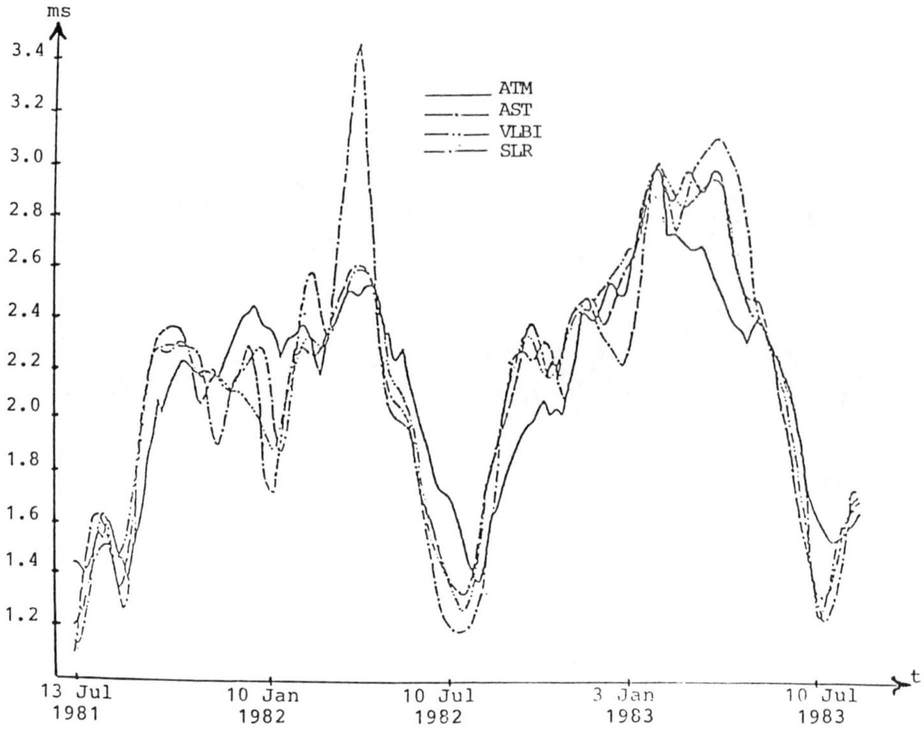

Figure 8. Comparison of LODR by three different techniques with Atmospheric Angular Momentum (A+M, unbroken line) (Zhao Ming, Gu Zhen Maian, 1985).

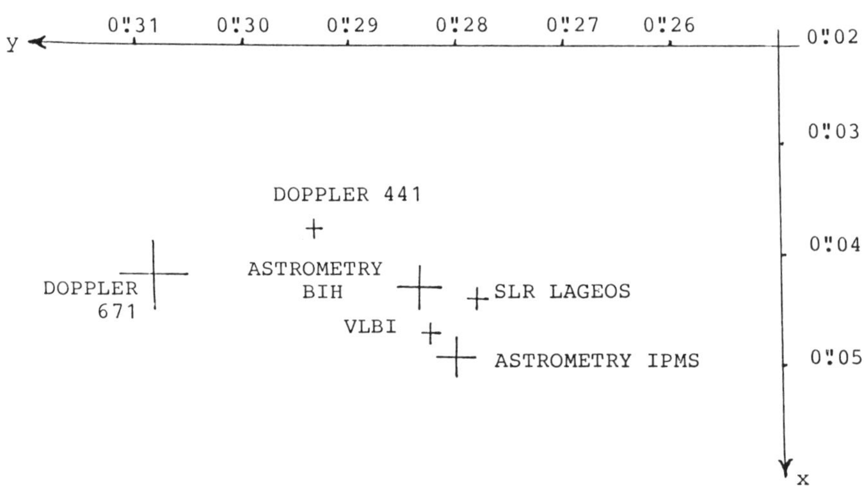

Figure 9. Relative polhode with their standard deviations (Mueller I.I. *et al.*, 1985)

TABLE 8. Biases and amplitudes of annual and semiannual terms of the ERP differences between individual series and the BIH global solution (Feissel, 1982).

TECHNIQUE ORG.	year	x (0"001)					y ("001)					UT (0.0001)				
		A	B	C	D	E	A	B	C	D	E	A	B	C	D	E
CIRCULAR D	78-82	0	-4	-1	1	-3	0	1	1	-4	-1	2	0	-13	0	4
ASTR/BIH	78-82	0	2	3	1	-1	1	1	1	-1	0	1	-1	-1	0	1
ASTR/IPMS	78-82	32	-3	2	8	-5	-18	10	4	-5	-1					
ILS-IPMS	79-81	129	-13	-10	-16	-10	-14	-3	7	-2	-16					
NNSS#92/IMA	79-82 78-79	-6	-2	-5	-3	2	-6	0	0	1	2					
SLR/CRS	78-82	1	2	0	0	0	-10	-3	0	0	0					
LLR/JPL	78-80											1	2	-2	-2	-3
VLBI/NGS	81-82	12	0	1	1	0						7	5	2	0	0
CERI		-34	1	-34	1	-12						35	20	39	-6	9

Abbreviations are explained in Table 5.

Constant biases and annual and semiannual terms of differences between the ERP determined by different techniques and raw data of the BIH global solution have been computed by the BIH by the use of the following formula:

$$\text{ERP (individual technique)} - \text{ERP(BIH)} =$$

$$= A + B \sin 2\pi t + C \cos 2\pi t + D \sin 4\pi t + E \cos 4\pi t$$

where t is in years.

The constant biases and the amplitudes of the annual and semiannual terms A-E were determined independently every year from the weighted differences at 0.05 years intervals. The constants A-E for chosen series are given in the Table 8. Note some large discrepancies in the biases as well as amplitudes of seasonal terms for instance in the case of the IPMS and the CERI. The best agreement is in the case of the SLR-CRS and VLBI-NGS series of the ERP.

5. PREDICTIONS OF THE ERP

The Earth polar motion and variations of the Earth rotation velocity, or length of a day, are complex phenomena. They depend on many changeable geophysical phenomena, variations of which have a wide spectrum range. The prediction of the ERP variations is of both theoretical and practical interest but it has not yet been solved satisfactorily.

The numerical prediction method used in practice by the U.S. Naval Observatory (NEOS, 1985-1987) is based on the simple model of the ERP adopting fixed periods of annual and Chandler wobbles. For instance:

$$x = A + B \cos(\Theta A) + C \sin(\Theta A) + D \cos(\Theta C) + D \sin(\Theta C)$$

Coefficients *A-D* are determined by the least square method from the previous ERP data. Accuracies of such predictions for 10-40 days are of the order of 0.″01 in pole coordinates and a few miliarcseconds in UT1. The model of pole motion containing wobbles with fixed periods was used also by the BIH (BIHa, 1968-1985). Comprehensive study of polar motion prediction is given by Zhu (1982) and Chao (1985).

REFERENCES

BIH, 1955-1967, *Bulletin Horaire*, Paris, France.

BIH, 1968-1986, *BIH Annual Reports* for the years 1968-1985, Paris, France.

BIHa, 1968-1985, *BIH Circulars*, A-E, Paris, France.

BIH, 1984-1986, *Merit Circular*, BIH, Paris, France.

Bock, Y., 1980, Rept. 298, Dept. of Geodet. Sci., Ohio State Univ. Columbus, USA.

Boucher, C., Altamani, Z., 1985, in *Proc. of the Intern. Conference on Earth Rotation and Terrestrial Reference Frame*, I.I. Mueller, ed., Dept. of Geod. Sci. Ohio State Univ. Columbus, Ohio, USA.

Boucher, C., Altamani, Z., 1986, *BIH Annual Report* for 1985.

Boucher, C. and Feissel, M., 1984, In '*Proc. of the Intern. Symposium on Space Techniques for Geodynamics*', Sopron, Hungary.

Cecchini, G., 1973, *Results of the ILS*, Vol XI, Firenze, Italy.

Chao, B.F., 1985, *Bulletin Géodesique*, 59, pp. 81-93.

Colquitt, E.S., Stern, W.L. and Anderle, R.J., 1984, '*Doppler motion measurements during MERIT*', report presented at the International Symposium on Space Techniques for Geodynamics, Sopron, Hungary, pp. 180-188.

Dickman, S.R., 1981, J, G. R., 86, 4904-4912.

Feissel, M., 1980, *Bulletin Géodesique*, 54, pp. 81-102.

Feissel, M., 1982, *In BIH Annual Report* for 1982, Paris, France.

Feissel, M., 1985a, The activities of the BIH in Earth Rotation, 1982-1984 Transactions of the XIX IAU General Assembly, New Delhi, India, Vol. XIX A, p.197.

Feissel, M., 1985b, Merit Campaigns. Report of the Coordinate Centre at BIH, XIX IAU General Assembly, New Delhi, India.

Feissel, M., 1985c, Catalogue of results on Earth rotation and reference systems. III part of the MERIT Final Report, BIH, Paris, France.

Guinot, B., 1978, in *'Proc. of Conference on Rotation of the Earth and Polar Motion Service'*, ed., I.I. Mueller, Dept. of Geod. Sci. Ohio State Univ. Columbus, USA.

IAU, 1955, *Transaction of the IX IAU General Assembly*, Resolutions, Reidel Publ. Company.

IAU, 1961, *Transaction of the XI IAU General Assembly*, Resolutions, Reidel Publ. Company.

IAU, 1979, *Transaction of the XVII IAU General Assembly*, Resolutions, Reidel Publ. Company.

IAU, 1985, *Transaction of the IXX IAU General Assembly*, Resolutions, Reidel Publ. Company.

ILS, 1903-1978, *Results of the ILS*, Vol. I-XII, The ILS.

IRIS, 1984-1987, *International Radio Interferometric Surveying Earth Rotation Bulletin*, Campbell J., Carter W., Kanejivi N., Ronnang B., Yeh Shu Hua, National Geodetic Survey, NOAA, Rockville, MD. USA.

IPMS, 1962-1984, *IPMS Annual Report*, Int. Latitude Obs., Mizusawa, Japan.

IPMSa, 1975-1987, *IPMS Monthly Notes*, Int. Latitude Obs., Mizusawa, Japan.

IUGG, 1960, *Travaux de l'IAG* edited after the IX IUGG General Assembly, Resolutions.

IUGG, 1979, *Travaux de l'IAG* edited after the XVI IUGG, General Assembly, Resolutions.

Lie Zheng-xin, 1986, *Bull. Géod.*, Vol. 60.

Lambeck K., 1971, *Bull. Géod.*, Vol. 101.

Lambeck K., 1980, *'The Earth Variable Rotation'*, Cambridge University Press.

Manabe S., Tanikawa K., Yokoyama K., 1987, *Simultameans determination of the ERP and coordinates of the optical astrometric stations.* Intern. Latitude Observatory, Mizusawa, Japan.

Markowitz W., 1968, in Proc. of the IAU Symposium No 32, eds. Wm. Markowitz and B. Guinot, Reidel Publ. Company.

Melbourne, W., *et al.* 1983, *'Project MERIT STANDARDS'*, U.S. Naval Obs. Circular No. 167, Washington, USA.

Moritz H., Mueller I.I., 1987, *'Earth Rotation'*; Theory and Observations, Wichman Publ. Company, USA.

Mueller, I.I., 1969, *'Spherical and Practical Astronomy'*, Frederick Ungar Publ. Co., New York.

Mueller, I.I., Ziqing Wei, 1985a, in *Proc. of the Intern. Conference on* 'Earth *rotation and terrestrial reference frame'*, I.I. Mueller (ed.).

NEOS, 1982-1987, *Earth Orientation Bulletin*, U.S. Naval Observatory, Washington.

Robertson D.S., Carter W.E., Tapley B.D., Schutz B.E., Eans R.J., 1985, Polar motion measurements; Subdecimeter accuracy verified by intercomparison.

Schutz, B.E., 1983-1987, *CSR Circular*, Analysis of LAGEOS laser ranging data, CSR, Texas State Univ., Austin, Texas, USA.

Tapley, B.D., Eans, R.J., Schutz B.E., 1985, in *'Proc. of the Intern. Conference on Earth Rotation and Terrestrial Reference Frame'*, Mueller I.I., ed., Dept. of Geod. Sci., Ohio State Univ., Columbus, USA.

Wilkins, 1980-1985, *Project MERIT Letters'*, No 1-8, Royal Greenwich Observatory, Herstmonceux.

Wilkins, G.A., 1986, *'Proc. of the Third MERIT Workshop and the joint MERIT-COTES Working Group meetings'* held in 1985 in Columbus. Part I of the MERIT Final Report.

Wilkins, G.A., Mueller I.I., 1986, Joint Summary Report of the IAU/IAG Working Groups MERIT and COTES., *Bull. Geod.* Vol. 60.

Wilkins, G.A., Babcock, 1987. *'Proc. of the IAU Symposium'*, No 128, Coolfont, USA.

Yumi, S., Yokoyama K., 1980. *Results of the International Latitude Service in a homogeneous System* 1899.0-1979, IPMS, Mizusawa, Japan.

Zhao Ming, Gu Zhen mian, 1985, in *'Proc. of Intern. Conference on Earth Rotation and Terrestrial Reference Frames'*, Mueller I.I., (ed.), Department of Geod. Sci., Ohio State Univ., Columbus USA.

Zhu S.Y., 1982, *Bulletin Géodesique*, 56, pp. 2548-273.

THE EARTH'S VARIABLE ROTATION: SOME GEOPHYSICAL CAUSES

KURT LAMBECK
Australian National University
Canberra, Australia

1. INTRODUCTION

The Earth's variable rotation, its departures from what it would be if it were a rigid body rotating in isolation, has occupied the interest of astronomers and geophysicists for more than 100 years. The reason for this is quite clear when one becomes aware of the range of processes that perturb the Earth from uniform rotation (Figure 1). A complete understanding of the driving mechanisms requires a study of the deformation of the solid Earth, of fluid motions in the core and of the magnetic field, of the mass redistributions and motions within the oceans and atmosphere, and of the interactions between the solid and fluid regions of the planet. The discussion of evidence for the variable rotation includes the examination of not only a variety of optical telescope evidence that goes back some 300 years, but also of historical records of lunar and solar eclipses, and planetary occultations and conjunctions for perhaps the past three millenia. The geological record, in the form of fossil growth rhythms in organisms such as corals, bivals or brachiopods or as cyclic organic growth and sediment sequences such as stromatolites or banded iron formations, extend, albeit with very considerable uncertainty, the record back through Phanerozoic time and into the Precambrian. To this variety of measurement techniques now has to be added the new methods derived from the space-oriented technological developments of the past few decades. One of the first geophysical conclusions drawn from the Earth's rotation was by W. Hopkins in 1839 who argued that the then widely held view by geologists, that the Earth was essentially fluid below the crust, was inconsistent with the astronomical observations of the precession constant. Observations of the Earth's irregular rotation also provide global constraints on the structure of the Earth; a consequence of the Earth being neither rigid nor an isolated system, of deformations and motions within the body, and of torques acting on it. Examples of mass redistribution include the rotational and tidal deformations of the planet, the seasonal and long period exchange of water mass between the oceans, atmosphere, ice sheets and ground water, and the large scale redistribution of mass associated with the stress cycle of large earthquakes. Examples of torques include the gravitational tidal torques, electromagnetic torques acting on the base of the mantle, and winds and ocean current acting on the surface.

Figure 1 illustrates schematically the variety of phenomena that perturb the Earth's rotation. The geophysical mechanisms may exhibit a variety of wavelengths but the Earth's response is primarily controlled by the global properties of the planet, properties that can often be measured

J. Kovalevsky et al. (eds.), Reference Frames, 241–284.

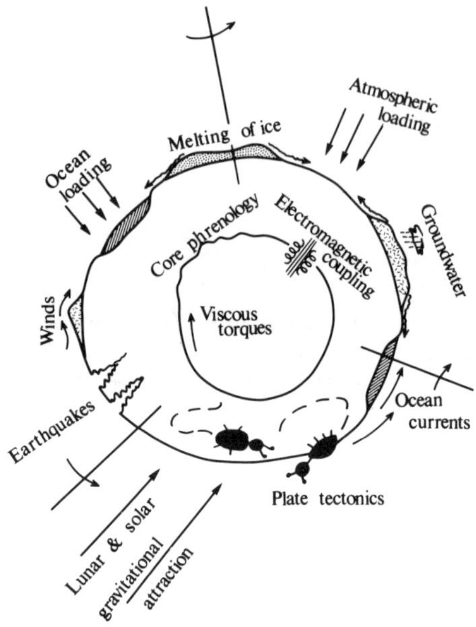

Figure 1 Schematic representation of the forces that perturb the Earth's rotation (Nature, 286, p. 104, 1980).

more directly and with greater precision by other geophysical techniques. Precise observations of the rotation nevertheless provide a number of useful pieces of geophysical information, including

(i) the frequency dependence of the Love number k_2 due to the dispersion effects and due to core resonances,

(ii) the Earth's specific dissipation function or Q at long periods,

(iii) the nature of energy dissipation in the oceans by providing estimates of the total rate of tidal energy dissipation in the ocean,

(iv) the acceleration of the Moon in its orbit and of dissipation of tidal energy within the Moon,

(v) constraint on electro-magnetic properties and processes at the core-mantle boundary.

The developments up to the late 1970's were examined and reviewed in 1980 (Lambeck, 1980), when developments in space science and technology were spurring new interest in the subject of the Earth's rotation. Precise tracking of satellites for gravity field studies, laser ranging for studying lunar motion, long-baseline radio interferometric observations for deciphering extra-galactic radio sources, and the precise manoeuvring of interplanetary flights all require a precise tracking of the motions of the Earth's rotation axis. At the same time, these new techniques permit the changes in rotation to be measured with an unprecedented precision and resolution (Lambeck, 1988). The geophysicist's signal is astrophysicist's and space engineer's noise. The most notable achievement with the new technology by that time had been the determination of polar motion from the Doppler tracking of satellites. In 1980, however, the impact of these new methods on the geophysical discussion had been minimal but the results were sufficiently promising to anticipate that new excitation functions would rise out above the measurement noise level.

I will limit the discussion in this Chapter to a few topics where either significant progress has been made since 1980 or where the new observations may lead to new insight into the physical processes, although the time span of the new data is too short, at most six Chandler wobble periods, to expect substantive developments in understanding the Earth's rotation that can be attributed solely to the result of the new methods. Much of the present discussion is, therefore, still based on the older observational data: an appropriate reminder that this earlier data will remain important for many years to come. Brief reviews, in which some of this progress is also discussed, are by Merriam (1983), Lambeck (1987), Rochester (1984), and Wahr (1985).

2. EQUATIONS OF MOTION

The fundamental equation describing the rotation of a non-rigid planet is the Eulerian equation in which both the angular momentum and inertia tensor of the planet and the torques L acting on it are time dependent. These equations are sometimes referred to as the Liouville equations. In most discussions of non-rigid rotation, the excursions from rigid body rotation are small and a perturbation form of the equations suffices. If the conventional terrestrial system, an "earth-fixed" cartesian coordinate system x_i (i=1,2,3), is introduced in which x_3 lies close to the mean position of the rotation axis and x_1 is directed towards the Greenwich meridian, it is convenient to define the rotation of these axes about themselves as

$$\omega_1 = \omega_0 m_1; \quad \omega_2 = \omega_0 m_2; \quad \omega_3 = \omega_0 (1+m_3); \tag{1}$$

where ω_0 is the mean angular velocity of the Earth. The m_i are small dimensionless quantities, of the order 10^{-6} or less. The m_1, m_2, $1+m_3$ represent the direction cosines of ω relative to x_3. In particular, m_1 and m_2 specify the position of the instantaneous rotation axis, the polar motion and, if the x_1 axis is directed towards the Greenwich meridian, m_1 is the component in the direction 90°W and m_2 is the component in the Greenwich meridian. The third component m_3, represents departures in the speed of rotation from the uniform value of ω_0. The change in length-of-day, $\Delta(\ell.o.d)$ is defined as $m_3 = -\Delta(\ell.o.d)/(\ell.o.d)$. The equations of rotation then take the form, with $j = (-1)^{\frac{1}{2}}$,

$$j\,\dot{m}/\sigma_r + m = \psi \tag{2a}$$

$$m_3 = \psi_3 , \tag{2b}$$

where

$$m = m_1 + j\ m_2 \tag{3a}$$

$$\psi = \psi_1 + j\ \upsilon_2 \tag{3b}$$

$$\sigma_r = \omega_0\ (C-A)/A\ . \tag{3c}$$

C and A are the mean polar and equatorial moments of inertia. The ψ_i are the excitation functions, characterizing the torques, relative motions and inertia tensor changes. They can be written as

$$\psi_i = \psi_i(\text{torque}) + \psi_i(\text{matter}) + \psi_i(\text{motion}) \tag{4}$$

with

$$\begin{aligned} \psi(\text{torque}) &= j\ L\ /\omega_0^2\ (C-A) \\ \psi(\text{matter}) &= \Delta I\ /(C-A) \\ \psi(\text{motion}) &= [\omega_0 h\ -\ j\ \Delta I\ -\ j\ h]/\omega_0^2(C-A)\ , \end{aligned} \tag{5a}$$

and

$$\begin{aligned} \psi_3(\text{torque}) &= (1/C\omega_0) \int_0^t L_3\ dt \\ \psi_3(\text{matter}) &= -\Delta I_{33}/C \\ \psi_3(\text{motion}) &= -h_3/\omega_0 C. \end{aligned} \tag{5b}$$

The inertia tensor I_{ij} of the Earth is written as

$$\begin{aligned} I_{11} &= A + \Delta I_{11}(t), \qquad I_{22} = A + \Delta I_{22}(t), \\ I_{33} &= C + \Delta I_{33}(t), \qquad I_{ij} = \Delta I_{ij}\ \ (i \neq j), \end{aligned} \tag{6a}$$

where the time dependent perturbations ΔI_{ij} are assumed to be small. Also,

$$\Delta I = \Delta I_{13} + j\ \Delta I_{23}; \quad h = h_1 + jh_2\ . \tag{6b}$$

The h_i are the angular momentum terms resulting from motion of particles relative to the x_i axes. Equations 2 define the motion of the rotation axis relative to the earth-fixed axes x_i. These equations clearly separate the geodetic problem of measuring the rotation (the left-hand-side of 2) from the geophysical problem of evaluating the mechanisms that drive the variable rotation, the excitation functions ψ_i. These equations also separate the polar motion (equation 2a and the definition 5a) from the length-of-day changes (equations 2b and the definition 5b), a separation that is convenient in view of different observational techniques that have been used in the past, to determine the two aspects of the rotational motion. The above equations are generally valid, provided that the excursions of the instantaneous rotation axis from its mean position are small. Otherwise, a more complete description is required in which the squares and products of small quantities must be retained but now the equations lose much of their convenience. At the present levels of accuracy of both the geodetic and geophysical data the first order equations, modified if necessary to include the small ellipticity of the polar motion path resulting from the axial asymetry of the Earth's density distribution ($I_{11} \neq I_{22}$), largely suffice. Equation (2) defines one part of the rotation problem. The other part is the motion of either the rotation axis or the earth-fixed axes in space. This latter motion can be expressed in several ways, for example by Euler's kinematic equations. If we define an inertial reference frame X_i, the position of x_i relative to it can be defined by the Euler angles

α_1 = inclination of the x_1-x_2 plane on the mean ecliptic,
α_2 = angle in the X_1-X_2 plane between X_1 and the ascending node of the x_1-x_2 plane on the ecliptic,
α_3 = angle in the x_1- x_2 plane between the ascending node and the x_1 axis.

The Euler equations then are

$$\begin{bmatrix} \dot{\alpha}_1 \\ \dot{\alpha}_2 \sin \alpha \\ \dot{\alpha}_3 + \dot{\alpha}_2 \cos \alpha_1 \end{bmatrix} = \begin{bmatrix} \cos \alpha_3 & -\sin \alpha_3 & 0 \\ \sin \alpha_3 & \cos \alpha_3 & 0 \\ 0 & 0 & 0 \end{bmatrix} \cdot \begin{bmatrix} \omega_1 \\ \omega_2 \\ \omega_3 \end{bmatrix} \qquad (7)$$

and they determine the motion of x_i relative to X_i, once the dynamic equations (2) have been solved for the ω_i. In equation (7), the angle α_3 is equal to $\omega_3 t$ and has a nearly diurnal period. This means that some nearly diurnal terms in the solutions for ω_i, generally of very small amplitude such as the lunar and solar torques on the earth, become the major terms in the Eulerian angles α_i. Large-amplitude long-period terms in the solutions for ω_i, such as the Chandler wobble, make only very minor contributions to the α_i. Certain geophysical quantities are therefore best studied by examining the m_i (the polar motion and the changes in length-of-day), while others are best studied by examining the α_i (the precession and nutation), always assuming that both astronomical quantities can be measured with comparable precision. In either case, the geophysical problem is one of evaluating the excitation functions; the variations with time of the torques L acting on the Earth, of the changes in the inertia tensor ΔI, and of the changes in relative momenta **h** caused by motions of particles making up the Earth relative to the terrestrial reference axes. An obvious, yet important point is that these excitations are integral quantities over the whole volume or surface of the Earth. The Earth's response is a globally integrated one to what may be either localized or global excitation functions. One example of this is the periodic change in the mass and angular momentum distribution within the ocean caused by the tides. The spatial spectrum of these changes is very broad, yet the Earth's rotation responds only to the small amplitude second degree component in the tide expansion. As such, the geophysical constraints imposed by the rotational data are often only of restricted value in understanding the process themselves.

3. THE FREE ROTATIONAL MODES

The Earth posseses a number of free rotational modes of which the Euler or

Chandler wobble and the nearly-diurnal wobble are geophysically significant. Their observation raises a number of questions. In a physical system such as the Earth, free oscillations are damped and cease to exist with time unless they are continually excited. The persistent occurrence of the Chandler wobble, therefore, raises the question of how this mode is maintained in the presence of dissipation. What is the nature of the excitation? Where is the energy dissipated? The same questions can be raised about the nearly-diurnal wobble except that it has not yet been observed with any certainty. A further question raised by observations of these modes concerns their period. Why does the observed Eulerian wobble period of about 430-435 days differ from that of the predicted values of about 300 days for a rigid Earth or about 270 days for a rigid Earth with fluid core? Part of the answer to these questions lies in the properties characterizing the deformational response of the planet to its rotation, and the study of the free oscillations may be expected to provide global constraints on the elasticity and anelasticity of the mantle and on the dynamics of core motions. Part of the answer also lies in the response of the oceans to the rotation and the study of the free oscillations may also yield some insight into the global ocean dynamics.

Of the two free oscillations discussed here, the Chandler wobble is well observed and its period is generally understood in terms of the departures from rigidity, exemplified by the fluidity of the core, the elasticity and anelasticity of the mantle, and the oceans. The nature of dissipation and excitation are more contentious matters, in part because the two processes are intimately linked. The nearly-diurnal oscillation has been a somewhat esoteric subject because it has remained largely unobserved, but this may no longer be so. Also, because it lies in a frequency band where there are considerable diurnal forced deformations of the Earth, resonance amplification of some of the forced motions may occur and this is anticipated to have observable effects.

3.1 THE CHANDLER WOBBLE

3.1.1 General formulation

A qualitative statement of the Chandler wobble problem can be found in most geophysics textbooks (eg Stacey, 1977). Consider a planet whose rotation axis, viewed from body fixed axes, rotates about the principal axis in accordance with Euler's equation. The potential V_c of the centrifugal force at a point P distant ℓ from the instantaneous rotation axis is

$$V_c = \tfrac{1}{2}\,\omega^2\,\ell^2$$

where

$$\ell^2 = \sum_i x_i^2 - \Big(\sum_i \omega_i\, x_i/\omega\Big)^2 \quad .$$

This potential can be written as

$$V_c = \frac{1}{3}\,\omega^2 r^2 + \Delta V_c$$

where $r^2=\Sigma_i x_i^2$ and $\omega^2=\Sigma_i\omega_i^2 \simeq \omega_0^2$. The second term ΔV_c can be expressed in a form

$$\Delta V_c = \frac{GM}{R}\left[\frac{r}{R}\right]^2 \sum_m (C^*_{2m}\cos m\lambda + S^*_{2m}\sin m\lambda)\, P_{2m}(\sin\varphi) \qquad (8a)$$

where

$$C^*_{20} = (R^3/6GM)\,(\omega_1^2 + \omega_2^2 - 2\omega_3^2)$$

$$C^*_{21} = -\,(R^3/3GM)\omega_1\omega_3, \quad S^*_{21} = -\,(R^3/3GM)\omega_1\omega_3 \qquad (8b)$$

$$C^*_{22} = (R^3/12GM)[\,(\omega_2^2 - \omega_1^2)\,], \quad S^*_{22} = -\,(R^3/6GM)\omega_1\omega_2 \quad .$$

The potential ΔV_c, harmonic in degree 2, deforms the Earth and, for an elastic response, produces an additional potential that can be defined with the aid of a rotational Love number k_2 as $k_2\Delta V_c(R)$ at $r=R$. The form (8a) is preserved outside this surface but the coefficients C^*_{2m}, S^*_{2m} are multiplied by $k_2(R/r)^3$. That is,

$$\Delta V_c = \frac{GM}{r}\left[\frac{R}{r}\right]^2 k_2\, (C^*_{2m}\cos m\lambda + S^*_{2m}\sin m\lambda)P_{2m}(\sin\varphi) \ . \qquad (8c)$$

It is this contribution that modifies the rotational motion of the planet from that described by the Eulerian solution. Equating the $k_2C^*_{2m}$ and $k_2S^*_{2m}$ with the appropriate elements in the second-degree inertia tensor gives the following time dependent elements $\Delta I_{ij}(t)$

$$\left.\begin{matrix}\Delta I_{13}\\ \Delta I_{23}\end{matrix}\right\} \simeq \frac{k_2R^5\omega_0{}^2}{3G}\left\{\begin{matrix}m_1\\ m_2\end{matrix}\right. \quad . \qquad (9)$$

The rotation excitation functions ψ_i defined by equation (5a) become

$$\psi_1 = \frac{k_2}{k_0}\left[m_1 + \frac{\dot m_2}{\omega_0}\right] \simeq \frac{k_2}{k_0}\, m_1$$

$$(10)$$

$$\psi_2 = \frac{k_2}{k_0}\left[m_2 - \frac{\dot m_1}{\omega_0}\right] \simeq \frac{k_2}{k_0}\, m_2 \quad ,$$

with

$$k_0 = 3(C-A)G/\omega_0^2 R^5 = 3GMC_{20}/\omega_0^2R^3 \simeq 0.942 \quad . \qquad (11)$$

With the definition

$$\sigma_0 = \sigma_r \ (1 - k_2/k_0), \tag{12}$$

where σ_r is defined by (3c), the equations of motion (2a) with (5a) and (9) are, in the absence of all other excitations,

$$j \ \dot{m} \ /\sigma_0 + m = 0 \quad , \tag{13}$$

and their solution is

$$m = m_0 \ e^{j(\sigma_0 t - \beta_0)} \quad . \tag{14}$$

The motion remains circular but the frequency is reduced from the Eulerian frequency σ_r to σ_0. Observations indicate that $2\pi/\sigma_0 \simeq 435$ days so that $k_2 \simeq 0.28$. This is close to the value of about 0.29-0.30 observed for the semi-diurnal body tide Love number and this analysis provides a qualitative explanation for the observed lengthening of the period from about 305 days to 435 days. The validity of equating the tidal and Chandler Love numbers is, however, not obvious and the agreement is largely fortuitous. In particular, the degree of coupling of core motions to the mantle is a significant factor in the Earth's overall rotational response but this is not so for the semi-diurnal tidal Love numbers. Also, the Chandler wobble Love number is a function of the ocean response to the centrifugal potential and a more quantitative evaluation, in terms of the planet's physical properties, is required.

If the Earth is subject to a forced excitation the total excitation ψ can be considered in two parts: the forcing function ψ_r evaluated as if the Earth were rigid and a second part ψ_D corresponding to the rotational deformation of the Earth. This is the contribution (10), or $\psi_D = (k_2/k_0)\mathbf{m}$ and the equations of motion now are

$$j \ \dot{m}/\sigma_0 + m = [k_0/(k_0 - k_2)] \ \psi_r \quad . \tag{15}$$

Thus the elastic yielding of the Earth modifies the amplitude of the excitation function by a factor $k_0/(k_0-k_2)$ and the effect of specific excitation functions can be evaluated as if the planet is rigid and using equation (15) in which σ_0 corresponds to the free oscillation frequency of the real Earth. This is appropriate for an excitation that does not load the Earth. If it does then ψ_r, must be replaced by $(1+k_2')\psi_r$, where k_2' is the second degree potential load Love number, and in general,

$$j \ \dot{m}/\sigma_0 + m = K \ \psi_r \quad , \tag{16}$$

where the wobble transfer function K is equal to $k_0/(k_0-k_2)$ when the excitation does not load the Earth and is equal to $(1+k_2')k_0/(k_0-k_2)$ when it does.

In the presence of linear damping the deformational bulge of the planet lags the potential of the centrifugal force by an angle ϵ such that (c.f. 9)

$$\begin{bmatrix}\Delta I_{13}\\ \Delta I_{23}\end{bmatrix} = \frac{k_2 R^5 \omega_0^2}{3G}\begin{bmatrix}\cos\epsilon & \sin\epsilon\\ -\sin\epsilon & \cos\epsilon\end{bmatrix}\begin{bmatrix}m_1\\ m_2\end{bmatrix} \quad . \tag{17}$$

For small ϵ, such that $\cos\epsilon \simeq 1$ and $\sin\epsilon \simeq Q_w^{-1}$,

$$\psi = (k_2/k_0)(1 - jQ_w^{-1})m \quad ,$$

leading to the equation of motion

$$\dot{m}/\sigma_r - j[1 - \frac{k_2}{k_0}(1 - jQ_w^{-1})]m = 0 \tag{18a}$$

whose solution is

$$m(t) = m_0\, e^{j(\sigma_0 + j\alpha)} = m_0 e^{j\sigma_0} \tag{18b}$$

with a dissipation time constant α^{-1} given by

$$\alpha = \sigma_0\, Q_w^{-1}\, k_2/(k_0 - k_2) \simeq \sigma_0/2Q_w \quad . \tag{18c}$$

The damping parameters α or Q_w serve, like the rotational Love number, as convenient functions to describe qualitatively the rotational motion but, without quantitative evaluations of specific energy sinks, they do not simplify the physics of the problem. Recent observations suggest that $50 \lesssim Q_w \lesssim 100$ (Okubo, 1982b) and the damping time constant is about 20-40 years.

In terms of energy dissipation, the wobble Q, is defined as

$$Q_w^{-1} = \frac{1}{2\pi E_w}\int \frac{dE}{dt}\,dt = \frac{1}{2\pi}\frac{\Delta E}{E_w} \tag{19a}$$

where E_w is the total wobble energy stored in the Earth that would, in the absence of any excitation, be dissipated in the time interval $0 < t < \infty$ and includes the kinetic strain, gravitational and ocean pole tide energies. ΔE is the amount of energy dissipated in one cycle of the motion. It should be noted that Q_w is not directly comparable with the shear mantle Q_μ as used in seismology. Several recent studies (Smith and Dahlen, 1981; Ben Menaham, 1982; and Okubo, 1982b) obtained

$$Q_w/Q_\mu \simeq 1.5\text{-}2.1 \quad . \tag{19b}$$

There are two ways in which the excitation mechanisms can be tested. One is to compute a theoretical m(t) for a specified $\psi(t)$; the other one is to deconvolve the observed **m**(t) to estimate an "observed" $\psi(t)$ and to compare this with the model value. The deconvolution process is one of estimating an "observed" $\psi(t)$, from the observed m(t) that contains a noise component. A simple process that does not attempt to separate noise from the signal m(t) is to write, for a discrete observation series, for time t

$$m(t) = e^{j\sigma_0 t}\,[m_0 - j\sigma_0 \sum_{0}^{t} \psi(\tau)e^{-j\sigma_0\tau}\Delta t]$$

and for an instant t-Δt

$$m(t-\Delta t) = e^{j\sigma_0(t-\Delta t)}[m_0 - j\sigma_0 \sum_{0}^{t-\Delta t} \psi(\tau)\, e^{-j\sigma_0\tau}\Delta t] \quad .$$

Then

$$m(t) = m(t-\Delta t)e^{j\sigma_0\Delta t} - j\sigma_0\psi(t)\Delta t \quad , \tag{20a}$$

and the inferred excitation is

$$\psi^0(t) \equiv \psi(t) = -\,(1/j\sigma_0\Delta t)[m(t) - e^{j\sigma_0\Delta t}\, m(t-\Delta t)] \tag{20b}$$

More sophisticated deconvolution filters have been proposed (see, for example, Smylie et al., 1970; Wilson and Haubrich, 1976).

The comparisons can also be made in the frequency domain. A series of step-function excitations will produce a spectrum of $\psi(t)$ that is "red", one of decreasing power with increasing frequency, whereas a series of delta-function excitations will produce a "white" or flat spectrum. It becomes possible, in theory at least, to gain insight into the forcing process by examining the excitation series derived by a deconvolution process from the polar motion process. Evaluation of the excitation spectra are generally suggestive of a red noise process, with power increasing towards the low frequencies, but in the neighbourhood of the Chandler wobble frequency the spectrum is essentially white and consistent with delta-function-like excitations (e.g. Wilson and Haubrich, 1976).

3.1.2 Observational evidence for wobble parameters

Numerous analyses of the Chandler wobble parameters have been made in recent years using different data sets and data lengths, different analytical techniques and different assumptions about the statistical nature of the noise of the data and of the excitation process. The estimation problem is one of matching the observed wobble spectrum, containing measurement noise, with the spectrum of the model for the wobble impulse response, usually assumed to be a damped linear oscillator, that has been subjected to an excitation process of certain assumed physical or statistical properties. In the presence of measurement noise $\epsilon_m(t)$, equation (20a) becomes

$$m(t) - e^{j\sigma_0\Delta t}\, m(t-\Delta t) = -\, j\sigma_0\psi(t)\Delta t + \epsilon_m(t) - \epsilon(t-\Delta t)e^{j\sigma_0\Delta t} . \tag{21}$$

If the spectra of both ψ and ϵ_m, denoted by $S_\psi(\sigma)$ and $E_m(\sigma)$ respectively, are assumed to be white, such that $S_\psi(\sigma)=\nu_\psi^2$ = constant and $E_m(\sigma) = \nu_m^2$ = constant, then the amplitude spectrum of **m** is

$$S_m(\sigma) = \frac{\pi}{2}\left[\nu_m^2 + \frac{(\sigma\Delta t)^2 \nu_\psi^2}{1 - 2e^{-\alpha\Delta t}\cos[(\sigma-\sigma_0)\Delta t] + e^{-2\alpha\Delta t}}\right] \qquad (22)$$

By matching this spectrum with the observed spectrum it becomes possible to estimate ν_m^2, ν_ψ^2, the damping constant α (or Q_w) and the free oscillation frequency σ_0. Equation (22) demonstrates clearly the interdependence between the various parameters. The damping parameter α is particularly sensitive to the assumptions made for the noise and excitation spectra. The various published solutions for these parameters differ essentially in the manner in which the observed spectrum is computed or in the manner in which the theoretical spectrum is matched to the observed spectrum.

Author	Data	Interval	Period	range (sid.d)	Q_w (sid.d)
Wilson and Vicente (1981)	BIH	1962-1979	429.87	(443.40 - 417.14)	
	IPMS	1967-1977	431.39	(446.65 - 416.19)	
	DMA	1969-1979	431.90	(451.05 - 413.84)	
	ILS*	1900-1979	434.46	(437.57 - 430.88)	>50(175)
Ooe (1978)	ILS	1900-1975	436.01	(434.00 - 438.05)	50-300
Okubo	ILS*, BIH & IPMS	1962-1980			50-100

Table 1 Summary of recent estimates of Chandler period and Q_w. The range of period estimates corresponds to ±1 standard deviation. The Q value preferred by Wilson and Vicente is 175 but values greater than 50 are consistent with their data and analysis technique. The asterisk denotes the revised ILS data set.

Table 1 summarizes some recent results for the wobble period and Q. The various estimates for this period are in reasonably good agreement. The data sources are the Bureau International de l'Heure (BIH), the International Polar Motion Service (IPMS) and the satellite Doppler solutions by the Defence Mapping Agency (DMA) are consistent (Wilson and Vicente, 1981) although the shorter data sets do result in rather large standard deviations for the period estimates. The revised International Latitude Service (ILS) data set (Yumi and Yokoyama, 1980) gives an estimate of considerably greater precision than the earlier solutions but the result is consistent with the analyses by Wilson and Haubrich (1976) and Ooe (1978) of an earlier ILS data set. The polar motion results from laser tracking of satellites or interferometric observations of radio sources do not yet cover a long enough time span to be useful.

Estimates of Q_w are more variable, being critically dependent on the choice of filter parameters used in estimating the power spectrum of the polar motion. Most analyses indicate that $50 \lesssim Q_w \lesssim 100$ (Okubo, 1982b; Wilson and Haubrich, 1976; Ooe, 1978) although Wilson and Vicente (1981) obtained $Q_w \simeq 175$ from an analysis of the revised ILS data set. Of some concern is the observation by Okubo (1982a) that an analysis of the ILS data for the years after 1963 produces quite different results than either the BIH or IPMS data sets for the corresponding period, placing doubt on Q_w values obtained from any part of the ILS data set. The estimates of the power in the neighbourhood of the Chandler wobble frequency required to excite the wobble is estimated by Wilson and Vicente (1980) as about (7-8) $(10^{-8}$ radian$)^2$/(cycles year) and by Ooe (1978) as about $21(10^{-8}$ radian$)^2$/(cycles/year), the difference being a consequence of different assumptions made about the measurement noise of $\mathbf{m}$.

3.1.3 Period

The quantitative evaluation of the Chandler wobble period involves the computation of the time dependence of the inertia tensor ΔI_{ij} and of the relative angular momentum vector h_i associated with the deformation and motion of the mantle, core and oceans. Thus if a particle at its equilibrium position x_i is moved to a position $x_i + d_i(x,t)$ in response to the rotation, the time dependent part of the inertia tensor (6a) is

$$\Delta I_{ij}(t) = \int_V \rho(x)[\,2x_k\, d_k(x,t)\delta_{ij} - x_i\, d_j(x,t) - d_i(x,t)\, x_j\,]dV \qquad (23a)$$

and the relative angular momenta are

$$h(t) = \int \rho(x)(x+d)\Lambda(\partial d/\partial t)dV \qquad (23b)$$

The elastic displacements within the mantle and core follow from the solution of equations that describe the deformation of a self-gravitating, elastic and, in this case, an ellipsoidally shaped rotating planet. An approximate solution follows from the Kelvin-Hough model for the core as defined by equations and from considering the motion relative to Tisserand axes such that h_i(mantle)=0. The mantle's deformational response to the variable rotation is assumed to be linear in m_i so that the displacements d_j of the mantle, and therefore the ΔI_{ij}, are also linear in m_i. In a most general form,

$$\Delta I_{ij}(t) = D_{ijk}(x,t)\ m_k(t),$$

where the D_{ijk} is a third order tensor and is a function of the mantle parameters. For an axially symmetric Earth, and ignoring the ellipticity of the planet, D_{ijk} is given simply by (9) or

$$D = k_2 R^5 \omega_0^2 / 3G \quad ,$$

where k_2 is the conventional static Love number of degree 2 (see also Sasao et al., 1977; Smith and Dahlen, 1981). The equations of motion can then be written as

$$\dot{m} - i\,\omega_0 \frac{(C-A)}{A_m}\left[1 - \frac{D}{C-A}\right] m - i\,\omega_0 \epsilon_c \frac{A}{A_m}\, n = 0$$

$$\dot{n} + i\,\omega_0 \frac{(C-A)}{A_m}\left[1 - \frac{D}{C-A}\right] m + i\,\omega_0 \left[1 + \frac{A}{A_m}\epsilon_c\right] n = 0 \tag{24}$$

and the corresponding eigenfrequencies are

$$\sigma_1 \simeq \frac{(C-A-D)}{A_m}\,\omega_0(1+\epsilon_c)\left[1 - \frac{A}{A_m}\epsilon_c - \frac{(C-A-D)}{A_m}\right] = \sigma_e$$

$$\sigma_2 \simeq -\,\omega_0\left[1 + \frac{A}{A_m}\epsilon_c - \frac{(C-A-D)}{A_m}\epsilon_c\right] \simeq \omega_0\left(1 + \frac{A}{A_m}\epsilon_c\right) \tag{25a}$$

The first is the Chandler wobble frequency, corrected for the mantle elasticity and the ellipsoidal core. The second is a nearly-diurnal wobble. The radial density structure of the entire Earth is used to compute the elastic yielding factor D but for evaluating the fluid core effect the core has been assumed to be of constant density. These solutions do not specifically allow for the elastic deformation of the core-mantle boundary, a neglect that is most severe for the second oscillation of frequency σ_2. More rigorous treatments are given by Smith (1977) and Sasao et al. (1980). These last authors obtained

$$\sigma_2 = -\omega_0\left[1 + \frac{A}{A_m}(\epsilon_c - \beta)\right] \quad , \tag{25b}$$

where $\beta \approx 6\times10^{-4}$ is a parameter defining the deformation of the mantle and core, and the ellipticity of the core-mantle boundary is effectively reduced by about 25%, because of the deformation.

Of the constants in (25a), C and A are based on observations of the precession constant H and the dynamic flattening and are known with sufficient precision, and

$$C = 8.0394\times10^{37}\ \text{kg m}^2$$

$$A = 8.0131\times10^{37}\ \text{kg m}^2\ . \tag{26a}$$

The core and mantle inertia elements are estimated from Earth density models based on the inversion of seismic data and on the assumption of hydrostatic equilibrium. For Earth model 1066A of Gilbert and Dziewonski (1975), Smith and Dahlen (1981) obtained

$$A_c = 9.1100\times10^{36} \text{ kg m}^2$$
$$C_c = 9.1332\times10^{36} \text{ kg m}^2 \quad (26b)$$
$$A_m = 7.1005\times10^{37} \text{ kg m}^2$$

Also, for the same model, $k_2 = 0.30088$. With these values the wobble period is 396.4 sidereal days. The more precise theory by Smith and Dahlen gives 396.9 days. Some uncertainty in this value results from the limitations of the Earth model, about ±0.5 days according to Smith and Dahlen (see also Sasao et al. 1980, Table 1, for estimates of this period for different Earth models). A further uncertainty arises from the assumption that the planet is in hydrostatic equilibrium, through the estimates of the core and mantle inertia elements, something that the planet is manifestly not.

Variations in the Earth's rotation induce small amplitude ocean tides. Of particular interest is the so-called "pole tide". This oceanographic curiosity is of more than passing geophysical interest because it significantly modifies the Earth's wobble from what it would be for an oceanless planet, and it may also account for the damping of the wobble. The amplitude of the pole tide is at most a few millimeters and only occasionally rises above the noise level of the sea level observations. Models of this tide are therefore, based on theoretical arguments, the simplest of which is that the tide follows an equilibrium theory. By the nature of the ocean-continent distribution the tide exhibits more regional variability than does the driving potential but only the components of degree 2 and order 1, those contributing to the inertia elements ΔI_{13} and ΔI_{23}, are relevant here. This abbreviated tide is given by

$$\xi = -(R^2\omega_0^2/6g)(1+k_2-h_2)[(A_1m_1+A_2m_2)\cos\lambda + (B_1m_1+B_2m_2)\sin\lambda]P_{21}(\sin\varphi), \quad (27)$$

where the A_i and B_i are functions of the coefficients in the ocean expansion (Lambeck, 1980, p.143). The time-dependent inertia tensor follows directly and the free oscillation frequency of the Earth is modified to, in a first approximation,

$$\sigma_0 \simeq \sigma_e\left[1 - \frac{1}{2}\omega_0\frac{\psi_0}{\sigma_e}(A_1+B_2)\right]. \quad (28a)$$

where σ_e is given by (25a). With $(A_1+B_2)\simeq3.05$ and with $\sigma_e\simeq2\pi/396.9$ days

$$\sigma_0 = 0.936\sigma_e. \quad (28b)$$

The Chandler motion is no longer circular but the ellipticity is small, the ratio of the two axes differing from unity by only about 0.3%. The effect of the equilibrium pole tide is to lengthen the wobble period by 27.2 days, from 396.9 days for the elastic body to 424.2 days for the elastic body with an equilibrium ocean tide. Smith and Dahlen (1981) obtained an increase of 29.8 days and Dickman and Steinberg (1986) obtained 30.9 days using more rigorous descriptions of the mantle deformation effects of the ocean tide.

Does the pole tide follow an equilibrium theory? The answer to this question is clearly of geophysical importance for if departures from equilibrium are significant, yet impossible to evaluate with precision, little or no information can be gleaned about the anelasticity of the Earth's mantle at the wobble frequency. On the other hand, if departures from equilibrium can be established then this contributes to the knowledge of the dynamics of long-period tides.

One consequence of a non-equilibrium tide is that the ocean surface deformation lags behind the rotation axis (c.f. equation 17) but for plausible values of this lag the period of the wobble is not affected by more than a fraction of a day. Alternatively, the tidal amplitudes could be enhanced, as would occur if the pole tide is near a resonance in the ocean and then $\sigma_0=\sigma_e\{1-\delta\Delta\sigma\}$ where $\Delta\sigma$ is the equilibrium ocean effect on the wobble frequency and δ is the amplification factor. The enhancement required to explain the discrepancy of about 8 days between the observed period of 435 days and the above model period of about 426-427 days is, from (28a) about 1.3.

Significant local enhancement of the pole tide has been known to occur but observational evidence for this tide in the open oceans remains scant. Certainly the requisite δ factor of 1.3 cannot be ruled out from the observational evidence. Theoretical arguments, however, run mostly in favour of a pole tide that closely follows equilibrium theory (see the recent studies by O'Connor and Starr, 1983; Dickman, 1985; Carton and Wahr, 1986; O'Connor, 1986; Wunsch, 1986). Wunsch argues that any departures from equilibrium in the open ocean will be too small to drive the observed pole tide in the North and Baltic Seas and that the observed enhancement must be the result of a near resonance of long-period waves in the basin with the pole tide potential. One exception to the above conclusion is the study by Molodenskiy (1985) who developed a more complete pole-tide model. Some of the previous authors developed this tide for either global oceans or for oceans with a zonal geometry and here the tide effectively follows an equilibrium theory but, according to Molodenskiy, for realistic ocean geometries the departures from equilibrium become significant. This conclusion does not follow from the results of Carton and Wahr (1986) and O'Connor (1986) for models of the tide in restricted ocean basins, and confirmation of Molodenskiy's results is desirable.

This uncertainty about the non-equilibrium pole tide is unfortunate for it does not permit strong conclusions to be drawn about the alternate explanation, that the discrepancy in period is a result of the anelasticity of the mantle. If it is, then σ_e in equation (28a) must be written as $(\sigma_e + \Delta\sigma)$, with the requirement that the σ_0 in this equation equals the observed value. With the Smith and Dahlen value for the equilibrium ocean tide lengthening of the period, the required decrease in frequency is $\Delta\sigma = - 0.0176\ \sigma_e$. Any such anelastic lengthening of the wobble period can be described by a modified Love number at frequency σ as

$$k_2(\sigma)=k_2 + \Delta k_2$$

where, k_2 is the elastic value. For the wobble model (5),

$$\Delta k_2 = -(k_0/\sigma_r)\Delta\sigma \simeq 0.013.$$

This is the increase in the Love number k_2 over the value based on seismic parameters. A popular Q model in $Q = Q_0(\text{frequency})^\gamma$, with $\gamma \leqslant 0.3$, where Q_0 is the Q value at the reference frequency, such as the frequency of the second degree free seismic oscillation (e.g. Anderson and Minster, 1979). With a mantle Q of between 350 and 600 at 54 minutes, $0.18<\gamma<0.30$. More precise models of this dispersion effect, based on realistic Earth models rather than on the Kelvin model, have been discussed by Smith and Dahlen (1981); Okubo (1982b) and Dehant (1987) with essentially similar results.

The ultimate estimates for the frequency dependence of attenuation in the Earth from the Chandler wobble period are only as reliable as is the assumption that the discrepancy in the wobble period of 7-8 days can wholly, or in a known part, be attributed to mantle anelasticity; that the pole tide can be adequately modelled. This question is an important one because while considerable information on the Earth's Q exists at seismic frequencies, the long period Q's are poorly constrained and the validity of extrapolating Q models beyond the seismic frequencies remains an open question.

3.1.4 Dissipation

The second aspect of the Chandler wobble is the dissipation process: where is the wobble energy dissipated? Is it in the oceans or mantle, or is it at the core-mantle boundary? This last possibility is usually discounted as estimates of viscous dissipation and electromagnetic damping appear to be wholly inadequate. The wobble dissipation function, or Q_w, is defined by equation (19a). Statistical analyses of the observations of the polar motion lead to estimates of Q_w and the geophysical problem is to evaluate both the total wobble energy stored in the Earth during one cycle of the oscillation and the rate at which this energy is being dissipated.

An order of magnitude of the rate of energy dissipation follows from considerations of the kinetic energy only. In its initial state the kinetic energy is

$$E_i \simeq \frac{1}{2}[A(\omega_1{}^2 + \omega_2{}^2) + C\omega_3{}^2]$$

and when the wobble is completely damped the final kinetic energy is

$$E_f = \frac{1}{2}C\omega^2$$

with

$$A^2(\omega_1{}^2 + \omega_2{}^2) + C^2\,\omega_3{}^2 = C^2\omega^2$$

because angular momentum is conserved. The loss of energy is, with (18b)

$$E = E_i - E_f = \frac{1}{2}\,A\,H\,\omega^2\,m_0{}^2\,e^{-2\sigma_0 t}$$

where H is the precession constant. The rate of energy dissipation is

$$\frac{dE}{dt} = \frac{\sigma_0}{2Q_w} A\, H\, \omega^2\, m_0{}^2\, e^{-\sigma_0 t/Q_w} \simeq 10^6 \text{ W}. \tag{29}$$

This is about six orders of magnitude less than the rate of energy dissipation in the semi-diurnal ocean tide.

Perhaps because of the insignificance of this energy loss, the sink has not been unambiguously identified. Ocean dissipation has generally been thought to be unimportant and this is perhaps surprising in view of the importance of the oceanic dissipation at the higher tidal frequencies. This wobble energy source can be evaluated in a similar way as the semi-diurnal and diurnal tides. The abreviated pole tide is given by (34) and the dissipation can be modelled by introducing a lag ϵ_0 in this response. The resulting equations yield a relaxation time constant of

$$\alpha = \omega_0(\sigma_0/\sigma_e)\epsilon_0\, \psi_0(A_1+B_2)/2 \tag{30a}$$

and

$$Q_w^{-1} = \epsilon_0\, \psi_0\, (A_1+B_2)\, (\omega_0/\sigma_e) \tag{30b}$$

For $Q_w = 1090$, $\epsilon_0 \simeq 4°$. Published information on the lag of the pole tide is even more sparse than the amplitude information. Values of a few tens of degrees have been reported from regions where the tide has been locally enhanced but global values cannot be established.

Wunsch (1974) assumes that the dissipation is by bottom friction in shallow seas and he estimates a dissipation of about 8×10^4 W for the North Sea although he cautions that this value is extremely model dependent and that this estimate could be excessive by an order of magnitude (see also Wunsch, 1986; Dickman and Preisig, 1986). Extrapolation to all of the world's shallow seas leads to a total dissipation of 4×10^6 W and possibly as low as 4×10^5 W, but, again, Wunsch cautions that these estimates may be excessive because some of the conditions that could have combined to produce the enhanced pole tide in the North Sea may not occur everywhere. If the bottom friction model is valid, then the pole tide currents in these restricted regions need to influence the global pole tide so as to produce the requisite lag angle. Perhaps this can be tested by numerical models. Other processes that lead to a more uniform dissipation in the oceans, either by dissipating energy along the coastlines, such as the Proudman model or internal friction, must also be considered.

If dissipation occurs primarily in the mantle then the shear Q_μ, corresponding to a wobble Q of 70-200, lies in the range 30-100 (equation 19b). This compares with a value of about 350-600 for the spheroidal seismic free oscillation mode of degree 2 whose period is about 54 minutes and this is suggestive of a strong frequency dependence. With the above frequency dependent Q law this leads to $0.18<\gamma<0.30$, (see also Smith and Dahlen, 1981; Molodenskiy and Zharkov, 1982; Okubo, 1982b) and this is consistent with the result obtained above from the analysis of the period elongation. However, the result is only as good as the assumption that dissipation occurs only in the mantle. The whole problem of dissipation remains wide open.

3.1.5 Excitation

The study of the excitation of the Chandler wobble by the atmosphere goes back to work by V. Volterra in 1895. The essential argument is that the atmospheric pressure variation over the Earth's surface is not strictly annual and that the wobble is maintained by the irregular variation in the elements in the excitation function. The problem is one of evaluating the excitation due to atmospheric motion and mass transport and to compare either the motion of the pole driven by this excitation with the observed motion, or the power in the excitation spectrum with the power deduced from the astronomical observations. This atmospheric forcing function is discussed further below. It contains two parts, ψ(motion) and ψ(matter). The first requires an evaluation of the relative angular momenta (h_1, h_2) from global meridional winds at all altitudes. Alternatively, the wind effect can be evaluated by computing ψ(torque) and this requires a knowledge of surface winds, of surface friction coefficients and of surface topography. ψ(motion) is generally believed to be small. This is indeed fortunate because measurements of the temporal and spatial variation of meridional winds are relatively sparse, and estimates of surface friction coefficients remain unreliable. The evaluation of ψ(mass) requires information of surface pressure only and it is usually easier to estimate this function than it is to estimate ψ(motion) from available meteorological observations.

A number of estimates of ψ(mass) have been made since Volterra's original suggestion. These include studies by Jeffreys in 1940, Munk and Hassan in 1961, Wilson and Haubrich (1976) and Wahr (1983). The general conclusion reached in these last three studies, based on relatively complete global monthly mean surface pressure variations for the same length of record as the astronomical time series, is that while there is significant power in the excitation spectrum it is insufficient, by a factor of 2 or 3, to maintain the wobble against damping. Wilson and Haubrich (1976) have argued that ψ(motion) from mountain torques does make a significant contribution but other studies do not support this conclusion.

In recent years, high resolution wind and surface pressure data has become available as part of global meteorological observing programs such as the Global Atmospheric Research Program and detailed excitation functions have been computed at daily intervals (Barnes et al., 1983). These data sets may lead to a revision of the atmospheric excitation hypothesis if changes in ψ occur on time scales that are much less than a month but, at present, these studies cover intervals that are too short compared with the damping time constant to estimate whether the motion can be maintained against dissipation of the wobble energy.

It has recently been suggested that variations in the ground water storage, including ground water, and water stored on the surface as ice and snow, may be sufficiently variable to contribute to the wobble excitation (Hinnov and Wilson, 1987). This hydrologic excitation function is somewhat smaller in magnitude than the seasonal atmospheric excitation of the polar motion (see Lambeck, 1980; p.154) but, because of the seasonal patterns of snow and ice coverage, the spectrum may contain significant power over a broad frequency band centred on the 12 month period (see also Chao et al., 1987). Global time series of hydrologic parameters are, however, incomplete and satisfactory tests of this hypothesis are difficult to make. Spectral analyses of the polar motion indicate that the power about the annual frequency is concentrated in a very narrow frequency band and this seems to argue against the hypothesis that major departures occur in the hydrologic excitation function from a strictly sinusoidal oscillation.

Soon after Chandler's discovery, J. Milne suggested that there may be some relation between polar motion and seismic activity. Later, in 1928, G. Cecchine also suggested such a relationship and the hypothesis that the wobble is excited by changes in the Earth's inertia tensor caused by large earthquakes has not been far from the forefront of discussions of the wobble ever since. If the earthquake modifies the inertia tensor according to a step-function at time t_s, no instantaneous change in the position of the rotation pole occurs and there is only a change in the direction of the pole path. A succession of earthquakes, associated with sufficiently large changes in the excitation function could then maintain the wobble and explain, at the same time, any secular drift in the pole position. That is (Lambeck, 1980)

$$m(t) = m_0 e^{j\sigma_0 t} + \frac{\omega}{\sigma_0 A} \sum_{t_s} \Delta I_{t_s} - e^{j\sigma_0 t} \frac{\omega}{\sigma_0 A} \sum_{t_s} \Delta I_{t_s} e^{-j\sigma_0 t} \qquad (31)$$

where the summation is carried out over all events occurring at times t_s. The second term is the shift in position of the excitation pole or of the mean pole of rotation while the third term represents the modified Chandler wobble about this new position of the mean pole. If the basic assumption of the step function model is valid, it remains to evaluate the ΔI from parameters characterizing the earthquake displacement field. Several authors have derived equations for this (Smylie and Mansinha, 1971; Dahlen, 1973; Israel et al., 1973; Mansinha et al. 1979) using elastic dislocation theory. The resulting expressions are of the form

$$\Delta I_{13} + j\Delta I_{23} = M_0 \sum_{i=1}^{3} \Gamma_i(r)(g_i + jh_i)$$

where M_0 refers to the seismic moment of the earthquake, Γ_i are functions of geocentric distance and the radial variations of elastic moduli and density and the g_i and h_i are functions of the earthquake coordinates and parameters defining the orientation of the fault plane and the direction of motion. A different approach has been adopted by O'Connell and Dziewonski (1976) who expressed the displacements in terms of a normal mode expansion but the results obtained by the two methods are consistent (see Table 8.2 of Lambeck, 1980). All of these studies indicate that very large earthquakes, such as the Chile (1960) and Alaska (1964) earthquakes, are able to shift the mean pole by amounts of 0".01 to 0".02.

The major difficulty with testing the seismic excitation hypothesis is the estimation of realistic seismic moments for most of the earthquakes. M_0 can be estimated from long-period spectra of the earthquake waves or from empirical relationships with the short-period seismic magnitude. Most of the moments of the older earthquakes, prior to about 1960, are based on the latter approach but different laws produce quite different moments once the magnitudes exceed about 7.5. There simply have not been enough very large earthquakes to permit a reliable average relationship to be established and considerable debate remains about the size of the moments of the large earthquakes.

The above shifts of 0".01 - 0".02 are for the two largest moment earthquakes of this century but there have not been enough comparable earthquakes before or after these two events to maintain a wobble that exhibits little evidence for major attenuation. Of earthquakes of recent years for which precise seismic moments are available, none have been adequate to maintain the motion against damping (Souriau and Cazenave, 1985; Gross, 1986) and just as the accuracy of the polar motion observations has increased significantly, so have the estimates of the seismic moments decreased! It has been argued that aseismic deformations, occurring at periods that are short compared with the wobble period but which lie outside the bandwidth of most seismometers, account for the missing excitation. Evidence for the aseismic deformations includes observations of slow (by seismology standards) deformation prior to the major shock. Studies of the after-shock area also provide evidence for post-seismic deformation: estimates of the fault plane areas based on after-shocks several months after the main shock are often much larger than the areas estimated from after-shocks immediately following the earthquake. The general discrepancies between estimates of seismic slip and plate motions along many plate boundaries, also suggests that significant pre-and post-seismic deformations occurs. Yet the observational evidence remains inadequate for evaluating quantitatively the role of aseismic deformation in exciting the Chandler wobble.

A third possible excitation mechanism is the action of electromagnetic torques on the base of the mantle. These torques arise from the penetration into the lower mantle of time-dependent magnetic fields self-generated within the liquid core, and they have often been invoked to explain the decade changes in the length-of-day (see below). It has been suggested by Runcorn (1982) that these torques also excite the Chandler wobble.

For the purpose of calculation, a stationary dipole magnetic field **B** is assumed to exist within the core and mantle. Differential motion between the highly conductive core and less conductive mantle is also assumed to occur. This relative movement produces an electric current density **J** that penetrates into the imperfectly insulating lower mantle. There it interacts with the field **B** to produce the Lorentz force **J** $\wedge$ **B** and a torque on the mantle of (Rochester, 1962)

$$\mathbf{L} = \int_{V_m} \mathbf{r} \wedge (\mathbf{J} \wedge \mathbf{B})\, dV \quad . \tag{32}$$

The integral is over the conducting part of the mantle. The geophysical problem lies in the evaluation of the current **J** circulating in the lower mantle. Dynamo theories generally predict the existence of a strong toroidal field in the core, produced by the differential movements at both the inner core - outer core and outer core - mantle boundaries, and it is the current produced by this part of the field that produces a net torque on the mantle. But this toroidal field is obscured from observations by the low electrical conductivity of much of the upper mantle and any estimates of the efficiency of the electromagnetic coupling are strongly model dependent. For these torques to excite the wobble they must occur over an interval of time that is short compared to the wobble period.

Two time constants control this coupling (Roberts, 1972). The first, τ_1, governs the time with which sudden changes in the toroidal field penetrate into the mantle. The second, τ_2, is the time constant of the electromechanical coupling. Neither τ_1 nor τ_2 alone is appropriate for characterizing the time constant for coupling but Roberts showed that the combination $\tau_3 = \tau_1{}^{1/3}\ \tau_2{}^{2/3}$ provides a reasonable estimate. A simple model is provided by Roberts in which

$$\tau_1 = \mu_0 \langle\lambda\rangle L^2 \quad , \quad \tau_2 = \frac{C_m C_c}{C} \cdot \frac{1}{4\pi\langle\lambda\rangle a_c^4 \, L \, B_r^2} \quad . \tag{33a}$$

In these equations μ_0 is the magnetic permeability and is nearly equal to that of a vacuum or $4\pi \times 10^{-7}$ H m^{-1}. γ is the electrical conductivity, and $\langle\lambda\rangle$ is the average value for of the lower mantle, of thickness L, defined as

$$\langle\lambda\rangle = L^{-1} \int_{a_c}^{a_c+L} \lambda \, dr \tag{33b}$$

where a_c is the radius of the core. B_r in (33a)) is the mean value of the vertical field at the core-mantle boundary. More complex models are discussed by Braginskiy and Fishman (1976) and LeMouel and Courtillot (1982) (see also Rochester, 1984). The evaluation of the models requires a knowledge of the lower mantle conductivity and of the strength of the field B_r at a_c. The latter can be approximately estimated by extrapolating the surface field downwards and $B_r(a_c) \simeq 3.5\times10^{-4}$ T. The conductivity of the lower mantle is poorly constrained (see, for example, Figure 8.5 of Stacey, 1977) and average values for the lower mantle range from about 100 Ω^{-1} m^{-1} with L = 100km (McDonald, 1957) to about 400 Ω^{-1} m^{-1} with L= 2000 km (Backus, 1983). Then τ_3 ranges from about 8 years for the upper value to about 15 years for the lower value. The more complex model of Braginskiy and Fishman (1976) gives $\tau_3 \simeq 8$ years for the conductivity model of Backus (Rochester, 1984) and all models give time constants that are more than an order of magnitude greater than required to excite the wobble. A principal limitation of these models is that they account for only the lowest-order terms in the magnetic field so that the actual field strength at the core-mantle boundary may be quite different than that assumed. Also, fluctuations with periods less than the screening time constants are not observed at the surface and, if they occur, they could significantly enhance the coupling (Hide, 1966).

There does appear to be another way to test the hypothesis. Any electromagnetic torques will also excite the length-of-day changes for there is no obvious reason why the torques should be primarily meridional. It is possible, therefore, to also test the hypothesis by examining the spectrum of length-of-day changes. Runcorn suggests that the impulse torque required to produce a change in the radius of the pole path of 0."01 is about equal to that required to produce the observed decade fluctuations, of $5x10^{-8}$, recorded in the proportional change in length-of-day or m_3. The problem is one of time scale: the m_3 power spectrum at periods of about 5 years and less is considerably smaller than at periods of the order of decades (Morrison, 1979), particularly when the m_3 have been corrected for the known atmospheric excitation (Lambeck and Hopgood, 1982) and it does not appear that such an impulse, occurring at intervals of about a decade or longer, can maintain the wobble against significant damping. The argument for exciting the Chandler wobble by these electromagnetic torques is, therefore, not strong.

The question of the nature of the excitation of the Chandler remains open. No single mechanism appears to be adequate to excite the wobble indefinitely and a combination of seismic and aseismic, atmospheric and hydrologic, and magnetic forcing functions is likely to be required. Yet not any one of these complementary contributions can be evaluated with sufficient confidence to permit the other mechanisms to be tested. What is required here is not so much improved astronomical or geodetic observations as improved geophysical models and observations to permit the excitation functions to be realistically estimated.

3.2 THE NEARLY-DIURNAL WOBBLE

The frequency of the nearly-diurnal wobble of the Kelvin-Hough model is given by

$$\sigma_2 = -\omega_0[(1 + \frac{A}{A_m}\epsilon_c) + O(\epsilon_c^2)].$$

where ϵ_c is the flattening of the core-mantle boundary. For an Earth in hydrostatic equilibrium, with the parameters (26), and $\epsilon_c \simeq 1/392.7$, $\sigma_2 \simeq -1.003$ (sidereal days)$^{-1}$ and the period is about 23 hr 52 m. The effect of the mantle elasticity and of realistic density distributions in the core is given by equation (25b) and these more realistic models give $\sigma_2 \simeq -1.0022$ (sidereal days)$^{-1}$ (Smith, 1977; Sasao et al. 1980). According to Sasao et al. the adoption of different core and mantle models modifies σ_2 by less than 10^{-4}. The corresponding period is 23 hr 52 m 55(±5)s. While this frequency is relatively insensitive to the details of the Earth model it is more sensitive to the choice of core-mantle boundary geometry. With

$$d\epsilon_c \simeq (da_c - dc_c)/a_c, \qquad (34)$$

where a_c, c_c are the equatorial and polar radii of the core. For a departure of this boundary from equilibrium by ±1 km $d\sigma_2 \simeq 4x10^{-4}\ \omega_0$ and the change in period is about ± 30 seconds.

The existence of the free wobble for the Kelvin-Hough model has long been recognized and the qualitative agreement between this result and the observed Chandler wobble suggests that the Earth should also possess the core mode, always provided that there is an adequate excitation mechanism.

Yet the existence of this oscillation has been a matter of much speculation. Numerous authors have claimed to have detected this motion with an amplitude of about 0".01-0".02 (see Table 1 of Rochester et al., 1974, for a summary of such claims). But such a retrograde wobble produces a long-period (about 460 days, for the complete theory) nutation of the rotation axis in space whose amplitude is some 450-460 times larger than the wobble amplitude, or about 4"-8" (see Toomre, 1974). Recently Capitaine and Xiao (1982) examined 14 years of Bureau International de l'Heure observations and noted a retrograde oscillation, referenced to inertial axes, of period of 434-444 days with an amplitude of only about 0".01. The 434-444 day period, compared with the predicted 460 days, is of potential interest for it implies that the factor $(\epsilon_c-\beta)A/A_m$ in equation (25b) is in error by $(8-12)10^{-5}$. A_m would have to be in error by 4-6% or β would have to be in error by a factor of nearly 2. More plausible is that it is a consequence of the departure of the core flattening from hydrostatic equilibrium. Then

$$\delta\sigma_2/\omega_0 \simeq a_c^{-1} \ (\delta a_c - \delta c_c)A/A_m \simeq (8-12)10^{-5} \quad ,$$

where δa_c and δc_c are the departures of the equatorial and polar radii of the core from its equilibrium shape. Then

$$(\delta a_c - \delta c_c) \simeq 200\text{-}300 \text{ m} \quad . \tag{35}$$

An alternative explanation could be that the free oscillation period is significantly modified by damping of the motion, but by analogy with the Chandler wobble, this damping would have to be severe.

The nearly-diurnal free oscillation will, like the Chandler wobble, be subject to damping and may also be excited. The efficiency of any damping mechanism will depend on values of the core viscosity for viscous damping, and on the electrical conductivity of the lower mantle and core and on the magnetic field strength at the core-mantle boundary for electromagnetic damping (see, for example, Sasao et al., 1980; Sasao and Wahr, 1981). These parameters are all poorly known, making any discussions of damping and excitation mechanisms of an as yet largely unobserved oscillation, little more than speculation. Yet if such damping can ultimately be observed it would contribute to an understanding of the physical properties in the neighbourhood of the core-mantle boundary. More effective will be the study of forced oscillations with periods close to that of the free-core wobble.

4. FORCED ROTATIONAL MODES OF THE EARTH

4.1 POLAR MOTION

The dominant term in the spectrum of polar motion observations, other than the Chandler wobble, is an annual oscillation, driven by seasonal distributions in the mass of the atmosphere, oceans and ground water. Recent estimates of the annual excitation, expressed in the form

$$\psi = \psi^{+}e^{j\sigma t} + \psi^{-}e^{-j\sigma t} \tag{36}$$

are given in Table 2. The first result is from Wilson and Vicente (1980) and is based on a preliminary revision of the ILS data. The second estimate, by Merriam (1982), is based on the revised ILS data set of Yumi and Yokoyama (1980) and the differences between the two solutions are insignificant. A third result, also from Merriam (1982), is based on an 18 year BIH data set (Feissel, 1980). The significant difference between this ψ and the ILS result reflects the general uncertainty of the retrograde component. According to Merriam, similar differences occur when the ILS and BIH data sets are compared over their common epoch and this suggests that the difference is more a consequence of the lower uncertainty of the retrograde term than of the year-to-year variability of the excitation function.

Earthquake		Wobble excitation Magnitude (0".01)	Wobble excitation Direction (Degree)	Author
Chile 1960	(1)	2.12	114	M.L. Smith (1977)
	(2)	2.80	11	
	(1)	2.56	109	O'Connell and Dziewonski (1976)
	(1)	2.2	101	Mansinha *et al.* (1979)
Alaska 1964		0.72	201	M.L. Smith (1977)
		0.73	202	Dahlen (1973)
		1.11	203	O'Connell and Dziewonski (1976)

Table 2 Comparison of estimates of the shift of the excitation axis due to the 1960 Chile and 1964 Alaska earthquakes. (1) is the main shock and (2) is the fore-shock of the Chile event.

Analyses of the seasonal excitation of the polar motion include the recent studies by Merriam (1982) and Wahr (1983). Both confirm the major role played by the longitudinal redistribution of atmospheric mass in exciting the annual change in the Earth's wobble (Table 2). The two studies also emphasize the importance of evaluating the ocean response to the variable surface pressure, although it does appear that the simple inverted barometer model is largely adequate. The agreement of the atmospheric excitation with the observed astronomical excitation remains poor (Figure 2), particularly in phase, and excitations other than the atmospheric mass must be investigated. Wilson and Haubrich (1976) suggested that mountain torques play an important role. Wahr (1983) concluded that both mountain torques and surface friction contribute about equally to the excitation, that neither contribution is well determined, and that each contribution represents less than about 10% of the pressure contribution. Table 2 gives the combined mountain torque and surface friction based on a general circulation model of

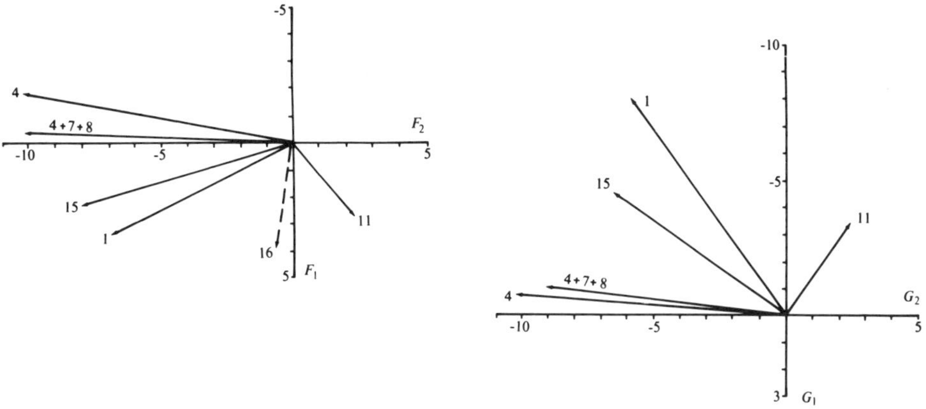

Figure 2 Seasonal excitation functions in polar motion expressed in the form (36) when $\psi^+ = F_1+jF_2$ and $\psi^- = G_1+jG_2$. The numbers correspond to the solutions given in Table 2.

the atmosphere. The surface winds over the oceans cause currents which contribute to the relative angular momenta h_i and which produce changes in pressure on the seafloor. In so far as the direct wind effects on the annual excitation are small, these indirect effects will also be small but, according to Wahr (1983), they are not entirely negligible (Table 2). These additional terms do not lead to a major improvement between the geophysical and astronomical estimates of the excitation function.

A more important contribution to the excitation function arises from the seasonal variations in the storage of ground and ocean water. The only global estimate is still the one by Van Hylckama (1970) (Table 2) and this amounts to about 30-40% of the atmospheric pressure contribution. The variation in ground-water storage must be balanced by variations in the water stored in the oceans and atmosphere. Table 2 gives this combined result and Figure 2 illustrates the combined term and it does move the geophysical excitation into closer agreement with the observed excitation. A preliminary re-evaluation of the ground-water excitation has been made by Hinnov and Wilson (1987) who used northern hemisphere records of monthly precipitation and temperature. Their result, and their water balance term, differ significantly from the Van Hylckama result. Possibly their basic model assumptions lead to an overestimation of the function, or the southern continents play a more important role than has been assumed by Hinnov and Wilson.

Using Van Hylckama's ground-water term, the combined excitation results in good agreement with the astronomical estimate (Figure 2) and this is perhaps surprising and fortuitous in view of the variety of assumptions made, and the inadequacy of many of the geophysical data sets used, in reaching this estimate. Probably the single-greatest uncertainty arises from the ground-water term and its re-evaluation is urgently needed.

4.2 THE FORCED NUTATIONS OF THE EARTH

Astronomical observations of the principal nutations have revealed discrepancies between the observed and theoretical amplitudes and these have generally been attributed to the Earth's departures from rigidity (e.g. Jeffreys, 1948; Federov, 1963). However, these observations were generally insufficiently precise to extract useful geophysical information. Recent improvements in observational accuracies of some of the nutation terms have led to a renewed interest in examining the forced nutations of a non-rigid Earth.

The consequence of the fluid nature of the core can be evaluated using the Kelvin-Hough model of an elastic mantle and an ellipsoidally-shaped core. Consider a torque $L_1+jL_2=L_0e^{j\sigma t}$, $L_3=0$, acting on such a model. The frequency σ is expressed relative to the body fixed axes such that it is nearly diurnal and retrograde for the long period nutations referenced to the inertial frame. That is

$$\sigma = -\omega_0(1-\Delta\omega/\omega_0) \quad . \tag{37a}$$

The solution of the equations of motion contain two parts: the motion m of the rotation axis and the motion $n = n_1+jn_2$ of the core with respect to the mantle. The solutions are of the form

$$m = m_0e^{j\sigma t}, \qquad n = n_0e^{j\sigma t}$$

with

$$\begin{bmatrix} m_0 \\ n_0 \end{bmatrix} = \begin{bmatrix} -\sigma & -\omega_0(1+\epsilon_c) \\ \sigma & \end{bmatrix} jL_0A_c/\omega_0\Delta \tag{37b}$$

where

$$\Delta \simeq CA_c\ \omega_0^2[\epsilon_c+(\Delta\omega/\omega_0)\ (-1+A_c/C)] \quad .$$

The determinant Δ vanishes when

$$\Delta\omega = \omega_0\epsilon_c/(-1 + A_c/C) \quad , \tag{38}$$

or when the forcing function has the same frequency as the free core oscillation, or

$$\sigma = -\omega_0+\Delta\sigma_2 = -\omega_0 - \omega_0\epsilon_c\ A/A_m \ . \tag{39}$$

The corresponding solution for a wholly rigid Earth is,

$$m_0{}^r = -jL_0/\omega_0A\Delta^r$$

with

$$\Delta^r = \sigma-\omega_0(C-A)/A \ ,$$

and the solution (37) can then be written as

$$m_0 \simeq [\ 1 - \Delta\omega(A_c/A_m)/(\Delta\sigma_2 - \Delta\omega)]m_0^r \tag{40a}$$

$$n_0 \simeq -[(A/A_m)(\omega_0 - \Delta\omega)/(\Delta\sigma_2 - \Delta\omega)]m_0^r \tag{40b}$$

where $\Delta\sigma_2$ is given by (39). The precession term in the nutation series corresponds to $\Delta\omega$=0 and the Earth responds as if it were wholly rigid. For the principal nutation $\Delta\omega/\omega \simeq -1/6800$ and the ratio m_0/m_0^r is reduced to about 0.993 by the fluid core. As $\Delta\omega$ approaches $\Delta\sigma_2$ the ratio departs significantly from unity. At frequencies far away from the resonance this ratio approaches the value $(1-A_c/A_m)$ and here the core does not partake in the motion (Figure 3).

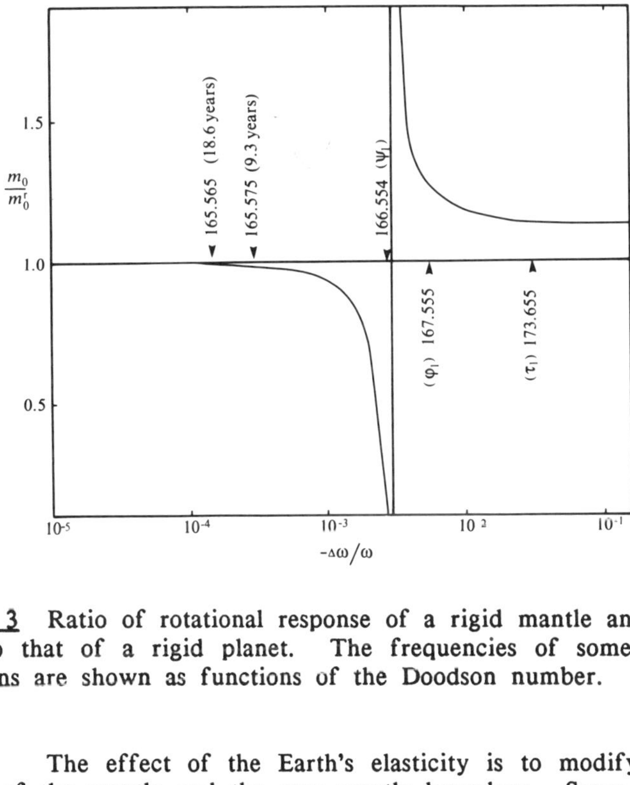

Figure 3 Ratio of rotational response of a rigid mantle and ellipsoidal fluid core to that of a rigid planet. The frequencies of some of the principal nutations are shown as functions of the Doodson number.

The effect of the Earth's elasticity is to modify both the inertia tensor of the mantle and the core-mantle boundary. Sasao et al. (1980) give a solution analogous to (40) as

$$m_0 = \left[1 - \frac{k_2}{k_0}\frac{\Delta\omega}{\omega_0}\right] m_0^r - \frac{A_c}{A_m}\frac{\Delta\omega}{\Delta\sigma_2 - \Delta\omega}\left[1 - \frac{\chi}{\epsilon}\left(1 - \frac{\Delta\omega}{\omega_0}\right) - \frac{k_2}{k_0}\frac{\Delta\omega}{\omega_0}\right] m_0^r \tag{41a}$$

$$n_0 = \frac{-A}{A_m}\frac{\omega_0 - \Delta\omega}{\Delta\sigma_2 - \Delta\omega}\left[1 - \frac{\chi}{\epsilon}\left[1 - \frac{\Delta\omega}{\omega_0}\right] - \frac{k_2}{k_0}\frac{\Delta\omega}{\omega_0}\right] m_0^r\ , \tag{41b}$$

where ϵ is the dynamical flattening of the Earth, χ is a function of the Earth's deformation, k_2 is the static Love number and k_0 is defined by (11). $\Delta\sigma_2$ now corresponds to the free oscillation frequency of the deformable Earth, that is, from (25b)

$$\Delta\sigma_2 \simeq - \frac{A}{A_m} (\epsilon_c - \beta)\omega_0 \tag{42}$$

The parameter χ in (41) is of the order 2×10^{-3} and values based on different Earth models vary by about 5% (Sasao et al., 1980). Now, for the principal 18.6 year nutation, $m_0/m_0^r \simeq 0.996$-0.997 and the effect of the mantle elasticity is opposite to the core effect.

Complete discussions of the nutations of a deformable Earth are given by Sasao et al. (1980) and Wahr (1981a,c) and differences for the two theories are less than 0."001 in amplitude. Table 3 summarizes the main results. The principal departures from the rigid Earth model occur at 18.6 years and 6 months for which the discrepancies are about 0."02-0."03. Not included in these theories are the effects of the ocean tides and Wahr and Sasao (1981) have shown that these can be important, about 0."001-0."002 for the 18.6 year nutation and about 0."0005-0."001 for the semi-annual nutation. It is also important to recognise that this oceanic contribution is likely to lead to the observed nutations being out of phase with the predicted values for the solid Earth.

McCarthy et al. (1980) have summarized, without critical comment, some published observations for the principal nutation terms. Estimates of the 18.6 year nutation range from 9."196 to 9."214 for the obliquity and from -6."819 to -6."858 for the longitude. More recently, Capitaine and Xiao (1982) examined 18 years of astrolabe observations of the Bureau International de l'Heure (the z-term in latitude and the w-term in universal time) (Table 4). For the 18.6 year nutation their second result in Table 4 is to be preferred and these values are in satisfactory agreement with estimates by Wako and Yokoyama and McCarthy (quoted in McCarthy et al., 1980). Overall these results are in agreement with the predicted amplitudes of Wahr and Sasao et al. but the observations are inadequate to distinguish between the theoretical models. Annual and semi-annual nutations are difficult to observe with precision because of systematic errors in the observations with a seasonal character and because of the need to precisely model the seasonal changes in polar motion and length-of-day. Estimates for the semi-annual nutation have been obtained by Capitaine (1980) and E.P. Federov (quoted in McCarthy et al. 1980) but the agreement in the obliquity is not satisfactory (see Table 4). Nor do the observations for this term agree well with the realistic model predictions. Likewise, estimates for the fortnightly nutations (Table 4) differ from one solution to another, although the Capitaine and Xiao (1982) values agree well with the model.

Recent results obtained by long baseline radio interferometry hold greater promise for improved nutation observations. Herring et al. (1986) analysed four and a half years of observations of a number of baselines to estimate the nutations with periods of 1 year and shorter (Table 4) (in addition to the terms tabulated here, these authors also observed the nutations at 122, 32, 28 and 9 days) (see also Herring, 1988). Agreement

	18.6 years		1 Year		6 Months		13.7 days	
	Obliquity	Longitude	Obliquity	Longitude	Obliquity	Longitude	Obliquity	Longitude
Rigid Earth (Kinoshita's theory)	9.2278	-6.8743	-0.0001	0.0499	0.5534	-0.5082	0.0949	-0.0881
Sasao *et al.* (Wang Earth model)	9.2018	-6.8407	0.0051	0.0565	0.5739	-0.5249	0.0977	-0.0904
Wahr (1066 A model)	9.2025	-6.8416	0.0054	0.0567	0.5736	-0.5245	0.0977	-0.0905

Table 3 Amplitudes (in arc seconds) of the nutations in obliquity and longitude for the principal terms in the nutation series according to the rigid Earth theory of Kinoshita (1977) and the elastic mantle-fluid core models of Sasao *et al.* (1980) and Wahr (1981). The nutation in longitude is defined here as $\alpha_2 \sin \alpha_1$.

	18.6 year		1 year		6 months		13.7 days	
	Obliquity	Longitude	Obliquity	Longitude	Obliquity	Longitude	Obliquity	Longitude
Capitaine and Xiao (1) (1982)	9.2032 ± 33	-6.8407 ± 14					0.0980 ± 10	-0.0905 ± 10
(2)	9.2095 ± 20	-6.8424 ± 30						
Wako and Yokoyama	9 214 ± 1	-6.838 ± 1						
McCarthy 1972	9.206 ± 4	-6.858 ± 4						
Capitaine (1980)					0.555 ± 6	-0.532 ± 8		
Federov					0.578 ± 4	-0.533 ± 4		
McCarthy (1976)							0.0922 ± 16	-0.0845 ± 15
Herring *et al.* 1986			0.0072 ± 2	0.0585 ± 2	0.5733 ± 2	-0.5240 ± 2	0.0979 ± 2	0.0909 ± 2

Table 4 Astronomical estimates of some of the principal nutations in obliquity and longitude. The first result by Capitaine and Xiao is based on analyses of 5-day values of the z term in 13 years of latitude data. The second value is based on analyses of 0.05 year values of both the z and w terms of 18 years of data. The results attributed to Federov, Wako and Yokoyama, and McCarthy, 1972, are taken from McCarthy *et al.* (1980). The results by Herring *et al.* (1986) are based on $4\frac{1}{2}$ years of long baseline radio interferometry observations.

with the theory is excellent at all periods except for the annual oscillation. The stated uncertainties of these nutation amplitudes are comparable or smaller than the differences predicted by the Sasao et al. (1980) and Wahr (1981c) theories, giving rise to an expectation that these observations may contribute to an improved understanding of the dynamics of the coupling of the core and mantle. In fact, Gwinn et al. (1986) interpret the above discrepancy between theory and observation at the annual frequency in terms of a departure of this boundary from the ellipticity predicted by the hydrostatic theory. The annual nutation term, referenced to the rotating frame, lies close to the frequency of the free core oscillation (Figure 3) and a small shift in this frequency caused, for example, by changing ϵ_c (equation 42) will modify the Earth's response without it significantly affecting the other nutations. Dissipation will also change the resonance frequency but if this is assumed to be small the requisite core flattening can be computed from (49) and (50). Gwinn et al. obtain $a_c - c_c \simeq 500$ m, similar to that deduced from the observed free oscillation period (equation 35).

5. LENGTH-OF-DAY

The non-uniform rotation of the Earth has traditionally been determined by measuring the time interval between successive transits of a star against an observer's meridian. For the present purpose the measured quantity can be considered to be the universal time UT1. The elapsed interval is measured against a uniform reference time R and the observed quantity is $\Pi = R$-UT1, the amount by which the Earth is slow after an interval ΔT. The measurement of Π therefore comprises two parts, the astronomic or geodetic part relating to the definition of the instants of transits of the celestial bodies, and the physics part relating to the establishment of a uniform time scale from frequency standards. Departures from uniform rotation are expressed in several ways, in terms of the m_3 defined by (1) or in terms of changes in the length-of-day. That is

$$m_3 = \frac{\omega_3 - \omega_0}{\omega_0} = -\frac{\Delta(\text{l.o.d})}{\text{l.o.d}} = -\frac{d\Pi}{dt} . \tag{43}$$

Modern methods of measuring the change in length-of-day, including radio-interferometric observations of stellar sources (VLBI) and the tracking of artificial satellites and the Moon, are essentially the same in which, for example, the diurnal orientation of the VLBI baseline is monitored relative to the stellar sources.

The observational record of Π is of variable accuracy and resolution but some observations go back to the seventeenth century (e.g. Morrison et al., 1982; McCarthy and Babcock, 1986). Until about 1955 the reference time was ephemeris time, determined from the astronomical observations themselves, and this proved to be the limiting precision of UT1 measurements. After 1955 the reference time is atomic time and the limiting factor is now the precision of the astronomical or geodetic observation but even for the past 30 years the data set is not homogeneous in accuracy and resolution.

Figure 4 illustrates the principal characteristics of the observations. The observations with respect to ephemeris time define the so-called decade fluctuations, changes in l.o.d that persist for about 10 years and longer, rise above the noise level for at least the period from 1850 to the present. Improvements in the ephemeris time determination after about 1920 resulted in fluctuations of about 5 year duration rising above the noise level. After 1955 seasonal oscillations became clearly evident and the tidal spectrum emerged. With the present VLBI observations, considerable high-frequency changes in length-of-day are observed.

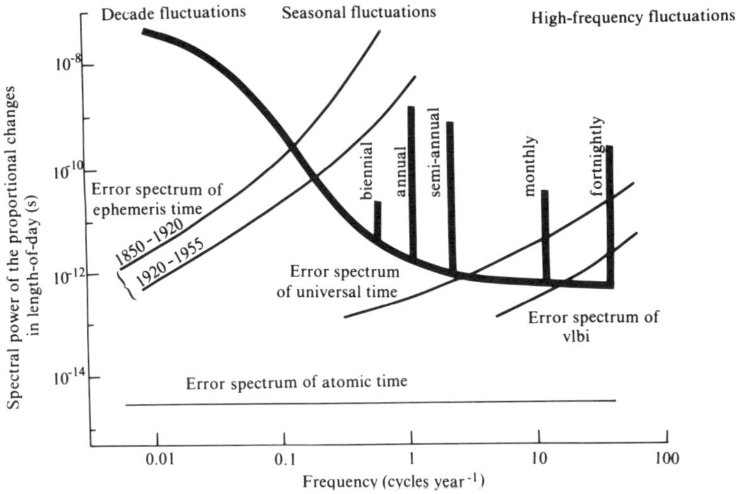

Figure 4 Schematic spectrum of the fluctuations in the length-of-day and the observational error spectra.

5.1 THE SECULAR CHANGE

The secular change in the Earth's rotation must be one of the most debated subjects in astronomy and geophysics for more than 200 years. Because of the decade-scale and longer fluctuations in the Earth's rotation, estimates of the true secular change are generally unreliable unless they are based on very long observational records but the longer the record the less reliable the data. The geophysical problem comprises two parts: one is the evaluation of the secular change resulting from the lunar and solar torques exerted on the Earth and the other is the evaluation of non-tidal contributions. The latter is generally the smaller amount, and the separation of the two remains uncertain. Causes for the non-tidal part include the Earth's adjustment to the melting of the ice sheets in Late Pleistocene time and possibly longer-period electromagnetic torques acting on the lower mantle. The subject is too vast to be discussed here and the reader is referred to the literature cited in the summaries by Lambeck (1980, 1988) (see also Rochester, 1984).

5.2 THE DECADE CHANGES

The decade changes in m_3 have reached 1 part in 10^7 over intervals of about 10 years and they are so large that their existence has been known for the best part of a century, yet their explanation remains obscure. Their magnitude is such that the atmosphere and oceans can only play a minor role in their excitation and the consensus is that the explanation must lie in differential motions between the core and mantle (Munk and MacDonald, 1960; Lambeck, 1980; Rochester, 1984). Where the disagreement occurs is the question of the mechanism(s) by which core motions are coupled to the mantle. Is it by viscous friction, by electromagnetic torques, or by the pressure of the core fluid as it flows past a non-spherical core-mantle boundary? Disagreement is not so much a matter of details about the mechanisms as one of the choice of appropriate physical parameters describing the core and lower mantle.

With a change in m_3 of 10^{-7}, as occurred from about 1870 to 1900, the torque required to act across the core-mantle boundary is about 10^{18} Nm. The strength of the coupling locally is defined by a tangential stress Σ_t at the boundary and the net torque exerted on the mantle is

$$L = \int_S \mathbf{R_b} \wedge \Sigma_t \, \mathbf{e_r} \, dS$$

where $\mathbf{e_r}$ is the normal to the core-mantle boundary R_b. The average tangential stress at R_b therefore has to be of the order 10^{-2} Nm^{-2}.

Surface friction gives rise to a stress

$$\Sigma_v = \rho\eta^* \, v/D$$

where ρ is the density, $\eta^*=\eta\rho^{-1}$ is the kinematic viscosity, v is the velocity of fluid past the boundary, and $D=(\eta^*/\omega)^{\frac{1}{2}}$ is the boundary layer thickness (Hide, 1977). An estimate of v is obtained from the westward drift of the magnetic field, at an average rate of about 0.2° a^{-1}, as this phenomenon is believed to be indicative of differential rotation between the core and mantle (but see Hide, 1966) so that $v \simeq 4 \times 10^{-4}$ m s^{-1}. Hence, for coupling to be effective, the core kinematic viscosity needs to exceed 0.1 m^2 s^{-1}. Few geophysical observations place strong constraints on this viscosity. Observations of the damping of seismic waves, for example, determine an upper limit of about $\eta^* < 10^4$ m^2 s^{-1}, while estimates based on the nutation observations requires that $\eta^* \lessdot 10$ m^2 s^{-1} (Toomre, 1974). Theoretical and physical estimates place much lower bounds on this parameter and Gubbins (1976) proposed that $\eta^* \simeq 4 \times 10^{-7}$ m^2 s^{-1} while Bukowinski and Knopoff (1976) suggested a value of about 10^{-4} m^2 s^{-1}. Based on these estimates, viscous coupling is generally believed to be unimportant.

Topographic coupling, by the flow of core-material past an irregular boundary has been proposed by Hide (1969,1977), and the mechanism is similar to the mountain torques exerted by the winds on the Earth's surface. The local stress is approximately (Hide, 1977)

$$\Sigma_r = C_t \rho\omega v(H-D)$$

where C_t is a drag coefficient, and H(>D) is the height of the irregular topography of the boundary. C_t is a function of H and wavelength of the bumps as well as of the magnetic field properties (Moffatt, 1978; Hassan and Eltayeb, 1982). Generally H>>D and the value of the viscosity of the core is unimportant in the evaluation of Σ_r. Hide suggested that $C_t \simeq 1$ but Hassan and Eltayeb proposed $C_t \simeq 10^{-3}$ and with the latter value H must approach 10 km for Σ_v to be significant but, from other geophysical evidence and arguments, this appears excessive. For $C_t \simeq 1$ the mechanism becomes a much more plausible candidate but clearly more work is required to establish appropriate limits on this coefficient.

Electromagnetic coupling has been introduced in section 3.1.5. The time constant for this coupling τ_3, defined by (33), is of the order 8-15 years and quite adequate to explain the decade-scale length-of-day changes. The local tangential stress at the core-mantle boundary

$$\Sigma_e \simeq C_e B_r B_t / \mu$$

where C_e is an electromagnetic "drag coefficient" and is approximately equal to unity, μ is the magnetic permeability and is not significantly different from the value for free space, and B_r, B_t are the radial and tangential components of the magnetic flux density **B** at the core-mantle boundary (Rochester 1962; Hide 1977). B_r can be estimated by extrapolating the surface field down to this boundary and more problematical is the field B_t, induced by the relative motion of the conductive core and lower mantle in the presence of the radial field, its value being a function of the motions within the core and of the conductivity of the lower mantle. In some of the early dynamo models B_t was found to be 100-200 times larger than B_r but in more recent models $B_r \simeq B_t$ (e.g. Busse 1975). With either case, most investigators agree that B_t will be adequate to produce the requisite stress at the core-mantle boundary (Hide 1977; Rochester 1984).

If electromagnetic coupling does occur, some correlation between fluctuations in magnetic field parameters and the planet's rotation may be anticipated. Several attempts have been made to determine such correlations between, for example, l.o.d. and changes in the drift rates of long wavelength components of the field or the variation in the intensity of the dipole term in the field (e.g. Yukutake 1973). Such correlations are particularly important for studying the conductivity structure of the lower mantle. The time constant τ_1 for a change in the magnetic field to diffuse through the lower mantle is given by equation (33a) so that a delay is expected to exist between the Earth's rotational response and any magnetic field changes at the surface, by an amount that is a function of the electrical conductivity. Evidence for these correlations is limited. First, the extrapolation of the field B_r to the core-mantle boundary is poorly constrained for the higher wavenumber components. Second, the component B_t is reduced to zero at the surface of the Earth because of the very low conductivity of the upper mantle and estimates of the magnitudes of these fluctuations are little more than speculations. Thus an absence of correlation cannot be used to infer the inadequacy of electromagnetic core-mantle coupling.

The coupling models examined so far generally consider only long wavelength terms in the magnetic field, although the analysis by Stix (1982) includes harmonics up to degree 12. Higher harmonics may be very effective in coupling core and mantle motions, particularly when the conductivity gradients are steep (Rochester, 1984), but they are attenuated rapidly as the Earth's surface is reached and at some wavelength, they cease to penetrate through the mantle so that they cannot be observed. If these higher harmonics are effective in producing high frequency coupling this should show up in the length-of-day spectrum although the latter contains little unexplained power at periods less than about 4-5 years and it does not appear that significant coupling occurs at these shorter periods.

5.3 METEOROLOGICAL EXCITATION

The principal contribution to the seasonal changes in the earth's rotation comes from the zonal circulation in the atmosphere and the evaluation of the excitation function is straight forward once this circulation is known over the globe and up to altitudes above the tropopause. From (5)

$$\psi_3 = \psi_3(\text{matter}) + \psi_3(\text{motion}) = -\Delta I_{33}/C - h_3/\omega_0 C$$

and the complete evaluation requires a knowledge of both the mass redistribution (ΔI_{33}) and the winds (h_3) in the atmosphere. Evaluation of the two terms has shown however that the relative motion term is up to an order of magnitude greater than the matter term and the principal task of evaluating ψ_3 is the measurement of the zonal winds.

The astronomically observed annual and semi-annual oscillations in l.o.d are well explained by the combination of wind excitation and tides (see below) (Lambeck and Hopgood, 1982) but nevertheless several limitations of the current wind data sets remain. These include the general lack of global wind data above 100 mbars for, as shown by Lambeck and Cazenave (1973) and Rosen and Salstein (1985), winds up to stratospheric heights are important. A second limitation is the paucity of data for the southern hemisphere. Apart from the annual and semi-annual terms, other oscillations at periods near 120 days, from 40 to 50 days, and shorter have been noted in the length-of-day spectrum, as have longer periods of 2-3 year duration. Most of these fluctuations are of meteorological origin (Rosen and Salstein, 1985).

Major efforts are underway at several centres to compute systematically the meteorological excitation functions, both ψ(motion) and ψ(matter), on a daily basis (e.g. Whysall et al., 1985; Naito and Yokoyama, 1985; Salstein and Rosen, 1985) and it appears that this excitation can now be removed with some confidence that the residual excitation has geophysical meaning. Unfortunately, within the present levels of accuracy of both ψ_3 and m_3, not much remains at these sub-seasonal periods other than the tidal signals (see Figure 5). The importance of being able to "correct" the astronomical data for the meteorological "noise" cannot be over emphasized for not only does it mask possible high-frequency solid-earth signals, it also contaminates the Earth's response to the lunar tide signals (see below). The interpretation of the decade changes in l.o.d is also complicated by the meteorological excitation. It is not suggested that the atmosphere makes a major contribution to the excitation but it may mask essential characteristics

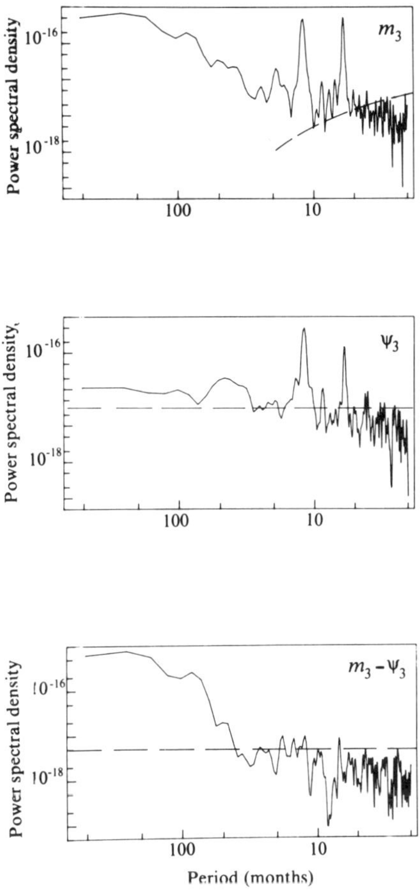

Figure 5 Power spectra of (a) m_3, (b) the zonal wind excitation ψ_3, and (c) m_3-ψ_3 less tidal signal. All spectra are based on the time interval from 1958 to 1980. The error spectra are indicated by the dashed lines (from Lambeck and Hopgood, 1982).

of these changes. For example, do these changes in length-of-day occur rapidly, driven by short-duration impulses in electromagnetic torques at the core-mantle boundary? Or do the changes occur more gradually, over periods of a few years. Because atmospheric contributions to the total excitation are significant at periods up to a few years this question cannot be answered with certainty unless the meteorological excitations are known.

5.4 TIDAL SIGNALS IN THE EARTH'S ROTATION

Jeffreys, in 1928, was the first to point out that the conservation of angular momentum of the earth implied that there must be periodic changes in the length-of-day. However, the semi-annual term in l.o.d was not observed until about 1953 and its early interpretation was complicated by errors in the FK3 star catalogue then in use, and by the recognition that seasonal factors, other than tides, could contribute. Observations of the monthly and fortnightly tidal terms were first reported by Markowitz in 1955, from an analysis of photo-zenith-tube observations. With the now available long series of length-of-day referenced to atomic time, the tidal terms rise clearly above the background measurement noise (e.g. Merriam, 1982; Yoder *et al.*, 1981).

Detailed theoretical calculations of the tidal perturbations were carried out by Woolard in 1959 for an elastic Earth. If a fluid spherical core is introduced with no coupling at the core-mantle boundary then the core will not take part in the rotation and the amplitudes for an elastic Earth will be reduced by $(A-A_m)/A_m$ or about 10%. The effect of the ocean tide is to increase the effective flattening of the Earth, assuming that this tide follows an equilibrium theory, and this increases the length-of-day changes, also by about 10%, so that the effective "whole Earth" Love number is actually close to that of an elastic Earth.

5.4.1 Solid Tides

A qualitative estimate of the changes induced in the Earth's rotation by the tidal deformation of the Earth follows readily from the Love number description of this deformation. The tide-raising potential of degree 2 and order m at r=R can be written as

$$V_2 = \sum_\nu V^{(\nu)} = Gm_c\left[\frac{R_e}{a_c}\right]^2 \sum_\nu \sum_{m=0}^{2} B_{2m}^{(\nu)} P_{2m}(\sin\varphi)\cos(\sigma_{2m}^{(\nu)}t+m\lambda+\beta_{2m}^{(\nu)}) \tag{44}$$

the frequencies, phase angles and amplitudes of these components. a_c and m_c are the semi-major axis of the lunar orbit and the lunar mass respectively. The corresponding rotational excitation functions are, neglecting the time derivative of the inertia tensor in the equations,

$$\begin{bmatrix}\psi_1\\ \psi_2\end{bmatrix} = m_c\left[\frac{R_e}{a_c}\right]^3 \frac{R_e^2 k_2}{(C-A)} B_{21}^{(\nu)} \begin{bmatrix}\cos(\sigma_{21}^{(\nu)}t+\beta_{21}^{(\nu)})\\ -\sin(\sigma_{21}^{(\nu)}t+\beta_{21}^{(\nu)})\end{bmatrix} \tag{45a}$$

$$\psi_3 = \frac{2}{3} m_c\left[\frac{R}{a_c}\right]^3 \frac{R_e^2}{C} k_2 B_{20}^{(\nu)} \cos(\sigma_{20}^{(\nu)}t+\beta_{20}^{(\nu)}) \quad . \tag{45b}$$

For m=0, the tidal potential has a zonal geometry and its time dependence is of long period. These tides contribute only to ψ_3 and m_3. For m=1 the periods group around 24 hours and the polar motion is subjected to small nearly-diurnal oscillations. There are no first-order contributions from the semi-diurnal (m=2) tides because the first order equations of rotational motion are independent of the inertia elements I_{12} and $(I_{11}-I_{12})$. The diurnal tides contribute significantly to ψ but the corresponding motion of the rotation axis has an amplitude that is approximately $\sigma_0/\sigma^{(\nu)}$ or 1/435 of

ψ and the diurnal tidal effect on the Earth's rotation becomes significant only at the very highest levels of precision and resolution of observations of the Earth's rotation.

The zonal tides play a considerably more important role. Lunar tides cause perturbations in length-of-day near 14 and 27 days and solar tides cause perturbations near 6 months and 12 months. Longer period perturbations occur near 8.8 years and 18.6 years, although their amplitudes are small when compared with the observed decade changes.

5.4.2 Core effect

In analogy with the Earth's rotational response to the lunar and solar torques, the tidal deformations of the fluid core may be decoupled from the mantle deformations. As noted by Merriam (1980) (see also Wahr *et al.*, 1981; Yoder *et al.*, 1981), this suggests that, instead of (5b) the rotational excitation function ψ_3 should be written as

$$\psi_3(\text{matter}) = -\Delta I^m_{33}/C_m \tag{46}$$

where ΔI^m_{33} refers to the change in the inertia tensor of the mantle only. The evaluation of these elements requires a solution of the equations of deformation for a planet with an ellipsoidally shaped fluid core that is acted upon by a body force potential of degree 2. The solution is (Wahr *et al.*, 1981)

$$\psi_3 = -\frac{\Delta I^m_{33}}{C_m} = -\frac{\Delta I_{33}}{C}(1-C_c\gamma/C\chi)/(1-C_c/C) \tag{47}$$

where $-\Delta I_{33}/C$ corresponds to the excitation function for the elastic Earth model, C_c in the polar moment of inertia of the core, γ is the non-dimensional deformation constant and $\chi=(k_2/k_0)(C-A)/A$. This indicates that the tidal response can then be written in the form (59) but with the elastic Love number replaced by the parameter

$$\kappa = k_2(1 - C_c\,\gamma/C\chi)/(1-C_c/C) \quad . \tag{48}$$

The parameters required to evaluate this equation are given by Sasao *et al.* (1980) for three different Earth models and κ ranges between about 0.266 and 0.268. The effect of the fluid core is therefore to reduce the elastic tidal response by about 10-11%.

If some degree of core-mantle coupling occurs, the effective Love number will lie between the above value for κ and the elastic value k_2. A lag angle will also be introduced into the response but this is unlikely to reach a magnitude where it can be observed (see Yoder *et al.*, 1981; Wahr *et al.*, 1981).

5.4.3 Ocean tides

The ocean tide contribution to the excitation can be evaluated in a manner that is similar to that used to examine the oceanic pole tide effects on the Earth's rotation. This contribution comprises two parts; the direct deformation of the ocean which, if the tide follows an equilibrium theory, will result in an increase in the tidal elements of the inertia tensor, and the indirect contribution, of opposite sign, arising from the Earth's deformation under the ocean tide load. The excitation function follows by expanding the ocean tide into spherical harmonics and the result is (Lambeck 1980)

$$\left.\begin{matrix}\psi_1\\ \psi_2\end{matrix}\right] = \frac{4\pi}{5}\rho_w \frac{R_e^4}{(C-A)}(1+k_2')D_{st}^{\pm(\nu)}\begin{bmatrix}\cos(\sigma^{(\nu)}t-\epsilon_{st}^{\pm(\nu)})\\ -\sin(\sigma^{(\nu)}t-\epsilon_{st}^{\pm(\nu)})\end{bmatrix} \tag{49a}$$

$$\psi_3 = \frac{8\pi}{15}\rho_w \frac{R_e^4}{C}(1+k_2')D_{st}^{+(\nu)}\cos(\sigma^{(\nu)}t-\epsilon_{20}^{+(\nu)}) \quad . \tag{49b}$$

where the $D_{st}^{\pm(\nu)}$ and $\epsilon_{20}^{\pm(\nu)}$ are parameters that define the tide with respect to the seafloor which itself is deformable. This latter response is summarized by the load Love numbers k_s'. It is important to note that the spectrum of the tidal perturbations in the planet's rotation will be more complex than for the elastic Earth model with or without a fluid core. The semi-diurnal ocean tide expansion, for example, contains terms harmonic in degree and order 2,0 and 2,1. Thus, while there are no solid Earth contributions from this tide potential (ignoring the second order terms in the equations of motion), there will be small semi-diurnal terms in both the polar motion and the length-of-day. Likewise, the long-period zonal tides contain harmonics in degree and order 2,1 and it could be anticipated that the polar motion contains small oscillations with the same periods as these long period tides. These excitation functions consider only the term ψ_i(matter) and ignore possible contributions from ψ_i(motion) arising from tidal currents, and this neglect warrants further investigation.

Consider the semi-diurnal M_2 tide. The amplitude D^+ of the 2,1 component is about 2.5 cm so that $|\psi|\simeq 2.8\times10^{-7}$ and $|m|\simeq 3\times10^{-10}$ and the semi-diurnal oscillation in polar motion is entirely negligible. The 2,0 coefficient in the M_2 tide has an amplitude of about 1 cm so that $|\psi_3| = |m_3| \simeq 2.5\times10^{-10}$ and the change in length- -of-day is about 0.02 msec. This is also below the present observational noise level. Likewise, the 2,0 coefficients in the diurnal tides produce negligible diurnal oscillations in the length-of-day.

The (2,1) harmonics in the diurnal ocean tide introduce perturbations in **m** that are of similar magnitude, or about 10% of the elastic tide effect, and these also are wholly negligible. But this is not true for the nutations in space for while the proportional change is the same, the amplitudes themselves are much larger: if the elastic deformation modifies the nutation amplitudes by about 30% then the ocean effect is about 3% of the rigid Earth nutation and there will be a small lag in the motion of the rotation axis with respect to the direct lunar and solar potential.

Whether or not the Mm and Mf ocean tides follow an equilibrium theory remains a matter of debate. Here it will be assumed that such a model is valid and Table 5 gives the corresponding perturbations in rotation. The largest perturbations in polar motion occur for the Mf tide although the amplitudes are small, with $|m|\simeq 2\times10^{-9}$ or less. The perturbations in m_3 are considerably more significant, about 10% of the corresponding solid tides.

Darwin notation	Doodson number	Period (sol. days)	m_3 (Elastic Earth) $\times 10^{-9}$	m_3 (Core+Ocean) $\times 10^{-9}$
	055.565	6817	1.654	1.730
	055.575	3409	-0.016	-0.017
Sa	056.554	366.3	-0.292	-0.305
	056.556	366.2	-0.015	-0.016
SSa	057.555	183.1	-1.840	-1.925
	057.565	178.3	0.045	0.047
	058.554	122.0	-0.107	-0.112
MSm	063.655	31.9	-0.399	-0.417
Mm	065.455	27.7	-2.084	-2.180
Msf	073.555	14.8	-0.346	0.362
Mf	075.555	13.7	-3.959	-4.141
	075.565	13.6	-1.642	-1.718
	085.455	9.1	-0.758	-0.793

Table 5 Amplitudes of perturbations in m_3 caused by the long period zonal tides. The fourth column corresponds to an elastic Earth with k_2=0.30 while the fifth column corresponds to an elastic mantle, fluid core and equilibrium ocean.

The tidal response of an elastic Earth with an equilibrium ocean tide is described by the effective Love number

$$\kappa = k_2 \ (1 + C \ \psi_3^{ocean} / k_2 \Delta I_{33}^*)$$

and equals about 1.125 k_2 - 1.129 k_2 for the above model (Agnew and Farrell, 1978; Merriam, 1980; Lambeck, 1980). If a non-equilibrium tide is used, κ will include an imaginary part, reflecting the lag of the combined elastic-ocean tide with respect to the tide raising potential. The total tidal excitation function is, with (47),

$$\psi = -k_2 \frac{\Delta I_{33}}{C} [\chi_c + C\psi_3^{ocean}/k_2 \Delta I_{33}] \equiv -\kappa \ \Delta I_{33}/C \qquad (50a)$$

where

$$\chi_c = [1 - C_c \ \gamma / C\chi] / (1 - C_c/C) \quad . \qquad (50b)$$

and for the above models κ = 1.03-1.04 k_2.

Numerous analysis of length-of-day observations for the tidal Love number κ have been made, but no consistent result appears. For example, Guinot (1974) and Lambeck and Cazenave (1974) found that $\kappa_{Mf} > \kappa_{Mm}$ although both estimates were found to be quite variable from data set to data set. Others, for example Yoder *et al.* (1981), found that $\kappa_{Mf} \simeq \kappa_{Mm}$. Capitaine and Guinot (1985) noted that κ_{Mf} exhibits considerable variability with time whereas κ_{Mm} appears to be relatively stable. This irregular type of behaviour led Lambeck and Cazenave to suggest that there is an increase in the power of the meteorological excitation about these tidal frequencies and the zonal wind excitation functions of Rosen and Salstein (1983) confirm this. Thus the length-of-day observations should first be corrected for this atmospheric excitation before Love numbers are estimated. Merriam (1984) has noted that some of the difference between the Mm and Mf results is also a consequence of the use of Woolard's nutation series instead of the Wahr (1981b) theory, although the Capitaine and Guinot results are based on this latter theory and the problem remains.

Merriam (1984) appears to be the only one who has estimated the Love numbers by first correcting the length-of-day data for the atmospheric excitation and he found that, within observational errors, κ_{Mf} equals κ_{Mm} and that the statistics of the atmospherically corrected results are better than those of the uncorrected length-of-day data set (see Table 6). With (50), the equivalent Love number for a static Earth model without oceans is about 0.309 and this is marginally greater than the value predicted from seismological data by an amount that is consistent with the frequency dependent Q law with $\gamma \simeq 0.2$. The uncertainties, however, are sufficiently large to permit a wide range of γ values, from 0 to more than 0.3. It should be noted that the load Love numbers will also exhibit dispersive behaviour so that the ocean loading term is frequency dependent and for detailed studies Merriam's formulation is preferred. Longer wind data sets are now available and a recalculation is worthwhile.

	BIH		Wind corrected BIH		
	κ	Phase (days)	κ	Phase (days)	Static Love number
Mf	0.331±0.023	+0.15	0.317±0.013	0.22	0.304
Mm	0.364±0.016	-0.09	0.327±0.011	-0.13	0.314
Mean			0.322		0.309

Table 6 Zonal Love number κ at the Mf and Mm frequencies before and after the removal of the atmospheric excitation from the BIH data from 1978-1982. A positive phase means that the observed signal leads the nominal tide signal. (From Merriam 1984). The static Love number is κ corrected for the equilibrium ocean tide (and loading) and the extent to which the core is not coupled to the motion of the mantle.

6. CONCLUSIONS

The very nature of the Earth's rotation requires observational records that extend over many years and even centuries. In consequence, the new measurement procedures based on space-age technologies have not yet made a major impact on the geophysical discussion of those motions. This is in part because the technologies themselves have evolved rapidly and homogeneous data sets do not yet exist for even a few years. Thus the Doppler satellite observations of the Earth's irregular rotation, which represented a major improvement over the classical astrometric observations, have themselves been supplanted by the laser range observations of satellites and the long baseline interferometric observations of stellar radio sources. Yet the geodetic developments of the past two decades have now reached the point where one can see geophysical signatures arising out of the noise from only relatively short series of observations and one example of this is the VLBI observations of the Earth's short-period nutations.

It would be hazardous to predict where our understanding of the deformations of the Earth will be when several decades of precise rotation observations are available. New responses to known driving forces are likely to be discovered and as yet unknown mechanisms will be postulated. One reason why such a prediction is hazardous is that the levels of precision of the observations are now such that they are much contaminated by meteorological excitations and what will be required in order to exploit fully the new geodetic measurements is a parallel program of the appropriate global atmospheric-oceanic-hydrologic parameters. Another reason why the prediction is hazardous is that there will be progress in complementary areas of geophysics as well. Advances in seismic tomography, for example, will undoubtedly lead to an improved understanding of the core-mantle boundary and to improved insight into core-mantle coupling processes.

REFERENCES

Agnew, D.C. and Farrell, W.E., 1978, Geophys.J., vol. 55, 171-178.

Anderson, D.L. and Minster, J.B., 1979, Geophys.J., vol. 58, 431-440.

Backus, G.E., 1983, Geophys.J., vol. 74, 713-746.

Barnes, R.T.H., Hide, R., White, A.A. and Wilson, C.A., 1983, Proc.R.Soc.Lond., vol. A387, 31-73.

Ben-Menahem, A., 1982, Geophys.J., vol. 70, 535-537.

Braginskii, S.I. and Fishman, V.M., 1976, Geomag.Aeron., vol. 16, 443-446.

Bukowinski, M.S. and Knopoff, L., 1976, Geophys.Res.Lett., vol. 3, 45-48.

Busse, F.H., 1975, Geophys.J., vol. 42, 437-459.

Capitaine, N., 1980, in "Nutation and the Earth's Rotation", E.P. Federov, M.L. Smith and P.L. Bender eds., Reidel, 87-94.

Capitaine, N. and Guinot, B., 1985, Geophys.J., vol. 81, 563-568.

Capitaine, N. and Xiao, N.Y., 1982, Geophys.J., vol. 68, 805-814.

Carton, J.A. and Wahr, J.M., 1986, Geophys.J., vol. 84, 121-137.

Chao, B.F. et al., 1987, J.Geophys.Res., vol. 92, 9415-9422.

Dahlen, F.A., 1973, Geophys.J., vol. 32, 203-217.

Dehant, V., 1987, in "The Earth's Rotation and Reference Frames for Geodesy and Geodynamics", A.K. Babcock and G.A. Wilkins eds., Reidel, 323-330.

Dickman, S.R., 1985, Geophys.J., vol. 81, 157-174.

Dickman, S.R. and Preisig, J.R., 1986, Geophys.J., vol. 87, 295-304.

Dickman, S.R. and Steinberg, D.J., 1986, Geophys.J., vol. 86, 515-529.

Feissel, M., 1980, Bull.Géod., vol. 54, 81-102.

Gilbert, F. and Dziewonski, A.M., 1975, Phil.Trans.R.Soc.Lond., vol. 278, 187-269.

Gross, R.S., 1986, Geophys.J., vol. 85, 161-177.

Gubbins, D., 1976, Geophys.J., vol. 47, 19-39.

Guinot, B., 1974, Astron.Astrophys., vol. 36, 1-4.

Gwinn, C.R., Herring, T.A. and Shapiro, I.I., 1986, J.Geophys.Res., vol. 91, 4755-4765.

Hassan, M.H.A. and Hayeb, I.A.E., 1982, Phys.Earth Planet.Int., vol. 28, 14-26.

Herring, T.A., 1988, in "The Impact of VLBI on Astrophysics and Geophysics", J.M. Moran and M.J. Reid eds. Reidel, in press.

Herring, T.A., Gwinn, C.R. and Shapiro, I.I., 1986, J.Geophys.Res., vol. 91, 4745-4754.

Hide, R., 1966, Phil.Trans.R.Soc.Lond., vol. A259, 615-650.

Hide, R., 1969, Nature, vol. 222, 1055-1056.

Hide, R., 1977, Phil.Trans.R.Soc.Lond., vol. A284, 547-554.

Hinnov, L.A. and Wilson, C.R., 1987, Geophys.J., vol. 88, 437-459.

Israel, M., Ben-Menahem, A. and Singh, S.J., 1973, Geophys.J., vol. 32, 219-247.

Lambeck, K., 1980, "The Earth's Variable Rotation", Cambridge University Press.

Lambeck, K., 1987, in "The Earth's Rotation and Reference Frames for Geodesy and Geodynamics", A.K. Babcock and G.A. Wilkins eds., Reidel, 1-20.

Lambeck, K., 1988, "Geophysical Geodesy: The Slow Deformations of the Earth", Oxford University Press.

Lambeck, K. and Cazenave, A., 1973, Geophys.J., vol. 32, 79-93.

Lambeck, K. and Cazenave, A., 1974, Geophys.J., vol. 38, 49-61.

Lambeck, K. and Hopgood, P., 1982, Geophys.J., vol. 71, 581-587.

Le Mouel, J.L. and Courtillot, V., 1982, J.Geophys.Res., vol. 87, 4103-4108.

Mansinha, L., Smylie, D.E. and Chapman, C.H., 1979, Geophys.J., vol. 59, 1-17.

McCarthy, D.D. and Babcock, A.K., 1986, Phys.Earth Planet.Int., vol. 44, 281-292.

McCarthy, D.D., Seidelmann, P.K. and Van Flandern, T.C., 1980, in "Nutation and the Earth's Rotation", E.P. Federov, M.L. Smith and P.L. Bender eds., Reidel, 117-124.

McDonald, K.L., 1957, J.Geophys.Res., vol. 62, 117-141.

Merriam, J.B., 1980, Geophys.J., vol. 62, 551-561.

Merriam, J.B., 1982, Geophys.J., vol. 70, 41-56.

Merriam, J.B., 1983, Sci.Prog.Oxf., vol. 68, 387-401.

Merriam, J.B., 1984, J.Geophys.Res., vol. 89, 10109-10114.

Moffet, H.K., 1978, Geophys.Astrophys.Fluid Dyn., vol. 9, 279-288.

Molodenskiy, S.M., 1985, Izv.Phys.Solid Earth, vol. 21, 167-175.

Molodenskiy, S.M. and Zharkov, V.N., 1982, Izv.Phys.Solid Earth, vol. 18, 245-254.

Morrison, L.V., 1979, Geophys.J., vol. 58, 349-360.

Morrison, L.V., Lukar, M.R. and Stephenson, F.R., 1982, Greenwich Obs.Bull., vol. 186.
Munk, W.H. and MacDonald, G.J.F., 1960, "The Rotation of the Earth", Cambridge University Press.
Naito, I., and Yokoyama, K., 1985, Proc.Int.Conf.Earth Rotation and their Terrestrial Frame, pp. 434-439, Depart.Geod.Sci., Ohio State University.
O'Connell, R.J. and Dziewonski, A.M., 1976, Nature, vol. 262, 259-262
O'Connor, W.P., 1986, Geophys.J., vol. 85, 1-11.
O'Connor, W.P. and Starr, T.B., 1983, Geophys.J., vol. 75, 397-405.
Okubo, S., 1982a, Geophys.J., vol. 71, 629-646.
Okubo, S., 1982b, Geophys.J., vol. 71, 647-657.
Ooe, M., 1978, Geophys.J., vol. 53, 445-457.
Roberts, P.H., 1972, J.Geomag.Geoelect., vol. 24, 231-259.
Rochester, M.G., 1962, J.Geophys.Res., vol. 67, 4833-4836.
Rochester, M.G., 1984, Phil.Trans.R.Soc.Lond., vol. A313, 95-105.
Rochester, M.G., Jensen, O.G. and Smylie, D.E., 1974, Geophys.J., vol. 38, 349-363.
Rosen, R.D. and Salstein, D.A., 1983, J.Geophys.Res., vol. 90, 8033-8041.
Runcorn, S.K., 1982, Phil.Trans.R.Soc.Lond., vol. A306, 261-270.
Salstein, D.A. and Rosen, R., 1985, Proc.Int.Conf. Earth Rotation and their Terrestrial Frame, pp. 440-449, Dept.Geod.Sci., Ohio State University.
Sasao, T., Okamoto, I. and Sakai, S., 1977, Publ.Astron.Soc.Jap., vol. 20, 83-105.
Sasao, T., Okuno, S. and Saito, M., 1980, in "Nutation and the Earth's Rotation", E.P. Federov, M.L. Smith and P.L. Bender eds., Reidel, 165-183.
Sasao, T. and Wahr, J.M., 1981, Geophys.J., vol. 64, 729-746.
Smith, M.L., 1977, Geophys.J., vol. 50, 103-140.
Smith, M.L. and Dahlen, F.A., 1981, Geophys.J., vol. 64, 223-281.
Smylie, D.E., Clarke, G.K.C. and Mansinha, L., 1970, in "Earthquake Displacement and Rotation of the Earth", L. Mansinha, D.E. Smylie and A.E. Beck eds., Reidel, 99-112.
Smylie, D.E. and Mansinha, L., 1971, Geophys.J., vol. 23, 329-354.
Souriau, A. and Cazenave, A., 1985, Earth Planet.Sci.Lett., vol. 75, 410-416.
Stacey. F.D., 1977, "Physics of the Earth" (second ed.), Wiley.
Stix, M., 1982, Geophys.Astrophys.Fluid Dyn., vol. 21, 303-313.
Toomre, A., 1974, Geophys.J., vol. 38, 335-348.
Wahr, J.M., 1981a, Geophys.J., vol. 64, 651-675.
Wahr, J.M., 1981b, Geophys.J., vol. 64, 677-703.
Wahr, J.M., 1981c, Geophys.J., vol. 64, 705-727.
Wahr, J.M., 1982, Geophys.J., vol. 70, 349-372.
Wahr, J.M., 1983, Geophys.J., vol. 74, 451-487.
Wahr, J.M., 1985, Scient.Am., Jan-Feb., 41-46.
Wahr, J.M., Sasao, T. and Smith, M.L., 1981, Geophys.J., vol. 64, 635-650.
Whysall, K.D.B., Hide, R. and Bell, M.S., 1985, Proc.Int.Conf. Earth Rotation and their Terrestrial Reference Frame, pp. 417-433, Dept.Geod.Sci., Ohio State University.
Wilson, C.R. and Haubrich, R.A., 1976, Geophys.J., vol. 46, 707-743.
Wilson, C.R. and Vicente, R.O., 1980, Geophys.J., vol. 62, 605-615.

Wilson, C.R. and Vicente, R.O., 1981, Astron.Nachr., vol. 302, 227-232.
Wunsch, C., 1974, Geophys.J., vol. 39, 539-550.
Wunsch, C., 1986, Geophys.J., vol. 87, 869-884.
Yoder, C.F., Williams, J.G., and Parke, M.E., 1981, J.Geophys.Res., vol. 86, 881-889.
Yakutake, J., 1973, J.Geomagn.Geoelectr., vol. 25, 195-212.
Yumi, S. and Yokoyama, K., 1980, "Results of the International Latitude Service in a homogeneous system", Mizusawa, 1899.0-1979.0.

PART 4

RELATIONSHIPS BETWEEN FRAMES

TRANSFORMATIONS BETWEEN CELESTIAL AND TERRESTRIAL REFERENCE FRAMES

IVAN I. MUELLER
Ohio State University, Columbus, Ohio

1. TRANSFORMATION FROM THE MEAN TO THE TRUE CELESTIAL REFERENCE FRAME

The rotation of a Cartesian frame of reference due to general (lunisolar and planetary) precession and astronomic nutation between the epochs T_0 and T is given by the matrix formula

$$\vec{e}_T = \mathbf{N}\,\mathbf{P}\,\vec{e}_M \tag{1}$$

where $\vec{e}_M$ and $\vec{e}_T$ denote unit vectors in the reference frame at the epoch T_0 (and affected by precession) and at the epoch T (and affected by precession and nutation) respectively. In the astronomical convention, the former is termed the 'mean' unit vector (at epoch T_0), while the latter is the 'true' unit vector (at epoch T). The terms mean and true are also used in the general sense, defining star positions, reference frames, celestial equator, equinox, pole positions, etc., depending on whether they are affected by precession only or also by nutation. Reference epoch designation must also follow. For the theory of precession and nutation, see Chapter 8.

The matrix **P** in eq. (1) represents the rotation due to general precession during the time interval $T-T_0$; thus it rotates the mean frame at T_0 to the mean frame at T. For equatorial right-handed celestial reference frames (i.e., first axis pointing to the vernal equinox, third axis perpendicular to the celestial equator (Fig. 1), it is given by the following combination of three rotations (Mueller, 1969, p. 65):

$$\mathbf{P} = \mathbf{R}_3(-z_A)\,\mathbf{R}_2(\theta_A)\,\mathbf{R}_3(-\zeta_A) \tag{2}$$

Here the rotation matrices are conventional ones, $\mathbf{R}_i$ denoting rotation about the x_i-axis, such that

J. Kovalevsky et al. (eds.), Reference Frames, 287–294.

$$\mathbf{R}_1(\alpha) = \begin{bmatrix} 1 & 0 & 0 \\ 0 & \cos\alpha & \sin\alpha \\ 0 & -\sin\alpha & \cos\alpha \end{bmatrix}, \mathbf{R}_2(\alpha) = \begin{bmatrix} \cos\alpha & 0 & -\sin\alpha \\ 0 & 1 & 0 \\ \sin\alpha & 0 & \cos\alpha \end{bmatrix}, \mathbf{R}_3(\alpha) = \begin{bmatrix} \cos\alpha & \sin\alpha & 0 \\ -\sin\alpha & \cos\alpha & 0 \\ 0 & 0 & 1 \end{bmatrix}$$

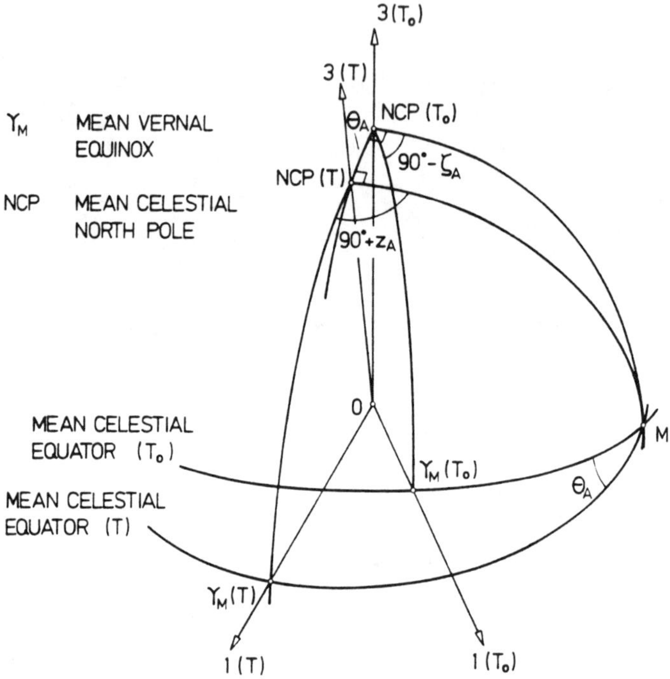

Figure 1. Relationship between mean celestial reference frames.

The equatorial precession parameters (Fig. 1) for precession between the starting epoch T_0 and ending epoch T are (Lieske, 1979; Lieske *et al.*, 1977):

$$\begin{aligned} \zeta_A &= (2306''.2181 + 1''.396\,56\, t_0 - 0''.000\,139\, t_0^2)\, t_1 + \\ &\quad + (0''.301\,88 - 0''.000\,344\, t_0)\, t_1^2 + 0''.017\,988\, t_1 \\ z_A &= (2306''.2181 + 1''.396\,56\, t_0 - 0''.000\,139\, t_0^2)\, t_1 + \\ &\quad + (1''.094\,68 + 0''.000\,066\, t_0)\, t_1^2 + 0''.018\,203\, t_1^3 \\ \theta_A &= (2004''.3109 - 0''.853\,30\, t_o - 0''.000\,217\, t_0^2)\, t_1 + \\ &\quad + (-0''.426\,65 - 0''.000\,217\, t_0)\, t_1^2 - 0''.041\,833\, t_1^3 \end{aligned} \tag{3}$$

where $t_0 = T_0 - J2000.0$, and $t_1 = T - T_0$ are in Julian centuries of TDB (see Chapter 15). The above equations are to be used from 1984 onward and are based on the IAU 1976 precessional constant of 5029."0966/Julian century at J2000.0. For pre-1984 values, see (Mueller, 1969, pp. 63-65).

The matrix **N** in eq. (1) represents the rotation due to astronomic nutation at the epoch T. For equatorial systems it is given by the following combination of rotations (Mueller, 1969, p. 75)

$$\mathbf{N} = \mathbf{R}_1(-\varepsilon - \Delta\varepsilon)\, \mathbf{R}_3(-\Delta\psi)\, \mathbf{R}_1(\varepsilon) \quad (4)$$

where ε is the mean obliquity of the ecliptic and $\Delta\psi$, $\Delta\varepsilon$ are the astronomic nutation components in longitude and in obliquity respectively (Fig. 2). Prior to 1984 these were computed from Woolard's rigid earth theory (Mueller, 1969, pp. 68-75; Woolard, 1953). For use from 1984 onward the IAU in 1980 adopted the theory of Wahr (1981) (see Chapter 8 and Seidelmann, 1982). The latter series referred to the mean equator and equinox of date may also be found in (*The Astronomical Almanac, 1984, Supplement*). Daily values of both parameters are tabulated in Section B of *The Astronomical Almanac* published annually.

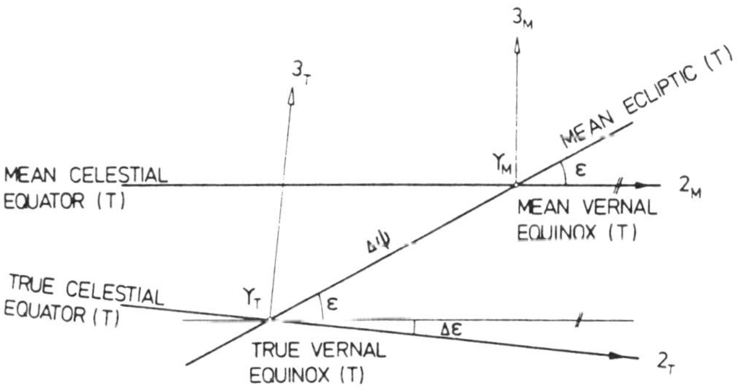

Figure 2. Relationship between the mean and true celestial reference frames.

The obliquity of the ecliptic in the above equation is calculated from

$$\varepsilon = 84381.''448 - 46.''8150\, t - 0.''000\,59\, t^2 + 0.''001\,813\, t^3 \quad (5)$$

where $t = T - J2000.0 = t_0 + t_1$ (of eq. 3) is in Julian centuries.

With eqs. (2) and (4), expression (1) thus provides the needed transformation between the celestial frames defined by the mean equator and equinox of date T_0 and by the true equator and equinox of date T. For case when T_0 = J2000.0, *The Astronomical Almanac* tabulates the nine elements of the matrix

$$\mathbf{R} = \mathbf{N}\,\mathbf{P}$$

on a daily basis in its Section B for the current year. The third axis of the true frame defined by the above transformation is the Celestial Ephemeris Pole (CEP), discussed in Chapter 8.

Now the geometric meaning of eq. (1) is completely clear: the precession matrix **P** transforms the mean frame ($\vec{e}_M$ may be any of the three unit vectors of the coordinate axes) at epoch T_0 to the mean frame at epoch T, and the nutation matrix **N** converts the latter to the true frame at the same epoch T.

2. TRANSFORMATION FROM THE TRUE CELESTIAL TO THE TERRESTRIAL REFERENCE FRAME

The coordinate systems involved are illustrated in Fig. 3 where the celestial reference frame (x,y,z) defined by the true equator and equinox of date T and the conventional terrestrial (u,v,w) reference frame (CTS) are shown together. The position of the axis z (e.g., the Celestial Ephemeris Pole, CEP) with respect to w (Conventional Terrestrial Pole, CTP) is given by the parameters of polar motion (x_p and y_p), while the angle between x (true vernal equinox of date) and u (Greenwich Mean Astronomical Meridian) is defined by the Greenwich hour angle of the true vernal equinox, the Greenwich Apparent Sidereal Time (GAST).

The terrestrial unit vector $\vec{e}$ may be obtained from that in the true equatorial frame of date, $\vec{e}_T$, by rotating the axis x about z (axis 3) with a positive GAST, then the axis z about the new x (axis 1) with a negative y_p, and finally the new axis z about the new y (axis 2) with a negative x_p. Thus, with the customary convention and notation (Mueller, 1969, p. 85),

$$\vec{e} = \mathbf{S}\,\vec{e}_T$$

where

$$\mathbf{S} = \mathbf{R}_2(-x_p)\,\mathbf{R}_1(-y_p)\,\mathbf{R}_3(\mathrm{GAST}) \qquad (6)$$

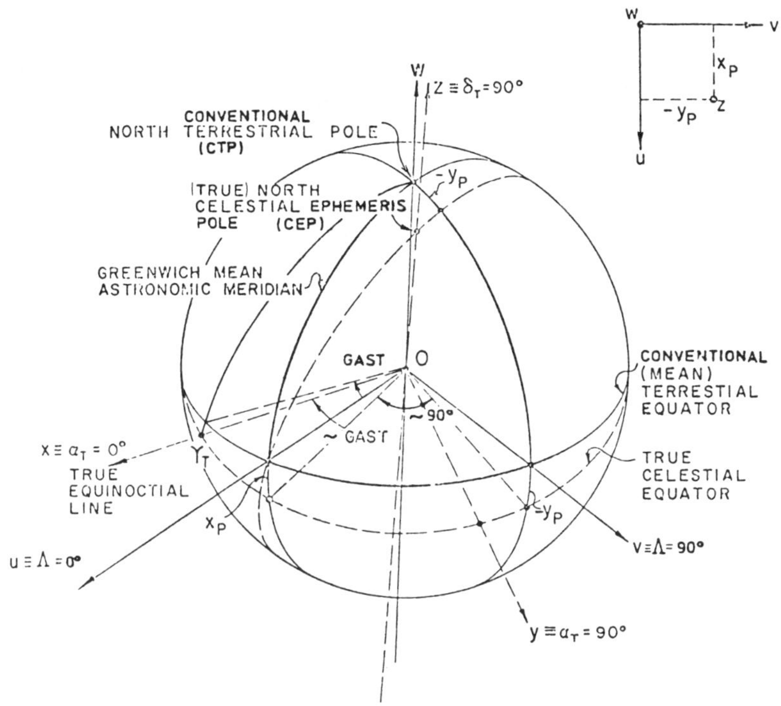

Figure 3. Relationship between the true celestial and the terrestrial reference frames.

This completes the transformation from the mean celestial reference frame of date T_0 (Conventional Inertial Reference Frame) to the Conventional Terrestrial Reference Frame at date T:

$$\vec{e} = \mathbf{S\,N\,P}\,\vec{e}_M \tag{7}$$

3. RECENT DEVELOPMENTS

3.1 EXPECTED CHANGES IN THE ADOPTED SERIES OF NUTATION

Recent analysis of modern highly accurate observations (e.g., VLBI) indicates significant departures from the IAU 1980 nutation series utilized in eq. (4). As pointed out at the end of Chapter 8, none of the existing theories based on various Earth models can adequately explain these departures from Wahr's model. Apparently more efforts are required both in theory and in observations to arrive at a resolution. In the interim, the corrections in Table 1, based on (Herring *et al.*, 1986), are being recommended as 'standardized working numbers' by the IAU Working Group on the Theory of Nutation until such time when adequate theoretical coefficients can be determined.

Table 1 Proposed Standardized Working Corrections to Nutation

		$\Delta\psi$ (0."001)	$\Delta\varepsilon$ (0."001)
Annual Term			
	in phase	+5.23	+2.08
	out of phase	+0.61	-0.24
Semiannual Term			
	in phase	+1.02	-0.41
	out of phase	-1.18	-0.47

These corrections are to be presented for adoption to the IAU 20th General Assembly, Baltimore, Md., in August, 1988.

Assuming that the CTS is to be maintained unchanged, corrections to the nutation terms in longitude ($\delta\Delta\psi$) and obliquity ($\delta\Delta\varepsilon$) would necessarily change the polar motion components and GAST, utilized in the transformation equation (6), as follows (Zhu and Mueller, 1983):

$$\begin{aligned} \Delta x_p &= \delta\Delta\varepsilon \sin\theta + \delta\Delta\psi \sin\varepsilon \cos\theta \\ \Delta y_p &= -\delta\Delta\varepsilon \cos\theta + \delta\Delta\psi \sin\varepsilon \sin\theta \\ \Delta(\mathrm{GAST}) &= \delta\Delta\psi \cos\varepsilon \end{aligned} \tag{8}$$

where θ is the sidereal time. As it is seen, the effects on polar motion are diurnal terms ($\delta\Delta\psi$ and $\delta\Delta\varepsilon$ being long periodic).

3.2 EXPECTED CHANGE IN THE CONSTANT OF PRECESSION

Modern observations also indicate a possible correction of –0."3/Julian century to the IAU 1976 constant of precession.

Williams and Melbourne (1982) and Zhu and Mueller (1983) investigated the effects of such a change. The effect on polar motion is a diurnal periodic term with an amplitude increasing linearly in time; on the GAST it is a linear term.

3.3 INTERMEDIATE REFERENCE FRAME ISSUES

The complete transformation from the mean celestial frame of date T_0 to the terrestrial frame CTS at date T is given by eq. (7). In geodetic applications generally only the complete transformation **SNP** is needed. Changes in the 'intermediate' reference frame defined by the **NP** transformation must either by 'absorbed' in the **S** matrix by changing appropriately x_p, y_p and UT1, or the CTS must change its orientation. There are seven options to choose from, and they are a matter of preference (Zhu and Mueller, 1983). One of these which would neither change the CTS orientation nor the UT1 is probably preferred by geodesists. It would however change the definition of the Greenwich Mean Sidereal Time by referring it to a nonmoving point on the equator (instead of to the mean equinox slowly moving due to precession). A similar option has been advocated by Guinot (1979) during the past decade but for different reasons. A recent proposal by Capitaine and Guinot (1988) is based on the observation that the classical definition of GAST representing the rotation of the Earth (i.e., CTS) is not satisfactory mainly for two reasons:
(i)- It is referred to the true equinox of date which is an inadequate and unnecessary intermediate reference point because modern observations of the CTS's orientation in space (especially VLBI) are practically insensitive to the orientation of the ecliptic and consequently to the position of the equinox.
(ii)- The presently adopted expression converting GAST to UT1 (Aoki *et al.*, 1982) neglects some cross-terms between precession and nutation which are of the order of 0."001 and should now be considered.

The definition advocated would thus be better adapted to the new methods of observation and would provide an accuracy of the order of 0."0001. It would also result in a new definition of Universal Time which would remain valid even if the adopted model for the **NP** transformation is revised (see also (Capitaine *et al.*, 1986)). The proposal is not without its critics. See (Aoki and Kinoshita, 1983; Aoki, 1988).

Related to the above issue is the definition of the third axis of the intermediate frame as defined by the transformation model **NP**, specifically, by the adopted theory of nutation (see Chapter 8). This pole, the Celestial Ephemeris Pole (CEP), conceptually has no diurnal motion with respect to an Earth-fixed or a space-fixed reference frame. Some of the modern observational techniques, however, are not very sensitive to this axis and, in fact, on the level of 0."001 accuracy, define a variety of technique dependent conventional poles. Capitaine *et al.* (1985) and Capitaine (1986) point out that clarification of this issue is necessary in order to intercompare and interpret polar motion coordinates determined at the level of 0".001 accuracy, by means of a variety of techniques ranging from VLBI to superconducting gravimetry.

REFERENCES

Aoki, S., Guinot, B., Kaplan, G., Kinoshita, H., McCarthy, D. and Seidelmann, P., 1982, *Astron. Astrophys.*, **105**, 359.

Aoki, S. and Kinoshita, H., 1983, *Celes. Mechan.*, **29**, 335.

Aoki, S., 1988, Relation between the celestial reference system and the terrestrial reference system of a rigid Earth, *Celes. Mechan.*, in press.

Capitaine, N., 1986, *Astron. Astrophys.*, **162**, 323.

Capitaine, N., Guinot, B. and Souchay, J., 1986, *Celes. Mechan.*, **39**, 283.

Capitaine, N., Williams, J. and Seidelmann, P., 1985, *Astron. Astrophys.*, **146**, 381.

Capitaine, N. and Guinot, B., 1988, in *'The Earth's Rotation and Reference Frames for Geodesy and Geodynamics,'* G. Wilkins and A. Babcock (eds.), Reidel Publ., Dordrecht, 33.

Guinot, B., 1979, in *'Time and the Earth's Rotation,'* D.D. McCarthy and J.D.H. Pilkington (eds.), Reidel Publ., Dordrecht, 7.

Herring, T., Gwinn, C. and Shapiro, I., 1986, *J. Geophys. Res.*, ***91***, 4745.

Lieske, J., Lederle, T., Fricke, W. and Morando, B., 1977, *Astron. Astrophys.*, **58**, 1.

Lieske, J., 1979, *Astron. Astrophys.*, **73**, 282.

Mueller, I.I., 1969, *'Spherical and practical astronomy as applied to geodesy,'* F. Ungar Publ., New York.

Seidelmann, P., 1982, *Celes. Mech.*, **27**, 79.

Wahr, J., 1981, *Geophys. J. R. Astr. Soc.*, **64**, 705.

Williams, J. and Melbourne, W., 1982, in *'High-precision earth rotation and Earth-Moon dynamics,'* O. Calame (ed.), Reidel Publ., Dordrecht, 293.

Woolard, E., (1953, *Astron. J.*, **58**, 2; also *'Astronomical Papers Prepared for the Use of the American Ephemeris and Nautical Almanac,'* **XV**, Part I.

Zhu, S.-Y. and Mueller, I., 1983, *Bull. Geodes.*, **57**, 29.

INTERCOMPARISON OF CELESTIAL REFERENCE FRAMES – GENERAL PRINCIPLES

YE SHU-HUA
Shanghai Observatory, Academia Sinica
Shanghai, China

Before we introduce the intercomparison of celestial reference frames, let us first examine the structure of a reference frame, what are the causes of difference between them and how they can be compared.

1 - ELEMENTS OF A CELESTIAL REFERENCE FRAME

Let us recall how a celestial reference frame is defined. Firstly, we must choose the origin of the frame to which the coordinate axes are attached; secondly, we choose the coordinate axes; finally, we choose the zero point of the fundamental plane from which the positions of the celestial bodies are measured.

From what is mentioned above, it is easy to see that a frame is just a geometrical construction. The realization of such a frame must be materialized by some celestial bodies, whose positions and motions are known perfectly. An ideal celestial reference frame should be non-rotational and uniform in space. However, the practical case is just more or less an approximation to the ideal frame (see the Introduction). What we can do, is to adopt a reference frame which is as good as possible.We call the one adopted 'conventional celestial reference system' (CCRS). According to what frame one chooses as the materialization of a system, we have different CCRS, e.g. stellar CCRS, the radio-source CCRS, the lunar, the artificial satellite, the planetary CCRS, etc... Usually, when we observe different objects, we are likely to choose a CCRS closely related to the observations.

For example, in optical astrometry, the bright stars are the objects observed, and the CCRS is materialized by a star catalogue, such as the FK4 or FK5 (see Chapter 1). The assigned positions and proper motions of the stars give reference points in the sky for the epoch of observations.

J. Kovalevsky et al. (eds.), Reference Frames, 295–304.

However, for fast moving objects, such as the Moon, LAGEOS, GPS satellites, etc..., we need to know their very complex motions precisely, and their ephemerides, which give their positions at the time of observations and are the materialization of the CCRS they define.

In short, a CCRS is defined by the following three elements and the position (or motion) of some specified objects.

(i)- The origin of the system : the barycentre of the solar system or of the Galactic system, etc...

(ii)- The fundamental plane, i.e., the equator, the ecliptic, a lunar orbital plane, an artificial satellite orbital plane, the galactic plane, etc...

(iii)- The zero point of the fundamental plane, i.e., vernal equinox, ascending node of the orbital plane, etc...

Although there are many possible ways to choose the three elements mentioned above, for several following practical reasons, the equator is usually chosen as the fundamental plane, and the vernal equinox as the zero point. Firstly, since the observations are conducted on the Earth, for the convenience of observations, the equator is more favourable than the others to be a reference plane: many instruments still have an equatorial mounting. Secondly, in reducing the observations to a common epoch, we need to take into account the motion of the Earth in space, thus the obliquity of the ecliptic and the equinox, cannot be avoided. Thirdly, in celestial mechanics computation, the expression of the equation of motion is simpler when referred to the invariable or Laplace plane. For artificial satellites, the Laplace plane is the equator. Considering both the benifits in computation and observation, it is natural to take the vernal equinox, as the zero point of the fundamental plane.

Even for observations in space (e.g. HIPPARCOS), the catalogue obtained does not depend on planetary motions, but, in order to link it with the ground-based catalogues, the orientation of the HIPPARCOS catalogue should still point to a common adopted equinox.

At a given time, the α, δ of the fundamental stars or radio-sources, represent the reference points in space corresponding to their CCRS; while the α, δ or the equivalent rectangular coordinates (X, Y, Z) of the Moon, the LAGEOS, GPS, etc..., given by their ephemerides, represent the reference points of their CCRS.

2. DIFFERENCES BETWEEN CCRS

In principle, all the CCRS can be made identical and can be precisely reduced one to another, provided the elements are chosen carefully namely, that the relations between different elements, i.e., the origin, fundamental plane and zero point are known exactly and the positions and motions of the celestial bodies which materialize the CCRS are given perfectly.

In fact, owing to the following points, the CCRS from different sources are not identical:

(i)- Differences in the parameters, constants, formulae, force models, etc..., which are used to define the origin, fundamental plane and zero point of the system, and the positions and motions of the celestial bodies materializing the specified CCRS.

(ii)- Differences in the methods and techniques of observation, observing networks, the instrumental errors, atmospheric refraction, thermal and other effects of the environment, etc...

(iii)- Differences in methods of data processing, rejecting, smoothing, weighting, truncating, steps of numerical integration, etc...

2.1. DIFFERENCES IN DEFINITIONS AND FORMULATION

Although the equator and the equinox are adopted universally as the practical fundamental plane and zero point for most of the CCRS, there still exist several ways to define the ecliptic and the equinox. For instance consider the long-existing stellar CCRS, the FK4: it uses the ecliptic defined by Newcomb theory and the Wollard theory of nutation, in which the calculations were done by using the old system of Astronomical constants adopted in 1896. The new system of constants adopted in 1976, reflected the progress in the intervening eighty years, and more recently, MERIT Standards further upgraded the constants by using more accurate observations including those from space probes and lunar laser ranging data, as well as the Wahr nutation series. Each upgrading slightly changes the equinox and thus the realization of the system.

As for the reference points in space, stellar CCRS use fundamental stars as indicators of these reference points. The positions and proper motions assigned to these stars, given in a fundamental star catalogue, are subject to changes both in the number of stars, and in the values of the positions and proper motions. Tracing back the FK, NFK, FK3, FK4 and FK5, the evolution of the stellar CCRS can be easily shown. The positions and proper motions of a group of fundamental stars are the adoped parameters of the stellar CCRS. Increasing the number of stars means that the reference points in space become denser than before, and more precise values of positions and proper motions of the stars make the distortions of the reference system in different parts of sky become smaller than the previous one, and thus the CCRS becomes more uniform in different parts of space.

Within a few years, the HIPPARCOS catalogue might become the most precise and dense stellar CCRS: the number of stars will be 115 000 (for FK5, there are not more than 3 000 stars), with a precision of position of 1-2 mas, about one order of magnitude better than the FK5; the precision for proper motions, 2 mas/year, is about the same as in FK5 (see Chapter 1).

The most precise kinematic CCRS is the radio-source CCRS observed with VLBI technique. The CCRS is materialized by a group (presently not more than 250) quasars and other compact extragalactic sources. The remoteness of these sources means their proper motion can be almost neglected. The precision of observation may be better than 1 mas. This CCRS is to be taken as the primary CCRS, owing to its very high accuracy (uniform in space) but not dense enough for all-purpose usage (see Chapter 2).

Let us take the dynamical CCRS defined by LAGEOS as an example. The position of LAGEOS in space for a given time, denoted in its ephemeris, represents the reference point. To construct the precise ephemeris, we need to take into account all the forces which influence the motion of LAGEOS and one has to adopt the masses and orbital elements of the major planets and the Sun, the oblateness of the Earth, the atmospheric drag model, the solar radiation pressure and that reflected by the Earth, a refined gravity model of the Earth, etc...(see Chapter 4).

As in the case of stellar CCRS, different input parameters correspond to different stellar CCRS, different equations of motion, different models constants and parameters used in each terms, will give different LAGEOS ephemerides, and thus different corresponding realizations of the CCRS.

In summary :

(i)- Different choices of the three elements of the frame that is, the origin, the funda-mental plane and the zero point, define different CCRS.

(ii)- Even though the same set of elements, say, the center of mass of the Earth, the equator and the equinox will be chosen, if we use different system of constants, models or formulae, different CCRS will still result.

(iii)- Different input parameters (for example the assigned star positions and proper motions in a stellar system) give different CCRS of the same kind, such as the FK4, FK5 and the NGS radio catalogue, the JPL catalogue, etc...

(iv)- Different terms in the equations of motion, force model and constants used in constructing an ephemeris give different CCRS of the same kind, for example based on various LAGEOS and Lunar Ephemerides. Fortunatelly all these differences mentioned above can be reduced easily to each other provided definitions, models and constants used are clearly specified.

(v)- Different materializations of CCRS, say, star, satellite, Moon, etc...

cannot be reduced to each other exactly only by their definition. This must be done systematically by careful comparison.

The MERIT Standards or any other system of standards (see Chapter 18), which recommend a unified system of constants, models and coefficients for the computations to reduce the observations, may help to eliminate the major parts of differences in CCRS mentioned in this section.

2.2. DIFFERENCES IN OBSERVATIONS

The constants, coefficients and parameters used in the previous section, are usually determined by observations conducted during a long time period. The techniques of observation may be optical visual, photographic, photoelectric, radar echoes, laser ranging, Doppler tracking, connected element radio interferometry, VLBI, meridian transit, equal-altitude, occultations, etc... Each technique has its own merits and shortcomings, bias and precision, is sensitive or insensitive to this or that effect and relies on different networks of observation.

For example, observations of the Moon were first performed using meridian circle, dual-rate camera, radar echoes, and more recently, lunar ranging. The precision from the beginning of this century until now, due to the progress of observing techniques, has been upgraded by 3-4 orders of magnitude. For the modern technique of lunar laser ranging, the precision depends on such factors as the laser pulse width, the width of the filter, the raising time of the photoelectric multiplier tube, the calibration of the system, the time resolution of the event-timer, the power of the laser, the time synchronization, the transparency of the local atmosphere, etc... Similarly, the VLBI observation precision depends on the noise of the system, tracking and pointing accuracies of the radio telescope, the band-width of the recording terminal, the stability of the local oscillator, the atmospheric refraction especially for the wet component of the troposphere, good geometric configuration of the network of observations, etc... In the case of satellite laser ranging, the requirements are similar to that of lunar laser ranging, but geographic distribution of a network of observation is needed for a good orbit determination.

In summary, differences in observations come from the following :

(i)- The mode and technique of observation, such as the visual, photographic, ranging, interferometry, etc...

(ii)- Instrumentation parameters, such as laser pulse width, time resolution, time synchronization, response time of the photoelectric element, noise of the radio receiver, recording band-width, performance of the hydrogen maser clock, etc...

(iii)- Effects of the environment, such as the abnormal refraction which deflects the zenith direction in optical observations, the atmospheric refraction which influences the path delay in laser ranging as well as in VLBI observations, thermal conditions around the observing site which distort the shape of the radio and optical images, oceanic and Earth tide loading which disturbs the direction of the local vertical and the inclination of the local crust, etc...

(iv)- Effects of the observation network distribution, for example poor geometric distribution in VLBI and polar motion determination networks which leads to the degradation of the weighting factors and even results in instable solutions for the observables and the adjusted parameters. The non-uniform distribution of observing stations cannot average out the effects of abnormal refraction coming from a regional atmospheric circular process neither can it eliminate the effects of regional plate motions, the errors arising from the local gravity field, the distorsion of the CCRS in some parts of space, etc... Obvious biases may result from these unbalances.

In fact, differences in observations cannot be easily put in the form of algorithms or formulae with an adequate accuracy. Only by skillful comparison and analysis, one can take out various factors inside these differences. Thus, even with the same definitions and formulations, different reference frames will result from different observations.

2.3. DIFFERENCES IN DATA PROCESSING

Such differences depend on the skill and personal taste of scientists or the capability of the computers. Thus they are quite arbitrary. How to solve a numerical equation? How many parameters can be adjusted simultaneously? How many steps of integration and iteration will be adequate? What would be a good choice of filtering window to cut off the noise or unwanted fluctuations? What are the appropriate methods of data smoothing? How to set up criteria for the rejection of data and weighting of observations? How to decide the truncation for the terms of a series? How to determine the constraints or boundary conditions of a set of equations? What is a suitable time span for a normal point? How to interpolate the values when there are very few or even no observations in some intervals of a time series of observations? How to control the results at the beginning and the end of a period of data series? How to fit the residuals with a physically meaningful equation? How to adopt some initial values in order to force the solution to coincides with some conventional systems or values in a given epoch, etc...? There are really a lot of procedures leading to different results even if one uses the same definition of frame elements, the same system of recommended constants, models and formulae, the same objects to materialize the CCRS, the same sources of observation data. Thus we obtain different reference frames. For example the global solution of LAGEOS orbits, lunar ephemerides, VLBI radio-source catalogues, and star catalogues compiled by different authors are sometimes different mainly because of data processing. Moreover, this kind of frame difference, owing to the existence of various factors and arbitrariness, cannot be reduced from one to another through analytical formulations, but can be found only by careful comparisons.

In summary, the three categories of differences lead to different CCRS. Some of them, mostly of the first category, can be reduced to each other by using the same definitions and formulations, but for the other two categories, the differences can be determined only by comparison. After the comparison, we can express the differences in two CCRS by a constant, a linear drift representing the average of biases drifting slowly with time; a rotation about axes, representing the difference in orientation of the frames, and some analytical terms to represent the different distorsions of the CCRS in different areas in space.

3. INTERCOMPARISON OF CCRS

Keeping in mind the way a CCRS is defined and the causes of their differences, we come to the methods of comparison. There are two methods of comparison. The first is to compare the positions of a common object given by two different CCRS. This certainly can be done for all the common objects, and we call this a direct comparison. However because the CCRS are materialized by different objects, there is sometimes no common object. One must then compare the results of the observation of some phenomena referred to different CCRS, and then deduce the CCRS difference.

3.1. DIRECT COMPARISON VIA COMMON OBJECTS

The most method of intercomparison is to compare positions of common objects at a common epoch. For example, differences of the α and δ of common stars in various fundamental star catalogues, show the difference of these stellar CCRS. The averages of the $\Delta\alpha$, $\Delta\delta$ for different sky areas, give the systematic errors between star catalogues and usually are represented by spherical harmonics over the sky. Similarly, differences of α, δ of common radio-sources show the differences of radio CCRS. Because the number of common sources is small, the intercomparison cannot show detailed differences for different parts of the sky.

Similarly, comparing the various lunar ephemerides, we can obtain the differences between the lunar CCRS. The same can also be applied to the comparison between the LAGEOS CCRS, the GPS CCRS, etc...

This comparison, via common objects, can be used between different kinds of CCRS. For example, optical positions of radio stars and radio-sources, observed with meridian instruments, astrolabes, photographic refractors or Space Telescope, and then observed with radio telescopes, can be used to compare stellar and radio CCRS (see Chapters 1 and 2). Optical positions of major and minor planets referred to a stellar system, give the relation between planetary CCRS and the stellar ones. Meanwhile, radio positions of major and minor planets, observed with the VLA, give a link between planetary CCRS and the radio systems. Recently, through the positions given by VLBI of some short period pulsars and their positions deduced from the time of arrival of pulses which depends on the Earth's position on its orbit, precise link between radio and planetary CCRS has been demonstrated (Bartels et al., 1985).

Direct comparison can also be extended to coincidences in position. Solar eclipse observations mark the coincidence of the position of the Sun and the Moon, thus can be used to link planetary and lunar CCRS. Similarly, lunar occultations of stars or radio-sources, link the stellar or the radio CCRS with the lunar CCRS. Finally, differential VLBI observations of the satellites of planets and transmitters on the Moon, link the radio CCRS with the planetary and the lunar CCRS.

For a good comparison, both a sufficient number of common objects or coincidences and an even distribution in the celestial sphere are needed.

In summary, direct comparisons are as follows :

(i)- Comparison of position of :
- common stars for different stellar CCRS
- common radio-sources for different radio CCRS
- lunar ephemerides for different lunar CCRS
- satellite ephemerides for different satellite CCRS

(ii)- Comparison of different kinds of CCRS :
- Optical positions of major and minor planets, for stellar and planetary CCRS
- Radio positions of major and minor planets, for radio and planetary CCRS
- Radio positions of millisecond pulsars determined by VLBI and time of arrival of pulses for radio and planetary CCRS
- Optical positions of radio stars and radio-sources, for stellar and radio
CCRS

(iii)- Coincidence methods :

- solar eclipses for planetary and lunar CCRS
- lunar occultations of stars for stellar and lunar CCRS
- lunar occultations of radio-sources for radio and lunar CCRS

(iv)- Differential VLBI method :

- radio-sources and transmitters on the Moon for radio and lunar CCRS
- radio-sources and the satellites of planets for radio and planetary CCRS

3.2. INDIRECT COMPARISON

From the relation between CCRS, CTRS (Conventional Terrestrial Reference System) and the S.N.P (the Earth rotation parameters, nutation and precession matrix, see Chapter 11), that is:

$$(\mathrm{CTRS}) = (\mathrm{S.N.P})(\mathrm{CCRS})$$

we have :

$$A\delta_1\,(\mathrm{CTRS}) = B\delta_2(\mathrm{S.N.P}) + C\delta_3\,(\mathrm{CCRS})$$

where A, B, C are coefficients and δ_S indicates the differenciation of these matrices.

(i)- When using the same observing network, the different of CCRS can be determined from the corresponding ERP observed, if the nutation and precession are unchanged. That is :

$$\delta_1(\mathrm{CTRS}) = 0,\ \ \delta_2\,(\mathrm{S.N.P}) = \delta_2\,(\mathrm{S})$$

we have:

$$B\delta_2(\mathrm{S}) + C\delta_3(\mathrm{CCRS}) = 0$$

from the three ERP parameters, we can determine the difference between CCRS.

(ii)- If (SNP) is adopted, we determine the station coordinates through various CCRS, the different results of station coordinates can be used to compare the CCRS, that is :

$$\delta_2(\mathrm{S.N.P}) = 0$$

and

$$A\delta_1(\mathrm{CTRS}) = C\delta_3(\mathrm{CCRS})$$

The indirect method, though not as precise as direct comparisons, can be applied to the comparisons between all kinds of CCRS. Details of comparisons, direct and indirect, will be given in the next chapters.

REFERENCES

Bartel, N., Ratner, M.I., Shapiro, I.I., Cappallo, R.J., Rogers, A.E.E. and Whitney, A.R., 1985, *Astron. J.*, **90**, 318.

Duncombe, R.L., (ed.), 1979, *'Dynamics of the Solar System'*, IAU Symposium n°81.

Fricke, W. and Teleki G., (ed.), 1982, *'Sun and Planetary System'*.

Gaposhkin, E.M. and Kolaczek, B., (eds.), 1981, *'Reference Coordinate Systems for Earth Dynamics'*, IAU Colloquium n°56.

Kolaczek, B., and Weiffenbach, G., (eds.), 1975, *'Reference Coordinate Systems for Earth Dynamics'*, IAU Colloquium n°26.

Kovalevsky, J., 1985, *Bulletin Astronomique*, **10**, n°2, 87.

Mueller, I.I., (ed.),1985, *'The MERIT/COTES Report on Earth Rotation and Reference Frame'*.

Mueller, I.I., 1985, *Bulletin Géodésique*, **59**, 181-188.

Wilkins, G.A. and Babcock, A., (eds.), 1986, *'The Earth Rotation and Reference Frames for Geodesy and Geodynamics'*, IAU Symposium n°128.

INTERCOMPARISONS BETWEEN KINEMATIC AND DYNAMICAL SYSTEMS

J. O. DICKEY
Jet Propulsion Laboratory
California Institute of Technology
4800 Oak Grove Drive
Pasadena, CA 91109 USA

1. INTRODUCTION

The advent of modern space techniques has revolutionized position measurements. These new techniques include laser ranging to the Moon and to artificial satellites; very-long-baseline interferometry (VLBI) observations of galactic and extra-galactic radio sources and of spacecraft; radio tracking of satellites; radar ranging to planets; and spacecraft tracking during planetary encounters. The accuracy of these measurements has reached a new milestone requiring a re-examination of both the concepts and realizations of reference frames.

The need for reference frames is seen in a variety of disciplines, such as astronomy, geodesy, celestial mechanics and geodynamics. One area in which this is a prime concern is Earth orientation (Earth rotation and polar motion) studies; analysis of results from the recent MERIT campaign verifies that Earth orientation is being determined with an accuracy of ~ 5 cm. Naturally, the goal is to have the frames themselves determined well below this accuracy level in order to optimize and combine the information content from the various complementary techniques. Frame determination and ties have a large "user community" and are critical in a broad variety of efforts. Some examples are the planetary missions, Galileo and Magellan. Here, a strong requirement exists for an accurate radio-planetary tie to relate the radio frame, the frame used for precise navigation, to the ephemeris frame, the frame of the planets. Frame determination and connection are also required for the upcoming Topex/Poseidon Mission, which is designed to measure the surface topography of the global oceans at the subdecimeter

J. Kovalevsky et al. (eds.), Reference Frames, 305–326.

level (Stewart *et al.*, 1986). This mission will benefit from a number of the new space techniques (e.g. laser ranging, Doppler and GPS tracking) and will require optimal frame determination and ties to reach its goals.

The main thrust of this chapter is to examine the determination of the various reference frames and the need, the current status and future prospects for accurate ties between them. Section 2 discusses the reference frames and their determinations, while Section 3, the bulk of this chapter, highlights the various connections or "intercomparisons" that are currently available. Prospects for the future are outlined in Section 4; our concluding remarks are given in the last section of this chapter. A short paper presented at the International Astronomical Union Symposium 128 (Dickey *et al.*, 1988) was the starting point for this more extensive and broader work. The reader is referred to the many accompanying articles in this volume which complement this chapter.

2. REFERENCE FRAMES AND THEIR DETERMINATIONS

Each technique observing a particular class of objects can be expected to establish its own reference frame (See Table 1). Contemporary astronomy has led to the development of three principal celestial coordinate systems: the optical frame (e.g. FK4/FK5) based on the positions of galactic stars, planets, and the sun (e.g. Kovalevsky, this volume, chapter 1; de Vegt and Johnston, 1988); the planetary/lunar ephemeris frame based on the major celestial bodies of the solar system (Willams and Standish, this volume); and the radio or the VLBI frame constructed from observations of extragalactic sources /quasars (Fanselow *et al.*,1984; Ma *et al.*,1986; Robertson *et al*, 1986 and Ma, this volume, chapter 2). It should be noted that the radio and ephemeris frames generate complementary terrestrial frames as well. Other terrestrial frames are developed through the analysis of the data from Earth-orbiting satellites [e.g. GPS (Global Positioning System), Doppler, laser reflecting satellites such as LAGEOS (e.g. Smith *et al.*,1985 and Tapley *et al,* 1985)-see also chapters 7 and 11 of this book]. These frames must consider local deformations as well as tectonic motion; for example, many of the laser sites are moving at rates of several cm/year. The celestial and terrestrial coordinate systems from a single technique and class of target are related through adopted constants and definitions. Each frame is rotated with respect to the others; and this offset may be time variable (e.g. the radio vs the FK4 frame). Dynamical systems include planetary/lunar ephemeris systems and reference frames for artificial satellites; kinematic systems encompass stellar or optical frames as well as extragalactic or radio systems.

TABLE 1 Frame Determinations

Frame	Technique	Target
* Ephemeris	* Planetary/Spacecraft Ranging	* Planets
	* Lunar Laser Ranging	* Moon
* Optical	* Optical Astrometry	* Stars, Sun and Planets
* Radio	* Very - Long - Baseline Interferometry	* Quasars, Radio Stars, Pulsars, and Interplanetary Spacecraft
* Terrestrial	* Satellite Laser Ranging	* Earth-Orbiting Satellite
	* Doppler	* Transmitting Satellites
	* GPS	* GPS Satellites

3. FRAME TIES

Measurements are inherently simpler to make and generally more accurate in their "natural" frame and hence should always be reported as such. However, to benefit from the complementarity of the various techniques, knowledge of the frame interconnections (both the rotation and the time-variable offset) is essential; these are summarized in Fig. 1 and Table 2 (both from Dickey *et al*, 1988). A comparison of Fig. 1 with a similiar figure by Williams *et al.* (1983) indicates a recent increase of activity in this area (techniques less accurate than 0.05 arcseconds are not included in either study). For example, ten lines in the earlier study instead of our current fifteen connected the targets with the techniques, and radio stars were listed as prospects for the future. The lunar/planetary system, integrated in a joint ephemeris, is by its nature unified by the dynamics (Williams and Standish, this volume). The radio frame is tied to the ephemeris frame in several ways; one is via differential VLBI measurements of planet-orbiting spacecraft and angularly nearby quasars (Newhall *et al.*, 1986). Another is the determination of a pulsar's position in the ephemeris frame (via timing measurements) and the radio frame [via radio interferometry -see Backer *et al.*, 1985]. Very Large Array (VLA) observations of the outer planets (Jupi-

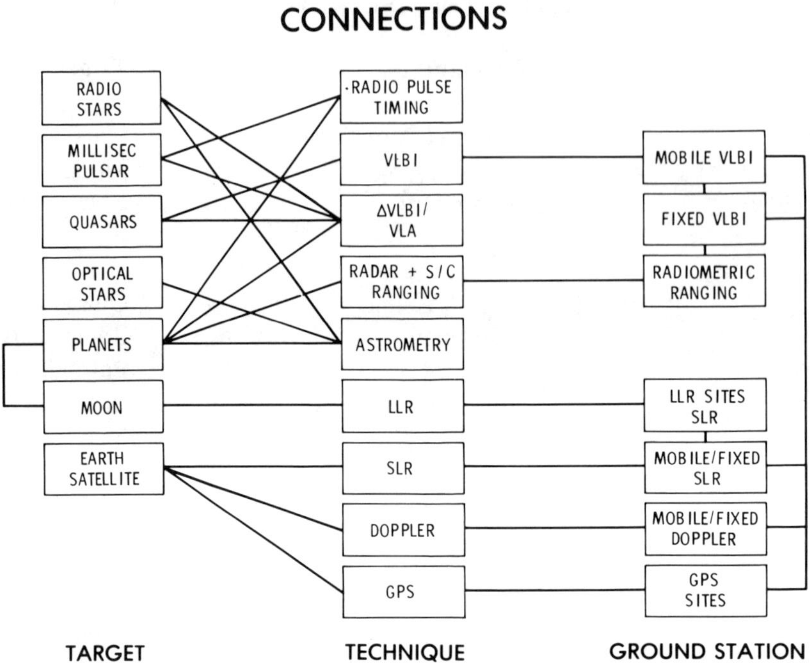

Figure 1 : Connections between Reference Systems

TABLE 2 : Frame Ties

Lunar Ephemeris/Planetary Ephemeris
* Coupled Dynamics
* Fundamental Zero Point: Dynamical Equinox

Radio/Ephemeris
* ΔVLBI
* Millisecond Pulsar
* VLA Measurements

Optical/Radio
* Optical-Radio Stars
* "Joint" Planetary Measurements

Optical/Ephemeris
* Occultations
* Optical Planetary Observation

Terrestrial/VLBI/LLR
* Collocation
* GPS Ties

ter, Saturn, Uranus and Neptune) or their satellites provide an additional tie between these two frames (Muhleman *et al.*, 1985). As for an optical-radio frame tie, a preliminary link has been established between the FK5 optical frame and the JPL radio reference frame via the differential VLBI measurement of optically bright radio stars and angularly nearby quasars coupled with comparisons of their optical positions (see Lestrade *et al.*, 1988), and also by the use of the optical positions of quasars (Purcell *et al.*, 1979). The optical and ephemeris frames are tied by optical observations of the planets. The remainder of this chapter will treat a few of the frame ties in some detail; for example, the connection between the radio and the ephemeris frames. In the some cases, such as the connections between the optical and radio frames, the highlights are given with reference given to a more detailed account.

3.1. CONNECTION BETWEEN THE RADIO AND THE EPHEMERIS FRAMES

3.1.1. Differential VLBI

The radio and ephemeris frames can be connected by measurements of objects belonging to each frame being observed by a single technique. A method of differential VLBI has been developed that allows essentially simultaneous radio observations of planets and quasars. A series of VLBI observations of planet orbiting spacecraft and angularly nearby quasars using the technique of differential VLBI has provided an estimate of the relative orientation of the JPL radio frame (Fanselow *et al.* 1984) and the ephemeris frame (Standish *et al.*, 1988; Williams and Standish, this volume). A strong motivator for these experiments is the accuracy requirements of the planetary encounters. Modern interplanetary navigation utilizes VLBI tracking (actually differential VLBI is used), while the natural frame of the planets themselves is that of the ephemeris. Therefore, an accurate frame tie between the ephemeris and radio frames is demanded. For a detailed account of the these analyses, the reader is referred to Newhall et al., 1986.

Between 1980 and 1983, eleven successful spacecraft-quasar differential VLBI observations were completed: eight using the Viking-Mars orbiter and three using the Pioneer-Venus orbiter. Figure 2 is a schematic illustration of the experiment. Single baselines between antennas of the Deep Space Network (DSN) were used. During any one observation, the two antennas (one at each end of an intercontinental baseline) simultaneously observe and receive signals from the same source, alternating from spacecraft and angularly nearby quasars. The received signals are digitized on magnetic tape; tapes are later analyzed by a digital correlator (see Ma, this volume, and Reid and Moran, 1988, for more details). The observations were from two to four hours in duration with a cycle time of five minutes.

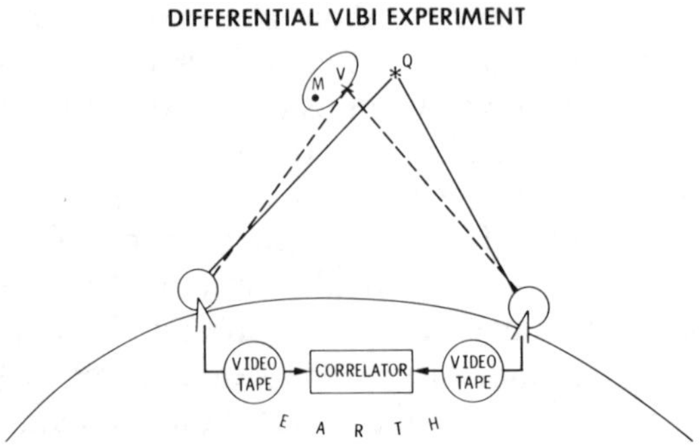

Figure 2: Schematic illustration of a differential VLBI experiment, where M refers to Mars, V to Viking and Q to the quasar.

The observations were made at S-band (2293 MHz).

Independently, the DSN stations received and recorded spacecraft Doppler tracking for a complete spacecraft revolution about the planet. An orbit was fit to these data by a weighted least squares after removal of a model containing the planet's gravity, the planetary ephemeris, solar radiation pressure, tracking station locations, Earth rotation (UT1) and polar motion. The resulting orbits had uncertainties at the few meter level with the principal source of error being the planet's gravity field. Thus, precise spacecraft orbits relative to the planet's center of mass were established allowing the spacecraft to be treated as a planet-fixed radio source.

After correlation, the phases from each source were separately connected to restore coherence lost by the cycle times within these observations. Data used in the estimation were time series of the observed phases of each source (spacecraft and quasar) throughout the entire observation. For each experiment, the coordinates of the quasar were determined in the planetary frame (DE200/LE200). An estimation program used a detailed model of physical and geometrical parameters affecting spacecraft and quasar phase observables. The time series of residual phase (observed minus computed) was produced for each source. These residuals were differenced with the differenced residual phase used to obtain a least-square estimate of the quasar position in the planetary frame. The strength of the differenced data is that the signature arising from common error sources, for example station location and equipment biases and atmospheric and ionospheric effects, will be cancelled out in large measure. The remaining signal will be mainly due to the need for a correction in the a priori value for the offset of the quasar and planet from each other. In these analyses, the a priori quasar coordinates were those of the radio source catalog. Hence, the a priori quasar radio-frame coordinates were subtracted from the calculated values in the planetary frame to obtain the offset of the two frames. Figure 3 shows the rms scatter ellipses for the Mars and Venus offsets separately; Figure 4 shows the scatter for the combined sets. The sizes and orientation of these ellipses in given in Table 3. The standard deviation ellipses include effects from uncertainties in orbit determination, planetary ephemerides, and radio-frame quasar positions.

TABLE 3 : Offset rms Scatter Ellipses

	Center (mas) RA	Center (mas) Dec	Semi-major Axis (mas)	Semi-minor Axis (mas)	Orientation Degrees
Mars	4.9	12.1	53.5	21.4	50
Venus	-11.2	-5.0	57.5	4.1	77
Combined	0.5	7.5	54.4	21.7	58

The two frames were found to be coincident in both right ascension and declination to less than 20 milliarcseconds (mas). The magnitude of the mean offset in right ascension for both the Mars and Venus observations sets is less than 12 mas with the two means agreeing to 16 mas. The calculated right ascension offset between the frames from all the data is very nearly zero (-0.5 mas); the uncertainty in this offset is estimated to be about 20 mas. The formal error ellipse for each experiment is significantly smaller than the scatter of the results indicating some error source that is not well modeled. The scatter is probably due to uncertainties in the spacecraft orbits, since orbits, determined from range rate data, provide marginal determination of their orientation about the Earth-planet line of sight.

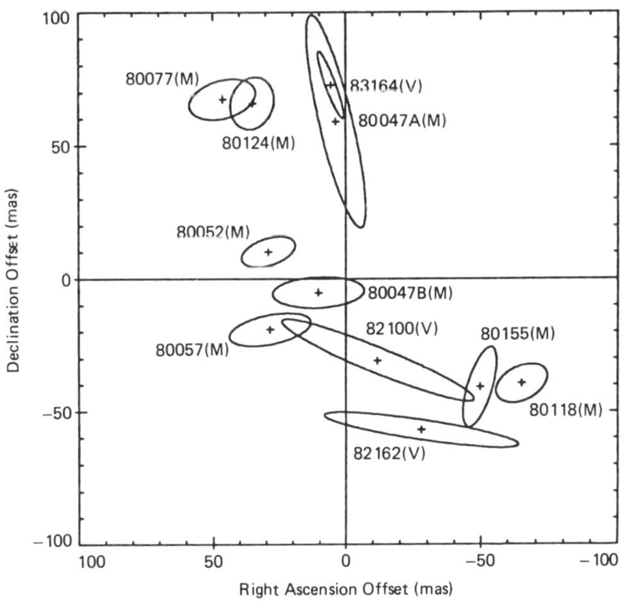

Figure 3: Quasar offsets and formal error ellipses for each of the eleven ΔVLBI experiments (adapted from Newhall *et al.*, 1986).

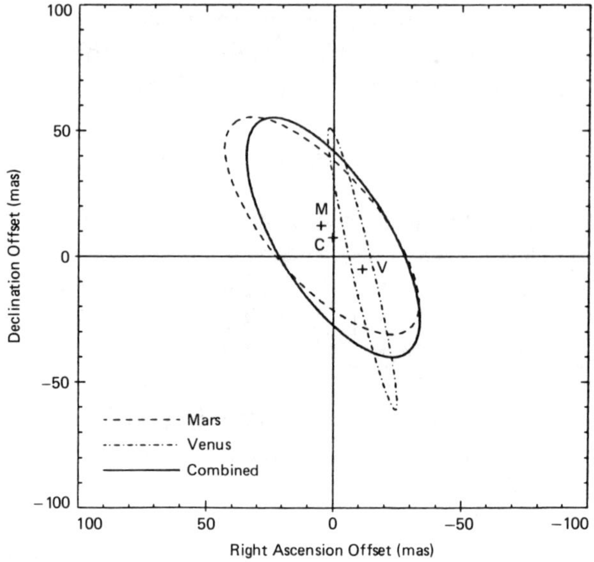

Figure 4: Mean and rms scatter ellipses for each of the ΔVLBI experiments for Mars (M), Venus (V) and for the combined set (C) [adapted from Newhall *et al.*, 1986].

For declination, the magnitude of the offset for both sets of observations is less than 13 mas, and the two means agree to 18 mas. Two quantities, an inclination and line of nodes, are needed to specify a declination difference between two coordinate systems. The scatter in the individual offsets precluded a firm determination of these values, but the mean absolute value of the declination difference is less than about 20 mas.

3.1.2. Very Large Array Measurements of the Outer Planets and Their Satellites

The radio and ephemeris frames can be connected by measurements of objects belonging to each frame being observed by a single technique. A method of differential VLBI was developed that allows essentially simultaneous radio observations of planets and quasars; this technique and results from experiments observing the orbiters of the inner planets, Mars and Venus, were discussed in the previous section (3.1.1). There have been no orbiting spacecraft about the outer planets (several missions are planned such as Galileo for the near future); hence, the technique as described previously cannot be applied. Very Large Array (VLA) measurements afford the best currently available method for such a tie. The motivation of the VLA measurements of the outer planets and in some cases their satellites are two-fold: first, a frame tie between the outer planets and the radio frame; and second, an improvement of the outer planet ephemerides.

The technique used consists of using the available antennas (usually 27) of the VLA in pairs (27 antennas would correspond to 351 pairs) to form this number of two-element interferometers. Each pair measures as a function of time the complex visibility, the phase of which is sensitive to the position of the source along the line of the instantaneous, projected baseline of that pair. In reality, the measurements are more complex because of many effects such as intrinsic errors in measuring weak signals and variations in troposphere delay along the line of sight at each telescope. The error sources can be considered as causing an instrumental phase error, which is a function of time for a given pair. This function is measured by periodically observing one or two quasars that are angularly nearby to the planet. The positions of the observed quasars are well known usually at the few milliarcsecond level or less (Fanselow *et al.*, 1984; Niell *et al.*, 1986; Sovers, private communication 1987). The measurement of the planet's position is achieved by "pointing" each two-element interferometer of the VLA at the ephemeris position of the satellite; if the ephemeris position were precisely correct the calibrated phases of the visibilities would be zero (to within the noise). A least-mean-square fit is used to fit the visibilities and estimate the offsets in the right ascension and declination of the apparent planet position relative to the ephemeris position.

Jupiter and Saturn have large natural satellites which emit sufficient radio flux for position measurements, while Uranus and Neptune are sufficiently small (angularly) that their positions can be accurately determined with radio-interferometry techniques. Measurements highlighted here were made with the Very Large Array (VLA) of the National Radio Astronomy Observatory near Socorro, New Mexico operating at a wavelength of 2 cm. The observations of the Jupiter system were made by G. Berge, A. Niell, and D. Muhleman; those of the Saturn system by G. Berge and D. Muhleman; those of Uranus by G. Berge and D. Jones. For a more detailed account of the Jupiter, Saturn and Uranus analyses, the reader is referred to Muhleman *et al.*, 1985. The Neptune results have been obtained directly from Dayton Jones, 1987, and Muhleman *et al.*, 1987.

The Jovian satellites, Europa, Ganymede and Callisto (see Fig. 5a) were observed on April 26 and May 1, 1983 in a repeated sequence with two cataloged quasars within ± 3 hours of the VLA meridian. A least-square procedure was used to fit the data and calculate the offsets of the true source positions of the planetary satellites from DE200. Measurements of planetary satellites must be done when the objects are near elongation with respect to the central planet so that the confusion flux from the planet itself can be minimized. Most of the flux from the central body is eliminated by interference effects when the data are combined over 6-hours. The satellite ephemerides are from the computations of J. Lieske and were combined with the Jupiter ephemeris by E. M. Standish. The results for Ganymede and Callisto are shown in Fig 6a.

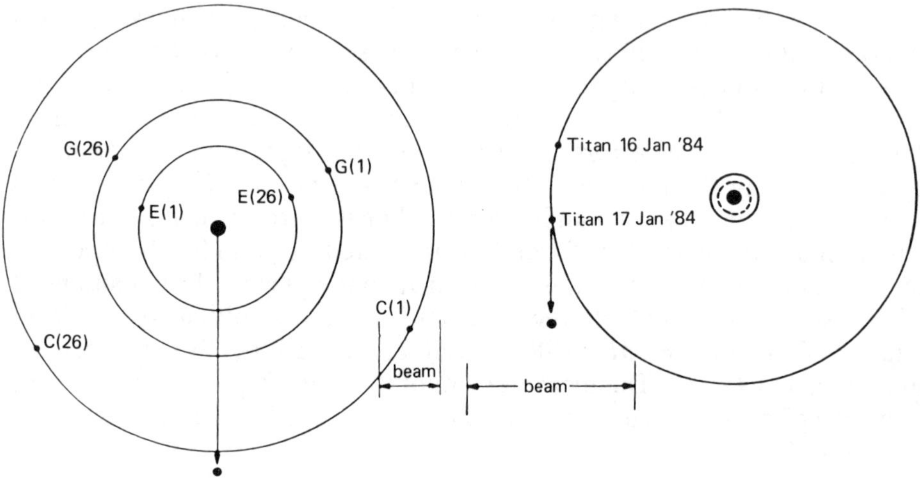

Figure 5 - a) Geometry of the Jovian Satellite System (left).
b) Geometry of the Saturn/Titan System (right)
(Muhleman et al, 1985)

The estimates of Ganymede are tightly grouped for the two days, while those for Callisto show considerable more scatter indicating an error in the orbit. Ganymede is expected to be determined better since its orbital period is considerably smaller allowing more accurate and a larger number of eclipses observations. Titan measurements were obtained January 16, 1984 using one quasar as a calibrator in a repeated sequence; the observing geometry of the Saturn/Titan system as well as the extent of the antenna beams (full widths at half maxima) for the individual telescopes are show in Fig. 5b. The Titan experiment is definitely more difficult and challenging than that of the Jovian satellites. The offsets of Titan with respect to two ephemeris systems is given in Fig. 6b.

Uranus was observed with VLA on April 30, 1985 in sequence with two quasars within ± 3 hours of its meridian crossing. The results for three separate solutions are displayed in Fig 6c. The filled circle indicates the least square fit for the center of the flux density; the plus symbol is the position after correction for the pole position; and the open circle represents the position based on the limb contours of the Uranus map (the formal error on the least-squares determination is approximately $\pm 0\farcs025$ on both dimensions). Observations of Neptune with the VLA were accomplished during three successive elongations of Triton in January 1987. The goals were here were threefold: a frame tie determination, a measurement of Neptune's position with respect to the DE200 and DE125 ephemeris predictions, and a detection of the motion of Neptune about the Neptune - Triton barycenter enabling an estimate of the Neptune / Triton mass ratio. These observations were hampered by poor weather conditions during all three elongations. Cloudiness is a source of variable attenuation and additional phase fluctuations at the wavelength used (2 cm); also large large-scale variations in the tropospheric water vapor can produce apparent shifts in the positions of radio sources. The offsets from DE200 ephemeris position are -0."42 ± 0."02 in right ascension and -0."06 ± 0."06 in declination. The offset with respect to more recent ephemeris DE125 is significantly better (see Fig. 6d) The Neptune position estimates have large scatter in declination, which is thought to be caused by variations in atmospheric refraction, already discussed.

These analyses were originally done with respect to JPL Planetary Ephemeris - DE200/LE200. Subsequent to the original analysis, offsets were determined with respect to the more recent JPL ephemeris DE125,

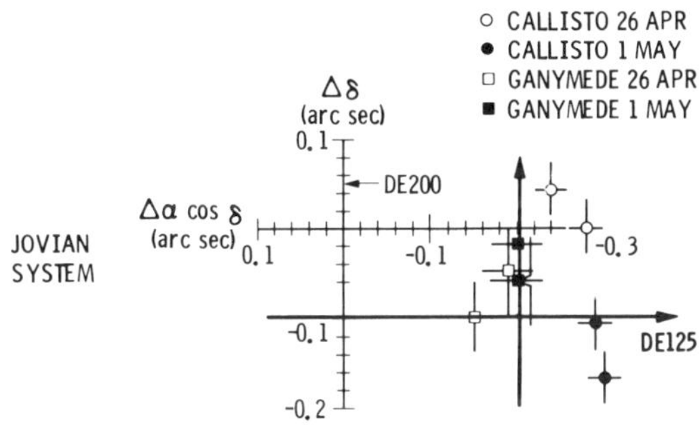

Figure 6a: Ephemeris offsets with respect to DE200 and DE125 of the Jovian System for 26 April and May, 1983 (Muhleman *et al*, 1985, and Standish, 1985).

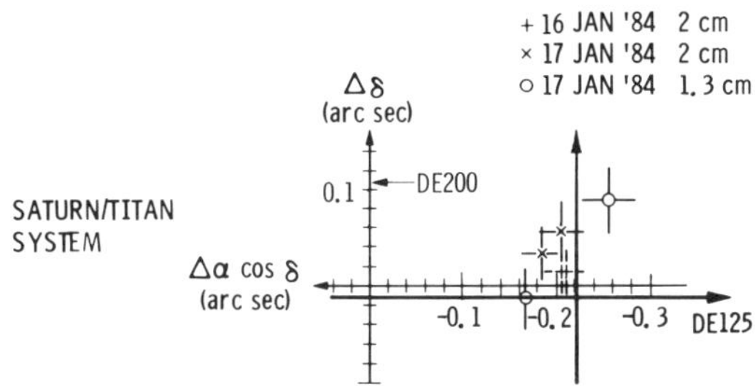

Figure 6b: Ephemeris offsets with respect to DE200 and DE125 for the Saturn/Titan system for 16, 17 January, 1985 (Muhleman *et al.*, 1985 and Standish, 1985).

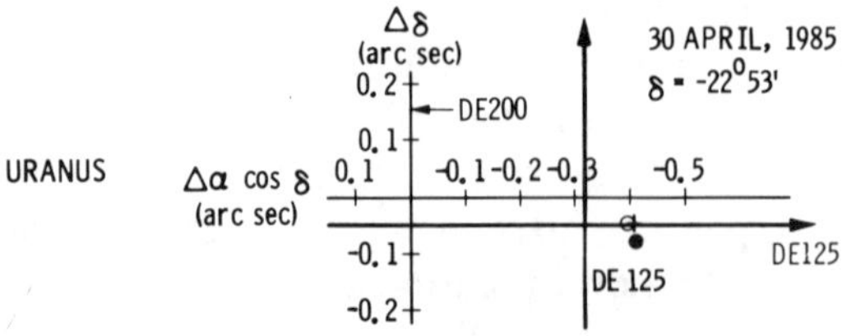

Figure 6c: Ephemeris offsets with respect to DE200 and DE125 of Uranus for 30 April, 1985. Filled circle is the least squares fit for the center of the flux density; the plus symbol indicates the position after correction for the pole position; and the open circle indicates the position based on the limb contours of the Uranus map (Muhleman *et al.*, 1985 and Standish, 1985).

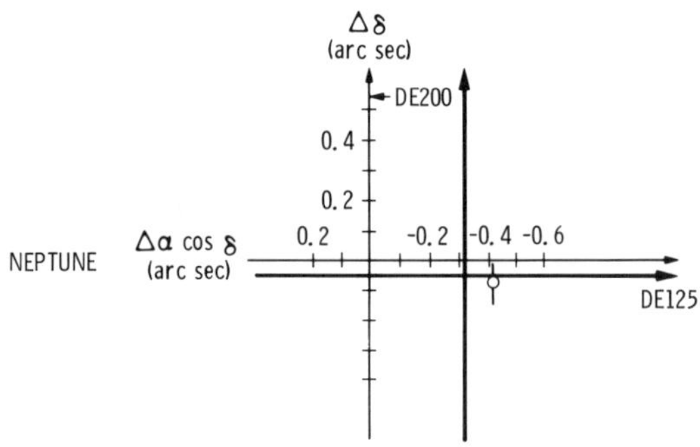

Figure 6d: Ephemeris offsets with respect to DE200 and DE125 of Neptune, for 6-12 January 1987 [Jones, (1987), Muhleman, et al., (1987), and Standish, M., (1985)].

which serves as the final Uranus Ephemeris Delivery for Voyager (Standish, 1985). These VLA measurements were not included in either DE200 or DE125 analyses and therefore can be used as a valuable independent check of the ephemerides. Comparison of results analyzed with respect to both frames are displayed in Figure 6. The DE200 results are very similar and suggest that the ephemeris reference frame for the outer planets is in error by roughly -0.2 arcseconds in right ascension. The agreement with DE125 is significantly better. For example, the residuals of the VLA Uranus observations with respect to DE125 are $-0.''06 \pm 0''03$ in right ascension and $0.''00 \pm 0.''04$ in declination. DE125 represents a major improvement to the ephemerides of the outer planets, which may be attributed to two major sources: a change in the data reduction procedure for the optical transit observations and the addition of new and more accurate data types, such as Voyager spacecraft tracking (Standish, 1985).

3.1.3. Millisecond Pulsar

Another method connecting the radio and the ephemeris frames is the measurement of millisecond pulsars, as joint "link" objects, in both of these systems. The pulsars are placed in the radio frame via radio interferometry measurements, while placement is done in the ephemeris system via timing measurements. For a detail account, the reader is referred to Backer *et al.*, 1985.

Accurate coordinates in the ephemeris frame can be achieved by the analysis of arrival times of pulses from pulsars. The timing positions for short period pulsars can be extremely accurate because pulsar-timing data tend to be a fixed fraction ($\sim 10^{-3}$) of the pulsar period. The positions of the two pulsars considered here, PSR 1937+21 and PSR 1913+61, can be determined with unusual precision because of their narrow pulse widths (~60 μs for the first and ~900 μs for each component of the double pulse of the latter) and because of their rotational stability. The microsecond accuracy of the Arecibo timing data PSR 1937+21 (Backer, Kulkarni and Taylor; 1983, Davis *et al.*, 1985) suggests that the position of this pulsar may be determined to a milliarcsecond accuracy with respect to the ephemeris frame. One begins with an ephemeris of the Earth's motion in an inertial frame; an incorrect position assumed for the pulsar will lead to a sinusoidal error in the predicted times with a period of about one year. Since we can determine both the amplitude and phase, position corrections in two coordinates can be computed. Ephemerides, DE118 and PEP740-R, were utilized in these analyses from two centers, JPL and the Center for Astrophysics (CfA), respectively. The reference frame defined by these ephemerides differ by a rigid rotation of ~ 0."4 (Bartel *et al.*, 1985). Since the DE118 reference frame is known to be in agreement with FK4 to within 0."05, the DE118 frame was adopted as fundamental and the results on PEP 740-R were rotated into this system. Positions were obtained in both the FK4 B1950.0 and the J2000 systems from the timing measurements that had uncertainties of 0.00003 seconds in right ascension and 0."0010 in declination for both ephemerides considered (see Table 4 on the following page).

The VLA position uncertainty (0."18) for PSR 1913+16 is several times larger than that for PSR 1937+21, which greatly limits the accuracy of the tie provided by these measurements. Since these results are consistent with but less precise than from PSR 1937+21, we will focus on frame tie results from PSR 1937+21. The millisecond pulsar 1937+21 was observed September 3-4, 1983 with the Very Large Array (VLA) antennas at two frequencies, which allow for the removal of possible ionospheric refraction. Just prior to the 1937+21 observations, five high-quality VLA calibration sources were observe for two hours to check and verify many parameters used in the data reductions such as station coordinates, Earth rotation and polar motion values. These calibration sources are known to better than 0.″05. The millisecond pulsar, 1937+21, was then observed for four hours with alternating observations made with three primary (which lie in a large triangle around the pulsar) and one nearby secondary (about 3 degrees away in the sky) calibrators. These data were used to determine a more accurate position for this pulsar and to estimate the overall systematic errors in the experiment. The positions of the primary calibrators were derived from previous astrometric runs at the VLA and are tied to the FK4 B1950.0. The error estimates for the source position are ± 0.″003 in right ascension and ± 0.″05 in declination. The error is a quadrature sum of the uncertainty in relating the pulsar position to the necessary secondary calibrator, the uncertainty of the position of the calibrator, and the uncertainty in the radio grid (see Table 4 below).

TABLE 4

Measured FK4 J2000 Positions for PSR 1937 + 21

	Right Ascension	Declination
VLA (1)	$19^h\,39^m\,38.^s5613 \pm 0.^s0030$	$+21^o\,34'\,59.''155 \pm 0.''050$
Timing, JPL(2)	$19^h\,39^m\,38.^s56025 \pm 0.^s00003$	$+21^o\,34'\,59.''1447 \pm 0.''0010$

(1) VLA observations epoch 1983.67
(2) JPL Ephemeris DE200

Comparison of the VLA and time measurement derived results are shown in Figure 7 in an inertial B1950.0 frame. The timing-based measurements using the CfA PEP740-R lie 0.″12 ± 0.″05 north of the VLA position; after rotation between the two frames, as was previously mentioned, the CfA agree with those from the JPL DE118 (0.″09 ± 0.″05 from the VLA positions). The B1950.0 offsets are consistent with the findings of Fomalont et al.(1984) for pulsars in the same region of the celestial sphere. The uncertainty of 0.″05 quoted here is clearly dominated by the VLA determination. Part of the 0.″09 discrepancy results from systematic differences between the B1950.0 reference frame in use at the VLA and in DE118. For example, the VLA frame has errors resulting from the FK4 constants of precession and nutation. When both results are transformed into the J2000 system, the positions agree well within 1 σ errors of the VLA measurements as shown in Table 4. The dominant uncertainty, at t his time,

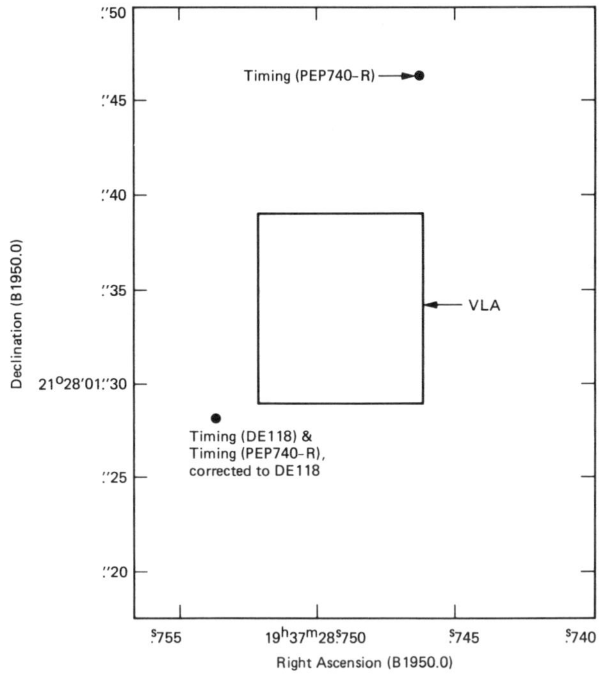

Figure 7: Relative positions for PSR 1937 + 21 from Table II. The box delineates the error of the VLA interferometer position. The dots indicate timing positions based on JPL ephemeris DE118 and CfA ephemeris PEP740-R (Backer *et al.*, 1985).

is that arising from the interferometric positions. Improvements in radio interferometric measurements could lead to a tighter link between these two frames. VLBI measurements of pulsars as opposed to VLA measurements will be central to a better tie; a challenge here is the weakness of the pulsar sources. Discovery of new millisecond pulsars would be welcomed; several such objects have recently been reported. Of key importance here are several factors: the pulsar's period, strength, rotational stability and distribution in the sky (see Section 4 for further discussion).

3.2. CONNECTION BETWEEN RADIO AND OPTICAL FRAMES

The optical and the radio reference frames can be connected by optical and radio observations of objects that are common to both frames; optical and radio positions of quasars (Purcell *et al.*, 1979) and radio stars (Lestrade *et al.*, 1988) provide such a connection. Such a frame tie determination would unify the optical and radio frames and would render the optical system non-rotating, by determining the possible angular rotation between the optical frame (which is subjected to proper motion) and the stable extragalactic radio reference frame. The reader is referred to Chapter 2 for a detailed account of the optical or stellar reference frames.

The stellar system, Algol, was the first star detected with a connected-element radio interferometer (Ryle and Elsmore, 1973) and then studied with the VLBI technique (Clark *et al,* 1975 and 1976). Since then, a few other radio stars have been investigated with VLBI: SS433 (Schilizzie *et al,* 1979; Walker *et al,* 1981; Niell *et al,* 1981) and CIRX-L (Preston *et al,* 1983). However, of these stars, only Algol is optically bright enough for HIPPARCOS observation. Currently, measurements are underway and a frame tie has been determined by utilizing optically bright radio stars as the link objects (Lestrade *et al,* 1988). A major motivation is the upcoming HIPPARCOS program. Here, the VLBI technique and the Bordeaux Automatic Meridian Circle were used to determine the positions of eight optically bright radio stars in the Jet Propulsion Laboratory (JPL) radio reference frame and in the FK4 system, respectively. The analysis of VLBI observations of these eight radio stars with angularly nearby quasars with a differential VLBI technique has resulted in measurement accuracies as fine as 2 milliarcseconds. These radio emitting stars have been positioned in the JPL-VLBI (J2000) reference frame following the IAU 1976 recommendations. The optical positions have been measured at the same epoch of the VLBI observations and are originally referred to the mean equator and equinox of B1950.0 with an expected precision of ~ 50 mas. To make the comparison possible, the original B1950.0 positions are transformed into the J2000 systems using the conversion matrix proposed by Aoki *et al.* (1983); this frame is in reality the preliminary FK5 frame. The positions were compared and the connection between the preliminary FK5 frame and the JPL radio frame was deduced. The mean difference (radio - optical) in right ascension and declination found are $+0''.02 \pm 0''.04$ and $-0''.02 \pm 0''.07$, respectively. Here, the standard deviations are quoted as realistic uncertainties; these uncertainties are consistent with the combined precision of the two techniques, VLBI ($0''.01$) and optical ($0''.05$). This indicates that the JPL radio reference frame and preliminary FK5 frame are aligned at the level of the precision of the optical measurements. These sources are all located in the Northern Hemisphere; a program is underway in Australia to provide accurate radio and optical positions for a grid of reference sources in the Southern Hemisphere (White *et al.*, 1986).

3.3. CONNECTIONS BETWEEN THE OPTICAL AND EPHEMERIS FRAMES

The determination of the optical or stellar reference frames is discussed in much detail in Chapter 2 (Stellar Reference Frames), while the reference frame of the ephemeris is covered in Chapter 3 (Dynamical Reference Frames in the Planetary and Earth-Moon Systems). The optical and ephemeris frames can be connected by optical observations of the planets and by occultations. Ground-based measurements of positions of planets with respect to stars have produced ties at the 0."05 level of accuracy; however, higher accuracies at the 0."01 level are unlikely (Williams *et al.*, 1983). Problems encountered here are the effects of the Earth's atmosphere, finite size of the planets and gradation of lighting at one limb. Lunar positions are affected by limb topography and by center of mass / center of figure offsets. Photoelectric observations of the occultations of stars by planets give very accurate angular differences between them, and therefore can provide a tie between the optical and planetary frames. However, in only a few cases, have the stars been bright enough to be included in the existing fundamental catalogue (FK4). Also, most of the recent observed occultations have involved the observation of the outer planets, which are not as well connected to the ephemeris frame as the inner four planets and the Moon. Astrometric measurements planned from Space Telescope and the HIPPARCOS satellite promise a new era of unprecedented accuracies in the optical measurements, which, in turn, will allow improved ties of the optical frame to other systems (See Section 4).

3.4. CONNECTIONS BETWEEN ARTIFICIAL SATELLITE FRAMES AND OTHER SYSTEMS

Artificial satellite terrestrial frames are developed through the analysis of the data from Earth-orbiting satellites [e.g. GPS (Global Positioning System), Doppler, laser reflecting satellites such as LAGEOS (e.g. Smith *et al.*, 1985 and Tapley *et al*, 1985)]. The determination of these reference frames is covered in much detail in Chapter 4 (Reference Frame for Artificial Satellites of the Earth). The concept will be outlined here for clarity in this chapter's presentation; the reader is referred to Chapter 11 (The Intercomparison of Terrestrial Frames) for a more detailed account.

It is important to connect the various frames of the artificial satellites to the terrestrial and celestial frames of the Ephemeris and Radio systems. Connections between terrestrial frames can be accomplished by collocation. Three alignments are of concern here. If the various techniques use data spans of a day or more and therefore are sensitive to the direction of the rotation axis, the alignment of the longitudes is the main issue. In principle, a minimum of one common observing site (or connection between sites) is required to align the longitudes of two geocentric techniques (such as satellite laser ranging-SLR and lunar laser ranging-LLR), while two connections are needed to align a baseline from non-geocentric techniques (e.g. VLBI) with a geocentric system (Williams *et al.*, 1983). Figure 1 summarizes the connections between the techniques. The line connecting all of the ground stations at the extreme right represents linking all terrestrial frames by using the mobile techniques. A separate line is used to indicate the connection through the joint SLR/LLR observing sites. The availability and high accuracy of mobile VLBI and SLR systems have made these connections realizable and have allowed for regular intercomparison of results (such as baselines) between the various techniques (see, for example, Ryan *et al.*, 1988). Collocation between the ephemeris frame and satellite frame as determined by laser ranging is automatic as several sites (McDonald Observatory - Texas, Haleakala - Hawaii, and Grasse - France) range to both the Moon and LAGEOS with either the same or adjacent telescopes. Comparisons between these two systems have been made by Williams *et al.*, (1985). Connection between the VLBI site at Fort Davis and the McDonald Laser Ranging System (only a few kilometers in separation) can be made by conventional surveying or via GPS. A GPS survey would be ideal for a tie between the SLR/LLR station at Haleakala, Maui and the VLBI site at Kauai. GPS furnishes an alternate method for terrestrial sites and its use is expected to increase (Section 4).

Care and caution must be exercised in the discussion of satellite coordinate systems in the context of an inertial celestial system. The even zonal harmonics of the Earth cause a satellite's node (orientation of the orbit in right ascension) to precess. Uncertainties in these harmonics cause drifts in right ascension, which restrict satellite results (Williams *et al.*, 1983 and Carter *et al.*, 1984) from being directly used for the highest accuracy inertial celestial system. Temporal variations of the zonal harmonics have been detected (Yoder *et al.*, 1983 and Rubincam, 1984). Therefore, in addition to connecting the systems by collocation, it is necessary for the satellite systems to use Earth rotation (UT1) values that do not exhibit offsets or drifts with respect to Earth rotation (UT1) derived by the inertial techniques (LLR and VLBI). Connecting the satellite frame to the inertial systems could provide an improved constraint for zonal harmonic determinations (Williams *et al.*, 1983).

4. PROSPECTS FOR THE FUTURE

The future is promising with on-going and planned efforts in several areas. Improved ephemeris-radio frame ties can be accomplished by VLBI observations of pulsars, additional VLA observations of the outer planets and satellites, and future differential VLBI experiments (such as that with orbiting spacecraft around Jupiter and Saturn). The millisecond pulsar - PSR1937+214, having a period of 1.6 msec, has exceptionally low timing noise. Its position in the ephemeris frame can be measured to ~ 1 mas. This will allow a radio-planetary frame tie limited only by the accuracy of an interferometric position measurement. Roughly, a factor of five improvement (down to 0."01) is expected here with the full implementation of VLBI observations (Linfield, private communication, 1986). An initial experiment of this type has been executed by R. Linfield and C. Gwinn (Linfield, private communication, 1987). For optical astrometry, HIPPARCOS will measure a network of stars over the entire sky with accuracies of ~2 mas (Kovalevsky, 1980), while the Space Telescope will measure small fields with similar differential accuracy. However, the Space Telescope can observe much fainter objects (Jefferys, 1980) and could observe the optical counterparts of extragalactic radio sources, all but possibly one of which are too faint for HIPPARCOS. A joint program would produce an accurate stellar network linked to the quasar radio frame by the Space Telescope. The occultations of stars by planets and planetary rings as measured by HIPPARCOS and the Space Telescope can provide an additional improved link between the optical and ephemeris frames. Also, optical interferometry offers exciting possibilities with the potential resolution being two or three orders of magnitude finer than that of VLBI (Reasenberg, 1986). Turning to the terrestrial-frame techniques, collocation provides direct links between the systems involved. GPS also furnishes an alternate method for terrestrial ties. Clearly in years to come Figure 1 will become even richer in the number and accuracies of connections.

5. CONCLUDING REMARKS

The concept and the realization of reference frames are clearly needed for a variety of efforts: astronomy, geodesy and geodynamics, to list a few. Reference frames are the standard scale against which measurements are made. The current space techniques have achieved remarkable accuracies; however, continued improvements and other new techniques may achieve even higher accuracy. The community must recognize the necessity of different frames, their transitory nature, and the need for improvements and the constant maintenance of their interconnections.

Acknowledgments: The author gratefully acknowledges interesting discussions and interactions with many colleagues: T. M. Eubanks, J. L. Fanselow, M. Feissel, R. Hellings, J. Hughes, K. Johnston, D. Jones, J. Kovalevsky, J.-F. Lestrade, R. P. Linfield, W. G. Melbourne, I. I. Mueller, D. O. Muhleman, X X Newhall, A. E. Niell, R. A. Preston, E. M. Standish, O. J. Sovers and J. G. Williams. This paper presents the results of one phase of research carried out at the Jet Propulsion Laboratory, California Institute of Technology sponsored by the National Aeronautics and Space Administration.

REFERENCES

Aoki, S., Soma, M., Kinoshita, H., and Inoue, K., 1983, *Astron. Astroph.*, **128**, p. 263.

Backer, D. C., Fomalont, E. B., Goss, W. M., Taylor, J. H. and Weisberg, J. M., 1985, 'Accurate Timing and Interferometric Positions for the Millisecond Pulsar 1937+21 and the Binary Pulsar 1913+16', *Astron. J.*, **90**, p. 2275.

Bartel, N., Capallo, R. J., Ratner, M. I., Rogers, A.E.E., Shapiro, I. I., and Whitney, A. R., 1985, *Astron. J.*, **90**, p. 318.

Carter, W. E., Robertson, D. S., Pettey, J. E., Tapley, B. D.., Schutz, B. E., Eanes, R. J., and Lufeng, M., 1984, 'Variations in the Rotation of the Earth', *Science*, **224**, p 957-961.

Clark, B. G., Kellermann, K. I., and Shaffer, D., 1975, *Ap. J.*, **198**, L123.

Clark, T. A., Hutton, L. K., Ma, C., Shapiro, I. I., Wittels, J. J., Robertson, D. S., Hinteregger, H. F., Knight, C. A., Rogers, A.E.E, Whitney, A. R., Niell, A. E., Resch, G. M., and Webster, W. J., 1976, *Ap. J.*, **206**, L107.

Davis, M. M., Taylor, J. H., Weisberg, J. M., and Backer, D. C., 1985, Nature **315**, 547.

de Vegt, C., and Johnston, K., 1988, 'Fundamental Reference Frames', Proceedings of the International Astronomical Union Symposium No. 128, The Earth's Rotation and Reference Frames for Geodesy and Geodynamics, eds. A. Babcock and G. Wilkins, D. Reidel,

Dickey, J. O., Fanselow, J. L., Melbourne, W.G., Newhall, X X, Standish, E.M., and Williams, J. G., 1988, 'Reference Frames: Determinations and Connections', Proceedings of the International Astronomical Union Symposium No. 128, The Earth's Rotation and Reference Frames for Geodesy and Geodynamics, eds. A. Babcock and G. Wilkins, D. Reidel, Boston.

Eubanks, T. M., Steppe, J. A., and Sovers, O. J., 1988, 'The Long Term Stability of Earth Orientation Measurements', to be published in the Proceedings of the International Astronomical Union Symposium Number 129, The Impact of VLBI on Astrophysics and Geophysics, eds. Reid, M., and Moran, J. M., D. Reidel, Boston, in press.

Fanselow, J. F., Sovers, O. J., Thomas, J. B., Purcell, G. H., Cohen, E. J., Rogstad, D.H., Skjerve, L. J., and Spitzmesser, D. J., 1984, 'Radio Interferometric Determination of Source Positions Utilizing Deep Space Antennas, 1971 to 1980', *Astron.J.*, **89**, p. 987-998.

Fomalont, E. B., Goss, W. M., Lyne, A. G., and Manchester, R. N., 1984, *Mon. Not. R. Astron. Soc.* **210**, 113.

Jefferys, W. H., 1980, 'Astrometry with the Space Telescope', *Celestial Mech.*, **22**, p. 175-181.

Jones, Dayton, 1987, 'VLA Astrometry of Neptune', Interoffice Memorandum, Jet Propulsion Laboratory, Pasadena, CA, USA.

Kovalevsky, J., 1980, 'Global Astrometry by Space Techniques', *Celestial Mech.*, **22**, 153-163.

Kovalevsky, J., 1985, 'Systèmes de Référence Terrestres et Célestes', *Bulletin Astronomique*, **10**, p. 87-93.

Lestrade, J. F., Requieme, Y., Rapaport, M., and Preston, R. A.., 1988, 'Preliminary Relationship between the FK5 and the JPL Radio Reference Frames', Proceedings of the International Astronomical Union Symposium 128, The Earth's Rotation and Reference Frames for Geodesy and Geodynamics', D. Reidel, A. Babcock and G. Wilkins, editors.

Ma, C., Clark, T. A., Ryan, J. W., Herring, T. A., Shapiro, I. I., Corey, B. E., Hinteregger, H. F., Rogers, A. E. E., Whitney, A. R., Knight, C. A., Lundquist, G. L., Shaffer, D. B.,

Vandenburg, N. R., Pigg, J. C., Schupler, B. R., and Ronnang, B. O., 1986, 'Radio-Source Positions from VLBI', *Astron.J.*, **92**, p. 1020-1029.

Muhleman, D. O., Berge, G. L., and Jones, D., 1987, 'VLA Observations of Neptune and the Mass of Triton', *Caltech Informal Report*.

Muhleman, D. O., Berge, G. L., Rudy, D. J., Niell, A. E., Linfield, R. P., and Standish, E. M., 1985, 'Precise Position Measurements of Jupiter, Saturn and Uranus Systems with the Very Large Array', *Celestial Mech.*, **37**, p. 329-337.

Mueller, I. I, 'Reference Coordinate Systems and Frames: Concepts and Realizations', *Bull. Geod.*, **59**, p. 181-188.

Newhall, X X, Preston, R. A., and Esposito, P. B., 1986, 'Relating the JPL VLBI Reference Frame and the Planetary Ephemerides', Proceedings of the International Astronomical Union Symposium 109: *Astrometric Techniques*, (Gainesville, Florida, 1984), H. K. Eichhorn and R. J. Leacock, eds., Reidel, Dordrecht-Holland, p. 789-794.

Niell, A. E., Lockhart, T. G., and Preston, R. A., 1981, *Ap. J.*, **250**, p. 248.

Niell, A. E., Fanselow, J. L., Sovers, O. J., Thomas, J. B., Liewer, K. M., Treuhaft, R. N., and Wallace, K. S., 1986, 'Accurate Positions of 120 Radio Sources with Declinations Above -45 Degrees', Proceedings IAU Symposium No. 109: *Astrometric Techniques*, (Gainesville, Florida, 1984), H. K. Eichhorn and R. J. Leacock, eds., Reidel, Dordrecht-Holland.

Preston, R. A., Morabito, D. D., Wehrle, A. E., Jauncey, D. L., Batty, A. J., Haynes, R. F., Wright, A. E., and Nicolson, G. D., 1983, *Ap. J. (Letters)*, **268**, L23.

Purcell, G. H., Cohen, E. J., Fanselow, J. L., Rogstad, D. H., Skjerve, L. J., Spitzmesser, D. J., and Thomas, J. B., 1979, 'Current Results and Developments in Astrometric VLBI at the Jet Propulsion Laboratory', Proceedings of the International Astronomical Union Colloquium No. 48: *Modern Astrometry*, F. V. Prochazka and R. H. Tucker, eds., published by the University Observatory, Vienna; Herausgeber and Verleger: Institut fur Astronomie (Universitats-Sternwarte Wien), p. 185-194.

Ramaty, R., 1969, *Ap. J.*, **158**, p. 753.

Reasenberg, R. D., 1986, 'Microarcsecond Astrometric Interferometry', Proceedings of the International Astronomical Union Symposium 109: *Astrometric Techniques*, (Gainesville, Florida, 1984), H. K. Eichhorn and R. J. Leacock, eds., Reidel, Dordrecht-Holland.

Reid, M. and Moran, J. M., eds., 1988, Proceedings of the International Astronomical Union Symposium Number 129, The Impact of VLBI on Astrophysics and Geophysics, D. Reidel, Boston, in press.

Robertson, D. S., Fallon, F. W., and Carter, W. E., 1986, 'Celestial Reference Coordinate Systems - Submilliarcsec Precision Demonstrated with VLBI Observations', *Astron. J.*, **91**, p.

Rubincam, D., 1984, 'Postglacial Rebound Observed by LAGEOS and Effective Viscosity of the Lower Mantle', *J. Geophys Res.*, **89**, p. 1077-1087.

Ryan, J., Clark, T., Ma, C., and Gordon, D., 1988, 'Toward the Realization of a Unified Terrestrial Coordinate Frame for Geophysics', Proceedings of the International Astronomical Union Symposium No. 128, The Earth's Rotation and Reference Frames for Geodesy and Geodynamics, eds. A. Babcock and G. Wilkins, D. Reidel, Boston.

Ryle, M., and Elsmore, B., 1973, *M.N.R.A.S.*, **164**, p. 223.

Schilizzi, R. T., Norman, C. A., Van Breugel, W., and Hummel, E., 1979, *Astron. Astroph.*, **79**, L26.

Smith, D. E., Christodoulidis, D. C., Kolenkiewicz, R., Dunn, P. J., Klosko, S. M., Torrence, M. H., Fricke, S., and Blackwell, S., 1985, 'A Global Geodetic Reference Frame from LAGEOS Ranging', *J. Geophys. Res.*, **90**, B11, p. 9221-9234.

Stewart, R., Fu, L. L., and Lefebvre, M., 1986, 'Science Opportunities From the Topex/Poseidon Mission', *Jet Propulsion Laboratory Publication*, Pasadena, CA, USA, 86-18.

Standish, E. M., 1985, 'Planetary and Lunar Ephemerides, DE/25/LE125', Interoffice Memorandum 314.6-591, Jet Propulsion Laboratory, Pasadena, CA, USA.

Standish, E. M., Newhall, X X, Williams, J. G., and Dickey, J. O., 1988, 'The Reference Frame of the Ephemerides', Proceedings of the International Astronomical Union Symposium 128, The Earth's Rotation and Reference Frames for Geodesy and Geodynamics', 1986, D. Reidel, G. Wilkins and A. Babcock, editors.

Tapley, B.D., Schutz, B. E., and Eanes, R. J., 1985, 'Station Coordinates, Baselines, and Earth Rotation from LAGEOS Laser Ranging: 1976-1984', *J. Geophys. Res.*, **90**, B11, p. 9235-

Walker, R. C., Readhead, A.C.S., Seielstad, G. A., Preston, R. A., Niell, A. E., Resch, G. M., Crane, P. C., Shaffer, D. B., Geldzahler, B. J., Neff, S. G., Shapiro, I. I., Jauncey, D. L., and Nicolson, G. D., 1981, *Ap. J.*, **243**, p. 589.

White, G. L., Jauncey, D. L. and Preston, R. A., 1986, 'A Radio-Optical Frame Tie Programme in the Southern Hemisphere', Proceedings of a colloquium on the European Astrometry Satellite HIPPARCOS - *Scientific Aspects of the Input Catalogue Preparation*, Aussois, 3-7 June 1985 (ESA SP-234).

Williams, J. G., Newhall, X X, and Dickey, J. O., 1985, 'The Coordinate Frame of the Lunar Laser Ranging Network', Report on the MERIT-COTES Campaign on Earth Rotation and Reference Systems, Part II: *Proceedings of the International Conference on Earth Rotation and the Terrestrial Reference Frame*, ed. I. Mueller Ohio State University, **2**, p. 590-600.

Williams, J. G., Dickey, J. O., Melbourne, W. G., and Standish, E. M., 1983, 'Unification of Celestial and Terrestrial Coordinate Systems', *Proceedings of the International Association of Geodesy (IAG) Symposia*, International Union of Geodesy and Geophysics (IUGG) XVIIIth General Assembly, Hamburg, FRG, August 15-27, 1983, Ohio State University, Dept. of Geodetic Science and Surveying, Columbus, Ohio 43210, **2**, p. 12-27.

Yoder, C. F., Williams, J. G., Dickey, J. O., Schutz, B. E., Eanes, R. J., and Tapley, B. D., 1983, 'Secular Variation of Earth's Gravitational Harmonic J_2 from LAGEOS and the Nontidal Acceleration of Earth Rotation', *Nature*, **303**, p. 757-762.

CURRENT INTERCOMPARISONS BETWEEN CTS's

C. BOUCHER
I.G.N., Saint Mande, France

1. MODELS AND METHODS

Due to the very nature of terrestrial reference systems, there exists a large variety of them. Their number is regularly increasing through the multiplicity of measurement techniques and data analysis.

As defined in Chapters 6 and 7, a terrestrial system is the actual implementation, in a specific data analysis, of an ideal terrestrial system, which is a quasi-cartesian system with a given origin, scale and orientation. But in order to ensure a proper realization of such a reference system, one selects a reference frame characterized by a finite set of points for which one provides coordinates as a function of time, $x(t)$.

Several aspects must be specified for such a terrestrial reference frame:

- the type of point (either reference points located on the Earth's crust such as geodetic control marks or tracking instrument reference points; or moving points such as an artificial satellite);

- the type of coordinate system associated to the quasi-cartesian reference system, such as cartesian, spherical, cylindrical, geographical, astronomical, ...

- the quality of the estimation of x.

The current terrestrial frames, as used in the geodetic or astronomical communities, are:

i)–Sets of tracking stations used in space geodesy, together with their cartesian (X, Y, Z) or geographical (λ, ϕ, h) coordinates, for which an ellipsoid has to be specified. The secular time variations

J. Kovalevsky et al. (eds.), Reference Frames, 327–348.

(as we shall discuss further below) have also to be given, either as time derivatives at a reference epoch t_0 (i.e. $\dot{X}$, $\dot{Y}$, $\dot{Z}$ or $\dot{\lambda}$, $\dot{\phi}$, $\dot{h}$), or as a sequence of estimations at various epochs t_k, usually regularly spaced - monthly or yearly values - (X_k, Y_k , Z_k or λ_k, ϕ_k, h_k).

ii)–Sets of optical instruments, with their astronomical coordinates (Λ, Φ) and their time variations.

iii)–Sets of geodetic marks with their geographical (λ, ϕ, h) or grid (E, N, h) coordinates. Here, h is obtained through geoid undulation N and orthometric height H:

$$h = N + H. \tag{1}$$

iv)–Sets of cartesian coordinates of the centre of mass of an artificial satellite, at regular epochs (ephemerides).

The case (i) is the most widely used in space geodesy. (ii) was used by the BIH up to 1983 for its system, and is also (implicitly) used by geodetic agencies for the orientation of terrestrial control networks using triangulation or traverses, through Laplace azimuths. The resulting frame is traditionally expressed as (iii). Finally, (iv) is also used, for instance, by Transit Doppler point positioning with Broadcast or Precise Ephemerides.

For all systems, the physical model adopted, either Newtonian or relativistic, enables us to define the concept of origin, scale and orientation. The general relationship between two quasi-cartesian systems $(X_k^{(i)})$ $k = 1,2,3$ and $i = 1,2$ is therefore

$$X_k^{(2)} = X_k^{(1)} + T_k^{(1,2)} + Q_{k,l}^{(1,2)} X_l^{(1)} + \delta X_k^{(1,2)} \tag{2}$$

where

$T_k^{(1,2)}$ are the translation components,

$$Q_{k,l}^{(1,2)} = \begin{bmatrix} D^{(1,2)} & -\varepsilon_3^{(1,2)} & \varepsilon_2^{(1,2)} \\ \varepsilon_3^{(1,2)} & D^{(1,2)} & -\varepsilon_1^{(1,2)} \\ -\varepsilon_2^{(1,2)} & \varepsilon_1^{(1,2)} & D^{(1,2)} \end{bmatrix} \tag{3}$$

$D^{(1,2)}$ is the scale factor,

$\varepsilon_k^{(1,2)}$ are the rotation angles around axis (k),

$\delta X_k^{(1,2)}$ is a correction term to take non-linearities and relativistic terms into account.

Such a general relationship is valid for the transformation between two terrestrial reference systems, providing they are nearly geocentric and vaguely aligned on the equatorial system with a common zero meridian (so called 'Greenwich').

One has furthermore assumed a same direct orientation of the vector basis. Consequently, (2) is a first order expansion of a three-dimensional Euclidian affinity, which is the general relationship between two orthogonal, isotropic and direct affine frames.

The two systems can be:

- either two different systems at the same epoch,
- the same system at two epochs, or
- in general, two systems at two epochs.

Finally, the general relation between the two coordinate systems $x_k^{(1)}$ and $x_k^{(2)}$ will be :

$$x^{(1)} \xrightarrow{\text{(I)}} X^{(1)} \xrightarrow{\text{(II)}} X^{(2)} \xrightarrow{\text{(III)}} x^{(2)}$$

Step (I) converts the coordinate system $x^{(1)}$ into its corresponding cartesian one. (II) is the transformation seen just before and (III) converts the cartesian coordinates $X^{(2)}$ into the final coordinates.

When (I) and (III) are of the same type, for instance, geographical to cartesian, it is possible to derive a differential formula giving directly $Dx = x^{(2)} - x^{(1)}$. In the case of geographical coordinates, this is the Molodensky formula.

Depending on the way the various terrestrial systems and frames are determined, there are discrepancies in either origin, scale or orientation, which can be predicted directly from the adopted definitions.

It is important to actually determine these discrepancies, in order to compare them with the values estimated from a direct comparison of the frames, as it will be discussed below. This can show systematic errors in the estimated values, coming from bias in the coordinates of the frames and the usually non global coverage of the colocation stations used in the computation.

We have selected some examples:

a) In VLBI, the origin of the frame is defined by fixing coordinates of a station. Consequently, the origin shift of this frame, with regard to any other system, will be given by colocating an instrument belonging to the other system at this VLBI reference station, and comparing related coordinates.

b) Techniques like VLBI or LLR currently derive information from a solar system barycentric frame, in a relativistic framework. The terrestrial geocentric frame is usually obtained by a Lorentzian boost and a spatial rotation. When comparing with a local geocentric Lorentz frame, as used by dynamical satellite techniques, such as SLR, there remains a scale discrepancy coming mainly from the solar gravitation at the Earth's level, or

$$< \frac{U_\odot}{c^2} > = \frac{GM_\odot}{A c^2} = 1.5 \times 10^{-8}$$

c) In most techniques, such as VLBI or SLR, the scale is defined through the adoption of the velocity of light, c. This value is fortunately fixed in the new definition of the metre to :

$$c = 299\ 792\ 458 \text{ m s}^{-1} \qquad \text{(CIPM 1983)}$$

The previous recommended value (IUGG 1957) of 299 792 500 m s^{-1} created a scale discrepancy of

$$\frac{\Delta c}{c} = 1.40 \times 10^{-7}$$

It must be noticed that this is only true for dynamical techniques if GM is adjusted. If this parameter is held fixed, it also influences strongly the scale of the terrestrial system.

1.1. TIME DEPENDENT EFFECTS

Another important aspect has to be taken into account for a refined realization of a terrestrial system, namely the time dependent effects.

On a general basis, the cartesian coordinates of a point are functions of time $X(t)$. For the corresponding frame, it is then certainly interesting to model all or a part of the time variations of the coordinates. This is particularly feasible for geodetic or astronomical points lying on the crust. The major phenomena which provide such time variations are

- solid earth tides (30 cm),
- ocean loading (a few cm),

- tectonic plate motion (10 cm/year),
- land uplift (a cm/year),
- local deformations due to ground water changes,
- deformations due to atmospheric pression (a few cm),
- direct luni-solar effect on astronomical vertical.

Some of these effects can be modelled with a good accuracy. A review of current models can be found in Melbourne *et al.*, 1983 (MERIT Standards). Two models are of particular interest for terrestrial frames:

- solid earth tide correction for ground station positions, especially important for the vertical component. In particular (MERIT Standards p. A5-7),

$$\Delta h = -0.121 \left(\frac{3}{2} \sin^2\phi - \frac{1}{2} \right) \text{ m} \tag{4}$$

  is the permanent tidal deformation, where ϕ is the latitude of the station. It must be specified if this correction is applied.

- tectonic plate motion correction, acting on horizontal components. The usual ones, such as the series of Minster-Jordan models, are defined through a set of angular velocity vectors $\bar{\Omega}_p$, one for each plate, and expressed in the terrestrial system, so that the velocity of a point of coordinates $\bar{X}$ is:

$$\dot{\bar{X}} = \bar{\Omega}_p \wedge \bar{X} \tag{5}$$

Two absolute motion models are usually adopted in data analysis:

- AM0-2, derived from the RM-2 model by applying a no global rotation condition,

- AM1-2, which minimizes the motion of a set of hot spots, also derived from RM-2 (Minster and Jordan, 1978).

AM0-2 depends only on the adopted contour of plate boundaries, whereas AM1-2 depends on the selection of the hot spots which are more subject to uncertainties. On the other hand, AM0-2 corresponds to the type of law of evolution one wants to give to terrestrial frames (see below), and has been consequently adopted by MERIT Standards (Update 1, December 1985). Nevertheless, AM1-2 leads to a system linked to the mantle which is needed to express a geopotential model without secular variations due to a residual rotation of the system. It is therefore favoured by groups which perform dynamical analysis of satellite tracking data.

In a given system (i), a ground point has therefore the following expression:

$$\overline{X}^{(i)}_{(t)} = \overline{X}^{(i)}_{0} + \dot{\overline{X}}^{(i)}_{0} (t - t_0) + \overline{L}^{(i)} (t) + \overline{\Delta X}_{tid}(t) \tag{6}$$

where

$$\dot{\overline{X}}^{(i)}_{0} = \overline{\Omega}^{(i)}_{p} \wedge \overline{X}^{(i)}_{0} + \dot{h} \frac{\overline{X}^{(i)}_{0}}{|\overline{X}^{(i)}_{0}|} \tag{7}$$

This modelling, although arbitrary, seems a good compromise between the model and stochastic components. The modelled part takes into account:

- tidal variation
- plate motion
- secular vertical motion.

As both points on the crust and reference systems are moving, it is important to understand the problems related to time evolution.

From (2), we see that if (6) is true in (i), we get in (j):

$$\overline{X}^{(j)}_{r} = \overline{X}^{(i)}_{0_r} + \dot{\overline{X}}^{(i)}_{0_r} (t - t_0) + \overline{L}^{(i)}_{r} (t) + \overline{\Delta X}_{tid}(t) + T^{(i,j)}(t)$$

$$+ Q^{(i,j)}(t)\, \overline{X}^{(i)}_{0_r} + \delta X \tag{8}$$

The assumption that (6) is also valid for (j) means that T and Q can be expanded to first order with a sufficient accuracy, and:

$$\left\{ \begin{array}{lllll} \overline{X}^{(j)}_{0_r} & = \overline{X}^{(i)}_{0_r} & + T^{(i,j)}_{0} & + Q^{(i,j)}_{0} & \overline{X}^{(i)}_{0_r} \\ \dot{\overline{X}}^{(j)}_{0_r} & = \dot{\overline{X}}^{(i)}_{0_r} & + \dot{T}^{(i,j)}_{0} & + \dot{Q}^{(i,j)}_{0} & \overline{X}^{(i)}_{0_r} \\ \overline{L}^{(j)}_{r} & = \overline{L}^{(i)}_{r} & & & \end{array} \right. \tag{9}$$

The question of the definition of the terrestrial system is then divided into two aspects:

- to define it at a reference epoch (t_0), through seven conditions for

origin, scale and orientation;
- to assess an evolution law. For this, several options are conceivable.

For the origin, either a station gets assessed coordinates (and time variations), like in VLBI or terrestrial system, or the system is modelled as geocentric, like in LLR or dynamical satellite techniques. This holds also for time evolution.

For the scale, a choice of constants (c, GM) determines it and its time evolution (no evolution).

For the orientation, one usually fixes or constrains Earth rotation parameters to a specific *a priori* series, or fixed value at a reference epoch, and applies a no net rotation condition (see below).

One could also define a terrestrial system and its evolution through unambiguous statements, such as principal axes of momenta for the origin (e.g. selected at geocenter) ....

But these definitions are hardly operational and sensitive to the redistribution of masses within the Earth.

1.2. TISSERAND MODEL

A rather tempting approach is the concept of Tisserand axes. Such axes are defined by minimizing the kinetic energy. They are characterized by null linear and angular momentums.

Let $\vec{X}$ and $\vec{V}$ be position and velocity related to an external frame, and $\vec{x}$, $\vec{v}$ for a terrestrial frame:

$$\begin{cases} \vec{X} = \vec{r}_0 + \vec{x} \\ \vec{V} = \vec{V}_0 + \vec{v} + \vec{\omega} \wedge \vec{x} \end{cases} \qquad (10)$$

The kinetic energy is:

$$T = \frac{1}{2} \int_C v^2 \, dm \qquad (11)$$

whereas the linear momentum is

$$\vec{p} = \int_C \vec{v} \, dm \qquad (12)$$

and the angular momentum:

$$\vec{h} = \int_C \vec{x} \wedge \vec{v}\, dm \tag{13}$$

If we apply a variational expression to the terrestrial frame, we get:

$$\begin{cases} \vec{0} = \delta\vec{r}_0 + \delta\vec{x} \\ \dot{\vec{0}} = \delta\vec{V}_0 + \delta\vec{v} + \delta\vec{\omega} \wedge \vec{x} + \vec{\omega} \wedge \delta\vec{x} \end{cases} \tag{14}$$

whence:

$$\delta T = \int_C \vec{v} \bullet \delta\vec{v}\, dm = -\vec{p} \bullet \delta\vec{V}_0 - \vec{h} \bullet \delta\vec{\omega} + \vec{p} \bullet (\vec{\omega} \wedge \delta\vec{r}_0) \tag{15}$$

so that

$$\delta T = 0 \Leftrightarrow \vec{p} = \vec{h} = \vec{0}. \tag{16}$$

One has some freedom to select the domain of integration (C). Usually, one chooses the crust (Smith, 1980).

A Tisserand terrestrial system is defined by its position at a reference epoch and the following laws, from (6).

$$\int_C (\dot{\bar{X}}_0^{(i)} + \frac{d\bar{L}^{(i)}}{dt}) dm = \bar{O} \tag{17}$$

$$\int_C \bar{X}_0^{(i)} \wedge (\dot{\bar{X}}_0^{(i)} + \frac{d\bar{L}^{(i)}}{dt})\, dm = \bar{O} \tag{18}$$

the tidal terms vanishing.

If we apply (7), we get:

$$\int_C \bar{\Omega}_p \wedge \bar{X}_0^{(i)}\, dm + \int_C \dot{h} \frac{\bar{X}_0^{(i)}}{|\bar{X}_0^{(i)}|}\, dm + \int_C \frac{d\bar{L}^{(i)}}{dt}\, dm = \bar{O} \tag{19}$$

$$\int_C \bar{X}_0^{(i)} \wedge (\bar{\Omega}_p \wedge \bar{X}_0^{(i)})\, dm + \int_C \bar{X}_0^{(i)} \wedge \frac{d\bar{L}^{(i)}}{dt}\, dm = \bar{O} \tag{20}$$

1.3. PRACTICAL REALIZATION

Analysis centres which process various types of data (either one specific type, or a combination) not only have to select models and constants, but also the parameters which are solved for in the (usually least squares) estimator. A variety of choices has been done by these centres.

Among the numerous possibilities which exist regarding the adoption of such models for station positions, we shall only mention four possibilities which are actually used:

(i)–solved-for constant positions: $\bar{X}_0$

(ii)–solved-for position and velocity at a reference epoch t_0:

$$\bar{X}_0 + \dot{\bar{X}}_0 \, (t - t_0)$$

(iii)–solved-for position at t_0, plus correction models:

$$\bar{X}_0 + \bar{\Omega}_p \wedge \bar{X}_0(t - t_0) + \overline{\Delta X}_{tid}(t)$$

(iv)–values solved-for at regular epochs (dayly, monthly, yearly) as step functions:

$$\bar{X}_k = \bar{X}_0 + \bar{L}_k \qquad t \in [t_k, t_{k+1}] \qquad t_{k+1} = t_k + \Delta t$$

In the cases (i) to (iii), the system and its time evolution are fixed by constraints on origin, scale and orientation as seen before. In the case (iv), it can be done in the same way or through other constraints, such as a discretized Tisserand condition:

$$\left\{ \begin{array}{l} \sum\limits_r m_r \, (\bar{X}_{k+1,r} - \bar{X}_{k,r}) = 0 \\ \sum\limits_r m_r \, \bar{X}_{k,r} \wedge (\bar{X}_{k+1,r} - \bar{X}_{k,r}) = 0 \end{array} \right. \tag{21}$$

The main remaining problem is to estimate reliable and accurate transformation parameters between systems. Even if they can be totally or partially derived from a priori considerations, it will be desirable to compute through a (usually least squares) estimation the numerical value of the (usually) seven transformation parameters occuring into (2), taken as model.

This will lead to a comparison of two systems or a combination of several, including the estimation of their relationships (see Boucher,

TABLE 1

Observational techniques	Observables	Number of stations
VLBI	Delay or delay rates of radiowaves coming from extragalactic sources	about 30
SLR	Laser range data of such satellites as LAGEOS, Starlette, AJISAI	several tens
LLR	Laser range data of 4 lunar retro-reflectors	4
Doppler	Doppler shift of radioelectric signals of several NNSS or TRANSIT satellites and GEOS 3, SEASAT 1, GEOSAT	about 35
GPS	Phase measurements on radiowaves transmitted by 18 satellites	about 20

TABLE 2

Terrestrial system defined by space techniques

Origin	Geocentric defined by satellite and lunar ephemerides
Scale	Defined by adopting c and GM
Orientation	Defined by pole coordinates and UT at given epoch

Feissel, 1984) using available frames and colocation.

2. SOURCES OF DATA

Space and terrestrial techniques provide a large number of types of data for the analysis of which it is necessary to select a terrestrial system. Such an analysis can be performed on a single kind of data, or on a combination of several (see tables 1 and 2).

This part reviews the main techniques and some combination efforts, concentrating, for each of them, on the characteristics of the associated terrestrial systems, on the currently available frames as determined by various analysis centres (see table 3).

2.1. VERY LONG BASELINE INTERFEROMETRY

Very Long Baseline Interferometry (VLBI), in its geodetic use, provides measurements of relative propagation delays received by a network (at least two) of ground radiotelescopes, of a radio signal coming from a compact extragalactic radiosource.

In the complex analysis of the data, a terrestrial system has to be determined. Currently, the analysis centres apply the following choices:

- the origin is arbitrarily fixed by adopting coordinates for a specific radio-telescope. A good external knowledge of them enables the system to be closely geocentric (to a few metres). If no crustal motion correction is applied, the origin will be slightly moving, together with the adopted fixed station, with regard to the geocenter (horizontal and vertical motions);

- the scale is entirely fixed by the choice of the velocity in vacuum of electromagnetic waves, *c*. One must also notice that, depending on the application of the suitable relativistic correction or not, two current choices of scales exist (differing by about 1.5×10^{-8});

- the orientation is defined through Earth rotation parameters. If they are adjusted, values of the three parameters (pole and UT1-UTC) will be held fixed at a given epoch.

Many groups currently perform VLBI data analysis. The main organizations are:

Harvard Center for Astrophysics, Harvard, USA

TABLE 3

Centers of analysis of different types of observational data

Analysis Centers	VLBI	SLR	LLR	Dop.	GPS
Harvard Center for Astrophysics, Cambridge, USA	*				
Jet Propulsion Laboratory - JPL Pasadena, USA	*		*		*
NASA Goddard Space Flight Center - GSFC, Greenbelt, USA	*	*		*	
National Geodetic Survey - NGS, Rockville, USA	*				
Center for Space Research - CSR, Austin, USA		*			*
Deutsches Geodatisches Forschungs-institut - DGFI, Munich, FRG		*			
Delft University, Netherlands		*			
Centre d'Etudes et de Recherches Géodynamiques et Astrophysiques - CERGA, Grasse, France			*		
MIT, Cambridge, USA			*		*
Naval Surface Weapon Center, Dahlgren, USA - NSWC				*	*
Defence Mapping Agency Hydrogaphic Topographic Center, Washington, USA - DMAHTC				*	*
Groupe de Recherche de Géodésie Spatiale, Toulouse, France - GRGS				*	
DoD for "Broadcast Ephemerides"					*

National Geodetic Survey, Rockville, USA (NGS)
NASA Goddard Space Flight Center, Greenbelt, USA (GSFC)
Jet Propulsion Laboratory, Pasadena, USA (JPL)

NGS is leading the IRIS/POLARIS project, GSFC the Crustal Dynamics program, and JPL the DSN VLBI program (in particular TEMPO).

2.2. SATELLITE LASER RANGING

Satellite laser ranging (SLR) provides laser range data from several tens of ground systems to properly equipped satellites such as STARLETTE, AJIKAIE, and especially LAGEOS.

Like all dynamical analysis of satellite tracking data, the terrestrial system is defined so:

- the origin is at the geocenter, by putting to zero the three first coefficients of the spherical harmonic expansion of the geopotential;

- the scale is fixed by adopting c and the geogravitational constant GM (if not adjusted). Again, the selection of the relativistic scale (barycentric or geocentric) has to be done;

- the orientation is fixed by three conditions, depending on the adjusted parameters (two latitudes and one longitude of station, or pole coordinates and length of day at a given epoch ...).

Again, several groups currently analyze SLR data:

Center for Space Research, Austin, USA (CSP)
NASA GSFC, Greenbelt, USA
DGFI, Munich, FRG
Delft University, Netherlands.

2.3. LUNAR LASER RANGING

Some laser ranging ground instruments have the capability to reach the Moon where retroreflectors have been set up by US and USSR lunar missions.

In the analysis of such data, the terrestrial reference system has:
- its origin at the geocenter, defined by the adopted lunar ephemerides;
- its scale defined by c, and again the choice in relativistic corrections. (The ephemerides are currently in the solar system

barycentric frame);
- its orientation.

Still a few analysis centers exist:
JPL, Pasadena, USA
CERGA, Grasse, France
MIT, Cambridge, USA.

2.4. DOPPLER

Measurement of the Doppler shift on radioelectric signal has been used from the early sixties for the tracking of satellites.

The major work was done on a continuous basis from 1967 with the series of Transit satellites, or Navy Navigation Satellite System (currently five or six such satellites are available).

Both permanent tracking networks (OPNET, TRANET, MEDOC) and mobile ground receivers provide Doppler data on Transit satellites, and also on special satellites (such as GEOS-3, SEASAT-1 or GEOSAT).

Characteristics of the terrestrial system are fully identical to SLR.

Current data analysis centres are:
Naval Surface Weapon Center, Dahlgren, USA (NSWC)
Defence Mapping Agency Hydrographic Topographic Center, Washington, USA (DMAHTC)
Groupe de Recherche de Géodésie Spatiale, Toulouse, France (GRGS)

These centres are primarily involved in the analysis of Tranet or MEDOC data for Transit.

Many other groups have performed analysis of Doppler data on a temporary basis for mobile receivers. Usually, they use orbits derived by DMAHTC or GRGS to determine station coordinates. Such a process typically involves three steps of realizations for the same terrestrial system:
1) primary: coordinates of the tracking stations;
2) secondary: satellite ephemerides;
3) tertiary: coordinates of mobile receivers.

NSWC, GSFC or GRGS have also performed analysis related to GEOS-3 or SEASAT-1 satellites.

2.5. GLOBAL POSITIONING SYSTEM

The Global Positioning System is a set of 18 satellites to be fully operational in 1991. They will be used very widely for geodetic and geophysical positioning, and also for orbitography of low satellites (e.g. TOPEX).

The general comments concerning reference systems are similar to Doppler.

Several analysis centers already compute ephemerides:

- DoD for Broadcast Ephemerides
- NSWC, Dahlgren, USA
- CSR, Austin, USA.

2.6. FORTHCOMING SYSTEMS

New satellite systems are under development, mainly ERS-1 (with SLR and a German radiotracking system, PRARE) and TOPEX/POSEIDON (with SLR, Tranet, GPS and a French radiotracking system, DORIS). Both satellites will fly a radar altimeter. A well defined and realized terrestrial system is useful, in order to provide an accurate ephemeris and a set of ground points (tide gauges) with ellipsoidal height of mean sea level. Both types of data expressed in the same terrestrial system will enable to determine an accurate sea surface topography.

The knowledge of the relationship between the reference system underlying the various analysis of the various radar altimeter missions, including past ones (GEOS-3, SEASAT, GEOSAT) will also enable fruitful comparisons and combinations.

2.7. OPTICAL ASTROMETRY

The analysis of data issued from global networks of optical astrometric instruments, such as done by IPMS or BIH, also require a terrestrial system. The frames were characterized by astronomical coordinates.

In particular, the BIH has published mean and annual values in its Annual Report for many years (see Boucher and Feissel, 1984). Unfortunately, these values were strongly affected by systematic errors (catalogue, refraction, etc ...).

2.8. TERRESTRIAL DATUMS

From several centuries, geodesists have established, on a local, national or continental scale, horizontal control networks. They have computed coordinates more or less rigorously and progressively, using terrestrial measurements of angles or distances, and astronomical azimuths.

For each of them, there is an underlying terrestrial reference system. Its origin is fixed through the adoption of a fundamental station. The scale is fixed by invar and/or electronic distances. In fact, this is only a planimetric scale. If one wants to express terrestrial coordinates as three-dimensional, one has to estimate also the ellipsoidal height of each station, or equivalently the geometric distance of each station to the origin. If any systematic error exists in these quantities, it will create a scale bias. Consequently, the three-dimensional scale of a terrestrial datum is basically hybrid.

The orientation is ensured by a set of Laplace stations (see Chapter 6).

Examples of continental networks are :

North American Datum (NAD 27, NAD 83)
European Datum (ED 50, ED 79, ED 87)
Australian Datum (AGD 66, GMA 80).

2.9. COMBINATIONS

Some groups are performing combination of data of various types, for three purposes:

- global gravity field determination, typically by
  NASA/GSFC (GEM series)
  DGFI and GRGS (GRIM series)
- global geometric and physical geodetic model (World Geodetic System), by the US Department of Defence (DoD WGS 60, 66, 72, 84)
- global terrestrial system for monitoring Earth rotation, by BIH since 1984 (BTS)

2.10. LOCAL SURVEYS

For comparing and combining systems through colocations of instruments, it is necessary to have the three-dimensional vectors which connect these points.

This can be obtained by local surveys or by GPS in relative mode (other systems in the future could also provide such data). Not only the existence, but also the accuracy and reliability of such data are of a primary importance.

3. CURRENT RESULTS

Intercomparison activities have been developed according to several ways. The main aspects are:

- Relations between global systems, especially in the efforts to combine various sources (NGS - Hothem *et al.*, 1982, or BIH - Boucher, Feissel, 1984). Seven parameters transformations are usually considered.
- Relations between global and continental networks, such as NAD, ED. Here again seven parameters can be applied, although only three translation components are often used (see table 6). Except for most recent adjustments, the existing frames associated to continental systems are contaminated by large systematic errors (several tens of metres). Consequently, the value of the estimated transformation parameters (3 or 7) is strongly influenced by the selection of the points used for the least square adjustment.
- Relations between the global and local datums, usually defined by a translation. Such a formula is valid as an average of the distorsion over a given area, either the whole coverage of the local datum, or only a specific area, in order to have a better fit when distorsions are large.

Table 4 gives a summary of major systems involved in such activities. Current results are also presented in table 5.

Several agencies have established such values, especially DMA for WGS 72 and WGS 84. Such data are currently used to convert between Doppler or GPS derived positions and local coordinates. Examples are presented in Table 6.

TABLE 4

Name	Extension	Spheroid	Realizations
NSWC 9Z2	World	NWL 8	Tranet tracking; Transit Precise Ephemerides point positions
WGS 72	World	WGS 72	US military products; GPS
WGS 84	World	GRS 80	US military products; GPS
BTS 86	World	GRS 80	BIH frame
SSC (NGS) 87R01	World		NGS VLBI frame
SSC (JPL) 83R05	World		JPL VLBI frame
SSC (CSR) 86L01	World		CSR Lageos laser frame
SSC (JPL) 87M01	World		JPL Lunar laser frame
SSC (GSFC) 87R01	World		GSFC VLBI frame
SSC (DGFI) 87L02	World		DGFI Lageos laser frame
NAD 27	North America	Clarke 1866	Control network
NAD 83	North America	GRS 80	Control network
ED 50	Europe	International	Control network
ED 79	Europe	International	Control network

TABLE 5

1	System 2	TX m	TY m	TZ m	D 10^{-6}	ϵ_x "	ϵ_y "	ϵ_z "	
NWL 9D	WGS 72	0	0	0	-0.83	0	0	0.26	1
WGS 72	WGS 84	0	0	4.5	0.2198	0	0	0.554	1
NWL 9D	BTS 86	0.167	0.212	4.314	-0.571	-0.036	-0.008	0.795	2
BTS 86	SSC (NGS) 87R01	0.108	0.125	0.058	-0.024	-0.005	0.010	-0.002	2
BTS 86	SSC (JPL) 83R05	0.055	-0.276	-0.158	-0.016	-0.005	-0.003	0.005	2
BTS 86	SSC (CSR) 86L01	0.0	0.0	0.0	0.0	-0.005	0.003	-0.004	2
BTS 86	SSC (JPL) 87M01	0.0	0.0	0.0	-0.022	0.002	0.004	-0.011	2
BTS 86	SSC(GSFC) 87R01	1.535	-0.961	0.450	-0.027	-0.004	0.010	0.010	2
BTS 86	SSC(DGFI) 87L01	0.003	-0.033	0.062	0.013	0.008	-0.006	0.121	2
BTS 86	ED 79	87.7	91.9	119.4	-0.72	-0.35	-0.06	-0.19	3

Remarks

(1) DMA Source
(2) BIH Annual Report for 1986
(3) IGN test computation for RETRIG

TABLE 6

Current datum shift parameters from local datum to WGS 84
(DMA source, 1987)

Datum	TX(m)	TY(m)	TZ(m)
Adindan	-162	- 12	206
AFG	- 43	-163	45
Ain El Abd 1970	-150	-251	- 2
Anna 1 Astro 1965	-491	- 22	435
Arc 1950	-143	- 90	-294
Arc 1960	-160	- 8	-300
Ascension Island 1958	-207	107	52
Astro Beacon "E"	145	75	-272
Astro B4 Sor. Atoll	114	-116	-333
Astro Pos 71/4	-320	550	-494
Astronomic Station 1952	124	-234	- 25
Australian Geodetic 1966	-133	- 48	148
Australian Geodetic 1984	-134	- 48	149
Bellevue (IGN)	-127	-769	472
Bermuda 1957	- 73	213	296
Bogota Observatory	307	304	-318
Campo Inchauspe	-148	136	90
Canton Island 1966	298	-304	-375
Cape	-136	-108	-292
Cape Canaveral	- 2	150	181
Carthage	-263	6	431
Chatham 1971	175	- 38	113
Chua Astro	-134	229	- 29
Corrego Alegre	-206	172	- 6
Djakarta (Batavia)	-377	681	- 50
DOS 1968	230	-199	-752
Easter Island 1967	211	147	111
European 1950	- 87	- 98	-121
European 1979	- 86	- 98	-119
Gandajika Base	-133	-321	50
Geodetic Datum 1949	84	- 22	209
Guam 1963	-100	-248	259
Gux 1 Astro	252	-209	-751
Hjorsey 1955	- 73	46	- 86
Hong-Kong 1963	-156	-271	-189
Indian	214	836	303
Ireland 1965	506	-122	611
Ists 073 Astro 1969	208	-435	-229
Johnston Island 1961	191	- 77	-204
Kandawala	- 97	787	86
Kerguelen Island	145	-187	103
Kertau 1948	- 11	851	5
La Réunion	94	-948	-1262
L.C. 5 Astro	42	124	147
Liberia 1964	- 90	40	88
Luzon	-133	- 77	- 51
Mahe 1971	41	-220	-134
Marco Astro	-289	-124	60

Massawa	639	405	60
Merchich	31	146	47
Midway Astro 1961	912	- 58	1227
Minna	- 92	- 93	122
Nahrwan	-247	-148	369
Namibia	616	97	-251
Naparima, BWI	- 2	374	172
North American 1927	- 8	160	176
North American 1983	0	0	0
Observatorio 1966	-425	-169	81
Old Egyptian 1930	-130	110	- 13
Old Hawaiian	61	-285	-181
Oman	-346	- 1	224
Ordnance Surveay of Great Britain 1936	375	-111	431
Pico de las Nieves	-307	- 92	127
Pitcairn Astro 1967	185	165	42
Provisional South Chilean 1963	16	196	93
Provisional South American 1956	-288	175	-376
Puerto Rico	11	72	-101
Qatar National	-128	-283	22
Qornoq	164	138	-189
Rome 1940	-225	- 65	9
Santa Braz	-203	141	53
Santo (Dos)	170	42	84
Sapper Hill 1943	-355	16	74
South American 1969	- 57	1	- 41
South Asia	7	- 10	- 26
Southeast Base	-499	-249	314
Southwest Base	-104	167	- 38
Timbalai 1948	-689	691	- 46
Tokyo	-128	481	664
Tristan Astro 1968	-632	438	-609
Viti Levu 1916	51	391	- 36
Wake-Eniwetok 1960	101	52	- 39
Zanderij	-265	120	-358

REFERENCES

Boucher, C. and Feissel, M., 1984, Realization of the BIH terrestrial system - *International Symposium on Space Techniques for Geodynamics*, Sopron, Hungary, 9-13 July 1984

Dickey, J.O. *et al.*, 1986, Reference frames: determinations and connections - *IUA, Symp.* 128, Coolfont

Hothem, L.D., Vincenty, T., and Moose, R.E., 1982, Relationship between Doppler and other advanced geodetic system measurements based on global data - *International Geodetic Symposium on Satellite Doppler Positioning*, Las Cruces, USA, 1982

Hothem, L.D., Vincenty, T., and Hoyle, D.B., 1983, Analyses of Doppler, Satellite Laser, VLBI and terrestrial coordinate systems - *IUGG/IAG XVIIIth General Assembly*, Hamburg, FRG, August 1983

Melbourne, W. *et al.*, 1983, Project MERIT Standards - *USNO, Circular n°* 167, Washington

Minster, J.B. and Jordan, T.H., 1987, Present day plate motions - *J. Geophys. Res.* 83, 5331-5354

Smith, M.L., 1980, The theoretical description of the nutation of the Earth - *IAU Coll.* 56, Varsaw

PART 5

TIME

GENERAL PRINCIPLES OF THE MEASURE OF TIME: ASTRONOMICAL TIME

B. GUINOT
BIPM, Pavillon de Breteuil, 92310 Sèvres, France

1. INTRODUCTION

The realization of a conventional frame of reference in the time dimension, which will be called a time scale, requires the same phases as the reference frame in space: conception, definition, and realization.

1.1. CONCEPTION

The initial concept is expressed by the statement of a principle or by an axiom which indicates how *an ideal time scale* could be obtained. But we have first to know what is understood by an ideal time scale! To say that it is the uniform time is not satisfactory, because the intuitive notion of uniformity cannot be clarified without some pre-existing concept. Not only we do not know what time is, but the very foundation of the measure of time escapes rational explanation and is truly intuitive. The measure of time is nevertheless the most precise metrology and the success of the models of the physical world is a sufficient justification.

The common perception that the complete location of an event requires three space coordinates and one time coordinate is a convenient starting point for our activity on the references. In particular we admit that there is only one time dimension. By virtue of this preliminary concept, two ideal time scales based on different concepts may diverge, but they are not independent. There is a link between them including possible space and time coordinates, although this link may be unknown. Precisely, the comparison of time scales based on different concepts may offer the possibility of extending our cosmological knowledge.

J. Kovalevsky et al. (eds.), Reference Frames, 351–377.

1.2. DEFINITION

The definition of a time scale is the application of the underlying concept. It requires two sub-phases. First a particular physical process must be selected. Then the definition must contain all the informations needed so that no ambiguity remains when the time scale is realized from the observations of the physical process. In other terms, if there are no observational uncertainties, a good definition should lead to a unique time scale, whatever be the required accuracy.

The ideal time of the definition is often designated by the word *Time* with a qualifier: Universal Time, Ephemeris Time, etc. The usual acronyms UT1, ET, etc. should refer to these ideal times.

1.3. REALIZATION

The experiments which strictly conform to the requirements of the definition give *primary* realized time scales. However these realizations are not equivalent on account of experimental errors and possible gaps in the definition which may be filled arbitrarily by the observers. As it matters to have a unique time scale realizing the definition of a particular time, it is necessary to agree conventionally on a single realization (or a single synthesis of various realizations). A correct notation of the realization of the time T is T(xxx) where xxx is an indentifier of the measurement or of the synthesis. For instance the UT1 time scale derived by the International Earth Rotation Service should be UT1 (IERS). Such a notation is already used for the atomic times, in accordance with the recommendation of international bodies. We suggest extending it to other times.

2. GENERAL REMARKS ON TIME SCALES

2.1. STRUCTURE OF A TIME SCALE

A time scale, although it is considered as "realized", is immaterial. It consists in a procedure of assigning *dates* to events.

In most cases the procedure uses a concrete evolutionary system called a *clock* (in a broad sense), conceived or selected so that to each state of the system it is possible to associate a number which is the corresponding date. Thus it is the ensemble of states and associated dates which forms the time scale. This ensemble may be continuous, which happens when the clock is, for instance, an apparent celestial motion (then the date is associated to a dimensional measure, usually angular). The ensemble of dates may also be discontinuous: pulses of atomic clocks, of time signals, etc.; in that case one must use interpolation to date an event (intervallometer).

But there are also time scales which are the result of calculation and which are not associated to any real device. This is the case of synthesis of individual time scales, such as the averaged UT1(IERS) mentioned above or the International Atomic Time TAI. It is useful to assume that the scale is the output of a *fictitious clock*, the behaviour of which can be characterized (reliability, stability, accuracy), as it is usually done for a real clock. As the datation of an event requires a real clock, the time scale is available through the knowledge of corrections, *clock corrections*, to be added to the readings of this clock.

The notion of date introduces an important metrological criterion: the ultimate *accuracy of reading* of a time scale. It is characterized by the one-sigma uncertainty on the date of an instantaneous event, when using the best possible means.

2.2. THE CONCEPTS OF THE MEASURE OF TIME

There are two ways which have been used for producing scientific time scales:

(i)–the addition of standard time intervals leading to *integrated time scales*;

(ii)–the use of dynamical theories leading to the *dynamical time scales*.

2.2.1. Integrated time scales

We consider two well identified states of a physical process. We can conceive that the duration between these states remains unchanged when the physical process is repeated under the same conditions and that it provides us with a standard of time interval or duration standard. In order to obtain a time scale, the duration standard can be indefinitely added without dead time: thus we get an *integrated time scale*.

This method, which is often qualified as intuitive, empirical or natural, can be applied with any type of phenomenon. It is often employed with dynamical phenomena for which there are no precise dynamical models, but which are assumed to be reproducible or periodic, in the construction of artificial clocks: pendulum clocks, crystal clocks, etc. However, for the realization of reference time scales, only two applications must be considered:

(i)–the solar time, as long as the invariability of the mean duration of the day, then of the speed of rotation of the Earth, was a dogma, i.e. until Euler's theory (1736);

(ii)–the atomic time based on quantum physics.

The defects of the integrated times are of conceptual and practical nature.

The flaw in the concept is that it is never possible to reproduce completely an experiment under the same conditions: at least one condition has changed, the time has flown and the universe has pursued its evolution. Is there a general aging of all the laws of physics and of ourselves? The question matters only in so far as the aging can be different for different types of phenomena and as all the models of the physics, with their constants and their laws, cannot be translated in time. Although such a differential aging still escapes our measurements, we cannot admit a priori that it does not exist. For instance, we have to keep in mind that the time implied by quantum physics may diverge from the time of dynamics.

From the practical side, important properties of an integrated time scale derive from the fact that the basic data is a duration taken as a standard (or equivalently a frequency). The indefinite addition of the standard duration (or the frequency integration) which realizes the time scale makes free the choice of the *origin*. This origin is only realized in a practical way by the very existence of the realized time scale, i.e. by the rules adopted for assigning dates. There cannot be an ideal definition of the origin.

Setting aside the arbitrary choice of the origin, the ideal integrated time should result from perfect addition of strictly reproducible standard durations. As the errors of the realization of standard durations are also indefinitely added, the realized integrated time scale diverges indefinitely from the ideal integrated time. The integration is particularly unfavourable in case of bias of the realized standard duration which generates a drift of the time scale, and of long-period fluctuations. That is the reason why it was necessary to wait until it was possible to benefit from the high stability and reproducibility of the frequency associated to atomic transitions to build an integrated time scale with a quality sufficient to study the long term phenomena that are encountered in astronomy.

By its mode of establishment, an integrated time scale gives the possibility of measuring the duration of a time interval by the difference of dates of its end and its beginning, in the unit defined by the measure attributed to the duration taken as a standard. Thus the integrated time scale makes permanently available the unit of time interval (or *unit of time*), without degradation.

2.2.2. Dynamical time scales

A dynamical theory is a mathematical model which represents and predicts the successive configurations of a material system as a function of a time parameter t. The only condition that has to be fulfilled a priori by the parameter t is that it be associated in a univocal mode to the various configurations of the system. However, one wishes that the theory could be based on a small number of *principles* which are as simple as possible. Though this criterion of simplicity may pose some interesting and unsolved metaphysical questions about the appropriateness of mathematics to physics, it is nevertheless efficient. "How then time is defined? Time is defined so that motion looks simple!" (Missner *et al.*, 1973, p. 23). We call *dynamical time* this parameter t of "simple" theories. (The International Astronomical Union has defined a Terrestrial Dynamical Time and Barycentric Dynamical Time with different meanings for the adjective 'dynamical'; these times will be considered later.)

At this point, we can illustrate by an example how a dynamical time scale can be established. The law of inertia of classical mechanics (probably due to Descartes (1596-1650), but included as one of the three laws of motion of Newton (1642-1727)) states that:

> "If a particle is far removed from the influence of all other particles in the universe, the particle will move with a constant velocity with respect to an inertial frame."

This law establishes a linear relationship between space coordinates and the parameter t in an inertial frame. It provides the means of measuring the dynamical time t, if we can sufficiently well approximate the conditions stated in the law, for instance by measuring the displacement of a star due to its proper motion!

In a more realistic application of the concept of dynamical time, the phases of the definition of some particular dynamical time are

(i)–the choice of a theory (Newtonian mechanics, general relativity, etc.),
(ii)–the choice of a physical system to which the theory is applied,
(iii)–the analysis of interactions in the system,
(iv)–the application of the theory (mathematics) and the introduction of observed initial conditions, which lead to an *ephemeris*.

The ephemeris gives the observable configuration of the system (or space coordinates) as a function of the dynamical time implied by the theory. Thus the realization of this time, the dynamical time scale, consists in observing at some instant the configuration of the system. The ephemeris gives, for this instant, the reading of the dynamical time scale. We have transposed a measure of time into a dimensional measure.

The transit from the concept to the definition involves many sources of defects: imperfection of the theory, incomplete knowledge of the actions, truncation of integrations, errors in the initial conditions, etc.. However, when the definition is adopted, the error of the realization of the time scale depends only on the error of the dimensional measure. It is limited and usually decreases as a result of technical improvements, contrary to what happens for the integrated times.

A dynamical time appears in the first place as a system to assign a coordinate in the time dimension. Can it be used to define a time interval unit? By analogy with what has been done for the integrated times, we call the duration of an interval measured by a dynamical time scale the difference of dates of its end and beginning. But, when this evaluation of the duration is applied to an experiment identically repeated at different dates, it is not a priori evident that the duration will be found constant. The variation of the duration might be due to obvious reasons such as defects of realization of the time scale, but also to conceptual reasons. That is why the rigorous definition of a time (interval) unit derived from a dynamical time must be the duration *between two specified dates* of this time. We find an example of these subtleties with the Ephemeris Time. In contrast with the integrated times, the primary data of a dynamical time scale is the scale itself; the unit of time is obtained from a derivation in the mathematical sense.

A dynamical time scale may depend on the realization of the conventional reference system in space. But, although it has not been done in practice, it could also be realized by relative measurements independent of the space references, such as angular distances between selected celestial bodies.

2.2.3. Relations between concepts

The method of establishing integrated time scales is more convenient for laboratory work and thus to obtain proper times of the laboratories. The only dynamical systems which can lead to sufficiently good dynamical time scales are the planetary systems; the theory requires coordinate-times. These matters are considered in Chapter 17.

But these relativistic distinctions are not related to the concepts. When they have been taken into account, no divergence between the integrated time scales and the dynamical time scales has been found, so far, which cannot be explained by realization errors.

The fundamental reason of this agreement is the simplicity criterion of the dynamical theories which requires that these theories be invariant with time: the same dynamical process, when it is repeated, should have the same duration, as measured with the time scale implicit in the theory. Historically, the dynamical theories have been constructed so that they agree with the measure of time based on the recurrence of the days, which was already a high quality integrated time scale.

Another and more fundamental problem is the agreement of the time scales given by quantum physics and the dynamics time invariant theories. Is this agreement real? Do we have to introduce variations with time of some 'constants'? There are no universally accepted answers to these questions.

2.3. STABILITY OF TIME SCALES

We assume that there is an ideal integrated time scale. It can be considered as the output of a generator of a perfectly constant frequency, associated to an integration device (a clock). The frequency output of real oscillators has some fluctuations which are called *instabilities*. The characterization of the instabilities (or of the *stability* of the oscillator) is important for the users. Among the several methods of stability characterization, a simple one has emerged as a common language: it consists in the evaluation of the pair variance (or Allan variance) of frequency samples. By definition, the *frequency stability of a time scale* (often abbreviated as the *stability of the time scale*) is the frequency of the fictitious clock of which the time scale is the output. The notion of stability mainly developed for frequency generators can be extended to astronomical time scales; it becomes essential for the time scales provided by the pulsars.

An elementary description of the stability characterization is given thereafter. For a full treatment of the subject see, for instance, Allan (1966) and Rutman (1978).

2.3.1. Normalized frequency deviation

The oscillator generating an ideal time scale produces a purely periodic signal $V(t)$

$$V(t) = V_0 \sin 2\pi \nu_0 t$$

where V_0 and ν_0 are constants. In real cases,

$$V(t) = \left[V_0(t) + \varepsilon(t)\right] \sin \left[2\pi \nu_0 t + \phi(t)\right].$$

We will henceforth neglect the amplitude modulation $\varepsilon(t)$. $\phi(t)$ is a varying phase. The instantaneous frequency is then

$$\nu(t) = \frac{1}{2\pi}\frac{d}{dt}\left[2\pi\, \nu_{0t} + \phi(t)\right] = \nu_0 + \frac{1}{2\pi}\frac{d\phi(t)}{dt},$$

which is written as

$$\nu(t) = \nu_0 + \Delta\nu(t).$$

The frequency instability of the generator (and of the associated time scale) is described by the behaviour of $\Delta\nu(t)$. It is convenient to be free of ν_0 by defining the *normalized frequency deviation* (also called the relative frequency deviation) as

$$y(t) = \Delta\nu(t)/\nu_0.$$

This is a dimensionless parameter, usually much smaller than unity. The ideal time t in $y(t)$ can be replaced without inconvenience by a real measure of time.

The instantaneous frequency is a mathematical notion, in physics these are only mean frequencies. We assume that the $y(t)$ are averaged over very short time intervals.

The knowledge of $y(t)$ requires the reference to an ideal time scale in principle. In practice, a much more stable oscillator is convenient; if it does not exist, the comparison of several oscillators of the same quality usually provides enough information on $y(t)$.

In the behaviour of $y(t)$ as a function of t, one can identify:

(i)–some deterministic components: bias, drift, periodic variations, etc.

(ii)–a random component with null average over infinite duration.

It is often difficult to distinguish between the random components having long-term fluctuations (strongly correlated noises) and some deterministic components. One should consider a noise as deterministic only when there are good physical reasons to believe that it is such (drift of the period of a pulsar, for instance).

We will now restrict our discussion in assuming that $y(t)$ is a purely random noise, stationary, which we intend to characterize. To this end, two methods are currently used: the estimation of the spectral density of $y(t)$ (Fourier frequency domain), and the estimation of variances (time domain).

2.3.2. Spectral densities

In general, it has been found sufficient to represent the one-side spectral density of $y(t)$, $S_y(f)$ by the sum of terms

$$[S_y(f)]_\alpha = h_\alpha f^\alpha,$$

where h_α is a positive constant, f the Fourier frequency and α an integer with values ranging from −2 to +2. Due to the necessary frequency averaging, there is a Fourier frequency cut off f_h, and the power law model of the spectrum is

$$S_y(f) = \begin{cases} \sum_{\alpha=-2}^{+2} h_\alpha f^\alpha & 0 < f < f_h \\ 0 & f \geq f_h \end{cases}$$

The finite duration on which frequency measurements are made also impedes the knowledge of the low frequency part of the spectrum.

The noise corresponding to the integral powers of f have distinct physical origins and receive specific names.

The *white phase noise* (α=2) is usually due to noise added externally to the oscillator, for instance, uncertainties due to phase reading. This type of noise is found in the dynamical time scales, because, in general, the precision of their time reading is low.

The *flicker phase noise* (α=1) may be due to some non-identified and slowly variable bias in phase (or time) readings.

The *white frequency noise* (α=0), in man-made devices, is usually a noise internal to the oscillator, such as the thermal noise.

The *flicker frequency noise* (α=-1) appears in all quasi-periodic processes. Its origin is usually not well understood. It is sometimes the consequence of deterministic effects, not recognized as such, occurring sporadically.

The *random walk frequency noises* (α=-2) often arises from deterministic perturbations, not recognized as such, for instance from residual frequency drift.

2.3.3. Pair variance (Allan variance or two sample variance)

One consider samples of mean normalized frequencies computed over the duration τ

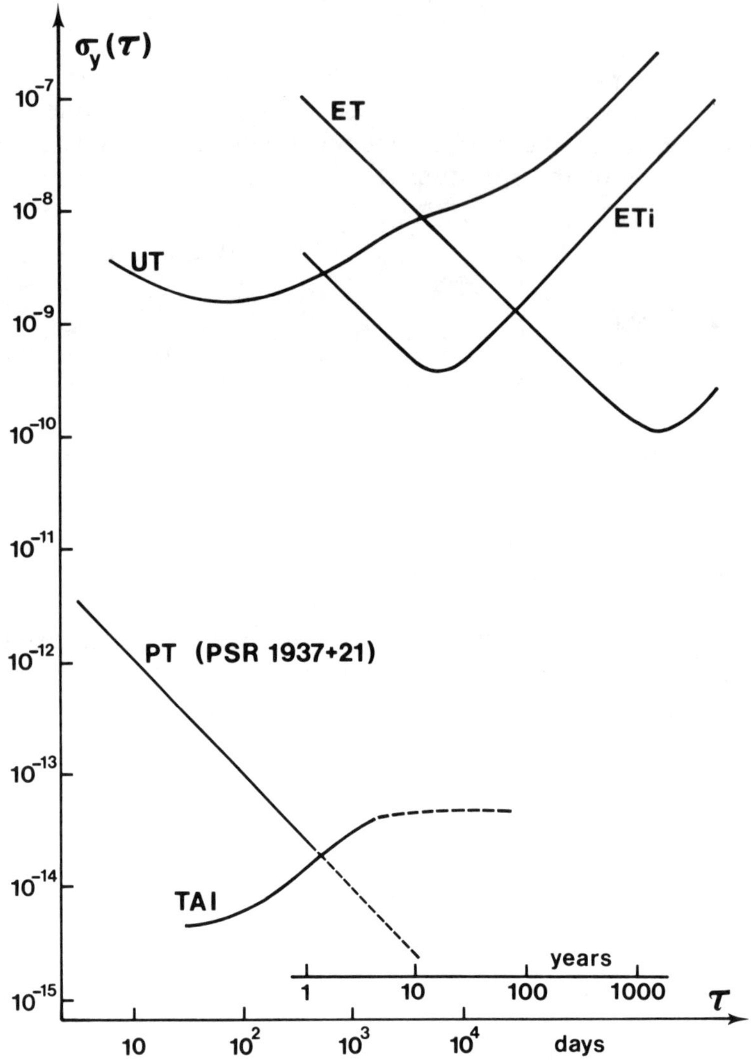

Figure 1. Stability of Universal Time UT1, Ephemeris Time ET (Sun) and ETi (Moon), Pulsar Time PT, and International Atomic Time TAI. See the text: UT1 in Section 3.1, ET and ETi in Section 3.2.5, PT in Section 3.3, TAI in Chapter 16, Section 7.

$$\bar{y}_k = \frac{1}{\tau}\int_{t_k}^{t_k+\tau} y(\theta)\, d\theta.$$

and the series of such samples, for consecutive interval s, without dead time. The pair variance is defined by

$$\sigma_y^2(\tau) = \tfrac{1}{2} < (\bar{y}_{k+1} - \bar{y}_k)^2 >$$

where < > represents a time average over infinite duration. In practice an estimator based on M samples is

$$\sigma_y^2\ (\tau) = \frac{1}{2\ (\text{M-1})} \sum_{i=1}^{M} (\bar{y}_{k+1} - \bar{y}_k)^2.$$

The pair variance is a function of τ. As in the frequency domain, the stability cannot be characterized by a single parameter.

Figure 1 and figure 1 of chapter 16 are examples of $\sigma_y(\tau)$ versus τ diagrams.

Many other types of variance have been invented in order to avoid some shortcomings of the pair variance. However, the pair variance, which is the simplest one, has the advantage of being universally used.

2.3.4. Relationship between the pair variance and the spectral density

One can demonstrate that in the regions where the spectral density can be modelled by

$$S = h_\alpha f^\alpha,$$

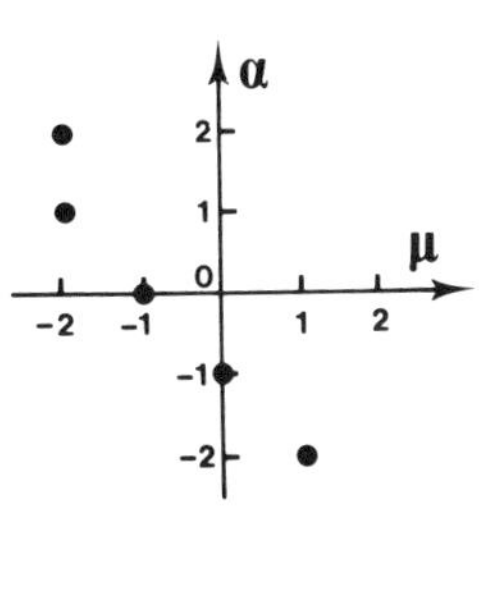

Figure 2. The α/μ diagram.

α being an integer, the variance $\sigma_y^2(\tau)$ can be represented by integral powers of τ:

$$\sigma_y^2(\tau) = K_\mu t^\mu.$$

The correspondence between μ and α is shown by figure 2. One notices that the pair variance fails to discriminate between the white phase noise and the frequency phase noise.

2.4. NOTATION OF DATES AND TIME DIFFERENCES

There is no unification of the way of expressing the date of an event. The unit of one day is often used in astronomy and space techniques; it is convenient since the readings of the various reference time scales, for the same event, differ by much less than a day so that no ambiguity is possible. The common calendar date and the julian day count (JD), with the subdivision of the day, although historically associated to the solar day, can be used with any time scale (Universal Time, Ephemeris Time, International Atomic Time), providing that the name of the time scale is clearly stated. A convenient form of the Julian day count is the Modified Julian Date:

$$\text{MJD} = \text{JD} - 2\ 400\ 000.5$$

which is widely used in time metrology and space techniques.

The same name or acronym can designate both a time scale (or a clock) and its reading at some instant. Thus one can write for an event *E*.

$$E \text{ occured at } R = \eta$$

where R is, for instance, UT1, TE, clock 872, ... and η a number.

This notation is extended to the expression of the difference of readings of time scales (or clocks) at some instant under the form

$$A - B = \xi \quad \text{at } R = \eta$$

where A and B are the designations of time scales (or clocks) and ξ a number which is the difference 'reading of A minus reading of B' at the instant η in the reference time scale R.

Example:

$$\text{UTC(USNO)} - \text{UTC(OP)} = 4.43\ \mu\text{s at UTC} = 1986 \text{ Nov. } 25,\ 0 \text{ h}.$$

TABLE I

Main dates of the development of time scales

Time scales based on the rotation of the Earth	Compromise time scale	Atomic time scales	Time scale based on the orbital motion of the Earth
Solar time **-500** The annual variation of the solar time is known ↓ Mean solar time **1884** Adoption of the Greenwich Mean Time **1925** The Universal Time UT is defined			**-400×10^{-6}** Informations on the duration of the day from paleosciences and eclipses **1630** Ephemeris Time established in retrospect from the study of old observations of the solar system
1955 Definition of UT1		**1955** First operational cesium standard. Beginnig of the atomic time scales	**1952** Definition of the Ephemeris Time by IAU
~1960 UT1 becomes an Earth rotation parameter	**1959** Coordination of some time signals using atomic frequencies		**1960** The SI second is the Ephemeris second until 1967
	1965 An approximation UTC (Coordinated Universal Time) to UT1 is linked to the BIH atomic time	**1967** The SI second becomes the atomic second	
	1972 Introduction of the leap seconds in the UTC system	**1971** The BIH atomic time becomes the official International Atomic Time TAI	**1976** Discussions on the future of Ephemeris Time. Definition of new time scales for the dynamics

The use of this so-called 'algebraic notation' is strongly recommended to avoid sign errors, especially when summing time differences: $(A-B) + (B-C) = A-C$.

2.5. MAIN DATES OF THE HISTORY OF THE MEASURE OF TIME

Some historical notes will be given with the time scales considered thereafter. Table I shows the parallel development of these time scales.

3. THE TIMES OF DYNAMICAL MODELS

3.1. THE MEAN SOLAR TIME AND THE UNIVERSAL TIME CONSIDERED AS A MEASURE OF TIME

3.1.1. Historical origin

The duration of the day, then of its annual mean were first considered a priori as a duration standard, and the associated time scales, true and mean solar times could be seen as integrated time scales.

The application by Euler, in 1736, of the principles of the Newtonian mechanics to the rotation of the Earth, by showing that the invariability of the speed of rotation was a consequence of these principles, conferred on the mean solar time the character of a dynamical time. However, even for many astronomers, the constancy of the duration of the day remained a dogma, and the second was simply defined as the duration of 1/86400 of the duration of the mean solar day, as if this duration was a standard.

In 1895 Newcomb gave a definition of the mean solar time which made it proportional to the rotation of the Earth with respect to an inertial system. Strictly speaking, the solar origin of the mean solar time was lost. But this offense to common sense was concealed by naming the 'fictitious mean Sun' the fiducial point on the equator, the hour angle of which is the mean solar time. The fictitious mean Sun does not correspond to a mean position of the Sun, on account to small secular terms. This terminology has been the source of much confusion.

D.H. Sadler (1978) has given the history of the adoption, for international use, of the mean solar time of Greenwich (GMT). The astronomical definition of the mean solar time places the beginning of the day at noon (0h, mean solar time). Complying at last with a wish expressed by the 1884 Conference on the Prime Meridian, the International Astronomical Union (IAU) decided in 1925 to match the astronomical and nautical day with the civil day beginning at midnight.

One must keep this change in mind when using data obtained prior to 1925. The confusion is all the greater since the acronym GMT was often kept for the mean solar time of the prime meridian, and for the new time advanced by 12 hours as well. The designation of the new time scale as 'Universal Time' suggested by the IAU slowly gained acceptance, at least in scientific work.

The difficulty of modelling the geophysical perturbations of the Earth rotation prevents the Universal Time from being a good realization of a dynamical time. Since the development of atomic time, it is the witness of the irregularities of the Earth rotation, a role which is emphasized by its modern definition (Aoki *et al.* 1982) which maintains, as Newcomb wished, the proportionality with the sidereal rotation of the Earth (see Part IV).

3.1.2. Forms of the Universal Time

The measurement of the Universal Time requires the knowledge of the coordinates of the pole of rotation in a terrestrial frame. When UT is referred to these variable coordinates, it is designated as UT1. Although only UT1 has a physical meaning, some other forms of UT have been defined for practical reasons: UT0 and UT2 (IAU, 1955) and UT1R (IAU, 1982).

UT0 is the form obtained with *fixed* coordinates of the pole computed from the observations when the coordinates of the pole are not known.

$$\mathrm{UT1} - \mathrm{UT0} = -(x \sin \lambda + y \cos \lambda) \tan \phi$$

ϕ being the astronomical latitude, λ the longitude (positive Eastward), x and y the coordinates of the pole (x towards the longitude 0, y towards the longitude 270°). UT0–UT1 reaches about 20 ms at mean latitude.

UT2 is a regularized form of UT1 by conventional correction of its seasonal variation, which has a peak to peak amplitude of about 60 ms.

UT1R is UT1 corrected for the effects of short period terms (up to 35 days) due to zonal tides; the correction can reach ± 3 ms.

3.1.3. Availability of Universal Time

When radio time signals appeared at the beginning of the 20th century, they were approximately disseminating the civil time of Greenwich (here called UT or UT1). A more precise access to UT was possible by using the UT times of reception of the time signals, based on observations regularly pursued in some observatories.

The need to agree on a single realization of UT led to the creation of the Bureau International de l'Heure (BIH) in 1911. Only in 1931 the BIH reference did appear based on the average of the astronomical measures of UT, under the name of 'heure définitive' which we will designate thereafter as UT(BIH) or UT1(BIH). This time scale is available as follows.

1931-1940 February, in BIH "Bulletin Horaire", Series IV, V, VI, VII, giving for some time signals
"Heure définitive des signaux horaires" ≡ UT(BIH) − emission time.

1940 February - 1955, in BIH "Bulletin Horaire", Series D, E, F, G (1 to 6), giving for most the international radio time signals
"Temps Universel − Heure définitive" ≡ UT(BIH) − emission time.

1956-1963, in BIH "Bulletin Horaire", Series G (7 to 24), H, giving for most international radio time signals
"Temps Universel 2 − Heure définitive" ≡ UT2(BIH) − emission time and UT1(BIH) − UT2(BIH).

1964-1967, in BIH "Bulletin Horaire", Series J giving
UT1(BIH) − UT2(BIH), and
(a) for the time signals following the UTC system
"TU2 déf − TUC" ≡ UT2(BIH) − UTC
E = UTC − emission time,
(b) for the other time signals
"TU2 déf − signal" ≡ UT2(BIH) − emission time.

1967-1987. UT1 should be considered as an angle representing the sidereal rotation of the Earth. Series of values referred to the atomic time are available in "BIH Annual Report" and "Circular D" (monthly)

Other realizations of UT1 by averaging were issued by the International Polar Motion Service (IPMS) and individual researchers. After 1988, the task of evaluating UT1 was given to a new service, the International Earth Rotation Service, replacing the BIH (for this domain of activity) and the IPMS.

3.1.4. The abandonment of the universal time as a measure of time

The first doubts about the invariability of the speed of rotation of the

Earth, of Kepler, Newton and Kant, were first of philosophical nature. In 1677, Flamsteed could not find any variation of the duration of the mean day by reference to pendulum clocks.

However the dogma of the uniform rotation resisted also the first scientific contradictions. As early as in 1864, Ferrel, and in 1865, Delaunay, explained the discrepancies between the lunar ephemerides and the observations by a lengthening of the duration of the day. However, in 1892, Tisserand still hesitated to change the basis of the measurement of time.

Finally, the work of Newcomb, then, during the first half of the 20th century, of Brown, de Sitter, Spencer Jones and Stoyko revealed beyond any doubt, by reference to the orbital motions and to artificial clocks, the main components of the irregularities of the Earth rotation and led to a measure of time based on the Earth orbital motion: the Ephemeris Time.

The definition of the second as the fraction 1/86400 of the mean solar day (which was traditional), was abandoned in 1960 (11th Conférence Générale des Poids et Mesures, CGPM).

The Universal Time, as a measure of uniform time, was abandoned in astronomy in the '50s. But UT remains the basis of public time in the form of the Coordinated Universal Time (UTC).

Figure 1 shows the frequency stability of UT1. In the long term the instability is dominated by the frequency drift due to the deceleration of the Earth rotation (2.3×10^{-8} per century in normalized frequency).

3.2. EPHEMERIS TIME

3.2.1. Concept and definition

The Ephemeris Time ET is a dynamical time in Newtonian mechanics. The theory is applied to the orbital motion of the Earth around the Sun, which is observable as the angular motion of the Sun in a celestial reference frame.

ET is thus defined by the conventional adoption of an ephemeris of the Sun (defining ephemeris) derived from a numerical expression of the positioning of the Sun in the Conventional Reference System (defining equation).

The defining equation is a numerical expression of the geometric mean longitude of the Sun L_0 given by Newcomb, adopted to this end (implicitly) by the IAU in 1952:

$$L_0 = 279^\circ\ 41'\ 48.''04 + 129\ 602\ 768.''13\ t_E + 1.089''\ t_E^2 \tag{1}$$

where t_E is measured in julian centuries of 36525 ephemeris days since the instant close to the beginning of 1900, when the geometric mean longitude of the Sun was 279° 41′ 48.″04. It is completed by the designation of the ET date in usual calendars, the date t_E=0 corresponding to

$$\begin{aligned} \text{ET} &= \text{1900 January 0.5 (exactly)} \\ \text{ET} &= \text{2415 020.0 in Julian Ephemeris Days (JED)} \end{aligned} \tag{2}$$

In (1) the numerical coefficients must be regarded as exact numbers (followed by noughts).

Historically, the expression of the mean longitude of the Sun was derived by Newcomb from observations made in the 19th century, which were referred to the mean solar time. The constant term A_0 defines the origin of ET; the UT and ET dates of an event are about the same in 1900-1905. The factor A_1 of t_E determines the duration of an ET day (or 'ephemeris day'); this duration is close to the average duration of a mean solar day during the 19th century.

The definition by (1) is incomplete because it does not relate t_E and ET to observable quantities. One has to know in addition
(a) the expression of the periodic terms of the Sun longitude,
(b) the value of the aberration.

Concerning (b), Newcomb had used a value of the aberration constant of 20.″50, different from the conventional value of 20.″47, in use from 1897 to 1967 and 20.″497 since 1968. Consequently, one finds sometimes the ET defining equation expressing the apparent mean solar longitude (which is better), with

$$A_0 = 279^\circ\ 41'\ 27.54''.$$

The space reference system is the ecliptic and the mean equinox. It must be understood that the realization of this reference by Newcomb has to be adopted.

3.2.2. Realization of the Ephemeris Time

In conformity with the definition (completed for its deficiencies), ET is realized by measuring the Sun longitude at some instant of date H read on an auxiliary clock. Solving (1) gives t_E, and TE at that instant, thus TE-H. The Universal Time has often been used as an auxiliary clock and the IAU has defined

$$\Delta\text{T} = \text{ET} - \text{UT1}.$$

The uncertainty of the measure of the Sun position is of the order of 0.″5. The corresponding uncertainty of ET amounts to about 10 s, which is exceedingly large: for instance, measuring the duration of an interval in the ET scale with a relative uncertainty of 10^{-8} would require that this interval exceeds 30 years.

The definition of a dynamical time scale by the longitude of the Moon, which varies 13 times faster than the longitude of the Sun, would have led to a better accuracy of reading. Unfortunately the deceleration of the mean motion of the Moon by tidal friction could be determined only by reference to a uniform time scale, until the implementation of lunar laser ranging (1971).

The motion of the Moon is nevertheless a good clock for providing a secondary realization of ET (denoted ETi, i being an index explained later) after calibration with respect to the 'primary' realization based on observations of the Sun. The calibration is made by correcting the constant, first and second order terms in time of lunar ephemerides, so that they give ETi in agreement with ET (as obtained from the Sun) over a long interval of the past.

The defining ephemerides of the secondary ETi are as follows

For ET0: The *Improved Lunar Ephemeris* which is based on the *Table of the Motion of the Moon* of Brown, after removal of an empirical term introduced by Brown (to compensate the effect of the irregularities of the mean solar time), addition of periodic terms to account for missing aberration corrections, and the addition of the 'calibrating' term.

For ET1: The *Improved Lunar Ephemeris* as for ET0, but corrected for
(a) the introduction of the 1968 IAU system of astronomical constants,
(b) an error in a coefficient of Brown's tables.

For ET2: The *Improved Lunar Ephemeris* as for ET1, with the substitution of Eckert's re-calculation of the solar perturbations instead of those in use.

3.2.3. Availability of ET

After the concept and the definition of ET were settled in 1952, it was possible to evaluate ΔT from ancient observations. Some values have been derived from solar and lunar eclipses as back as 700 B.C. Since 1620 increasingly numerous and better occultation observations give reliable values. The definition of ET has stimulated the observation of occultations by visual and photoelectric methods, and the construction of the Markowitz's Moon camera giving timed photographs of the Moon against the background of stars.

In modern observations, the measurement errors have been reduced, but other sources of error limit the *accuracy of reading* of ETi. These include irregularities of the Moon limb and erroneous positions of the stars. The uncertainties of the ΔTi on yearly averages are of the order of 0.1 s.

The Ephemeris Time ET (according to its primary definition by the Sun) should be highly uniform in the very long term, as will be seen in 3.2.5. But the calibration of the ETi is not perfect and additional long-term variations of the ETi normalized frequency may amount to several units of 10^{-9}.

The lunar laser ranging could have provided a better accuracy of reading of ETi, and also, in conjunction with the improvement of the theories and ephemerides, a better uniformity. But, before its implementation in 1971 it was already found more convenient to use the atomic time as a secondary clock for giving ET. The atomic time interval unit is based on frequency measurement made with respect to ET0 in 1955-1958 (Chapter 16). Therefore, in 1958, the difference between ET and atomic time was a mere time offset. Subsequent measurements have not shown a significant relative drift, so that a convenient realization of ET is

$$\mathrm{ET} = \mathrm{TAI} + 32.184 \text{ s},$$

TAI being the International Atomic Time (see 16.6).

For reasons which will be mentioned in 3.2.6, the IAU, in 1970, began to consider the need of a new definition of time scales for modern dynamics which should replace ET. The availability of Ephemeris Time is especially useful for interpreting observations prior to the existence of the atomic time (before 1955).

No service has been officially designated to analyze and combine individual measurements of ETi, and to issue a unique series of ΔTi conventionally accepted. Yearly averages of ΔTi are found in the main national ephemerides. A table of values of ΔT from 1630 to 1980 (every 5 years from 1630 to 1780, then every year) has been given by Stephenson and Morrison (1984).

3.2.4. The second of Ephemeris as time unit

The defining equation (1), which entirely defines the t_E datation system, could have been used to define the unit of time interval by stating that it is the duration for the increase of L_0 from $(L_0)_1$ to $(L_0)_2$, $(L_0)_1$ and $(L_0)_2$ being two specified numerical values of some year (on account of the 2π ambiguity). However, in the same spirit, a more compact definition has been devised.

The derivative

$$\frac{dL_0}{dt_E} = A_1 + 2A_2 t_E$$

is supposed to be equal to the rate of finite increments $\Delta L_0/\Delta t_E$. ΔL_0 is taken equal to 2π (the increase of L during a *tropical year*, by definition). Solving Δt_E gives the 'instantaneous' duration of the tropical year for date t_E, as measured in the t_E scale:

$$D = 31\ 556\ 925.9747\ ...\ \text{s} - 0.5303\ \text{s} \times t_E.$$

The definition of the time unit is given for t_E=0 and has been worded as follows (11th Conférence Générale des Poids et Mesures, 1960)

> "The second is the fraction 1/31 556 925.9747 of the tropical year for 1900 January 0 at 12 hours Ephemeris Time."

The non-astronomers were rather puzzled by the mention of the duration of a year at a given instant. Moreover, the definition does not provide the information needed to realize the second at any instant.

This unfortunate definition of the second of the International System of Units (SI) could have been avoided because atomic frequency standards were already well studied when it was adopted, and it was evident that the SI second would be soon based on an atomic transition. It was in fact done in 1967 and the definition of the ephemeris second as a SI unit was abrogated after only 7 years of existence.

3.2.5. Stability of ET and ETi

(i)-*Stability of the solar ET*

In the very long term (over centuries), the main cause of instability of ET is the frequency drift due to the error of the factor A_2 of t_E^2 of (1), and to the neglected higher order terms. Modern theories give

$$A_2 = 1.''093241\ cy^{-2}$$

(Bretagnon, 1982). However, the difference with Newcomb's value of A_2 is almost entirely due to the correction of the precession terms, the mean sidereal motions differing only by $0.''0003\ cy^{-2}$. Thus, if the observations of the Sun are all referred to the equinox realized by Newcomb's expression of the precession, the normalized frequency drift $\dot{y}$ of ET as given by (1) may correspond to an error on A_2 of $0.''0003\ cy^{-2}$:

$$\dot{y} = 1.5 \times 10^{-21}\ \text{s}^{-1} = 4.6 \times 10^{-12} \text{ per century.}$$

The stability curve of ET on figure 1 has been drawn with an assumed white noise of 2 s on the reading of a year average of ET and the drift of 4.6×10^{-12} per century.

(ii)-*Stability of the lunar ETi*
The main cause of frequency drift is the uncertainty of the adopted value of the lunar deceleration. In the calibration of ETi with respect to ET, Spencer Jones has taken

$$\dot{n} = -\ 1.09 \times 10^{-23}\ \text{rad s}^{-2}. \tag{3}$$

More recent studies give higher values. For instance, from lunar laser ranging (Dickey and Williams, 1982)

$$\dot{n} = (-\ 1.22 \pm 0.06) \times 10^{-23}\ \text{rad s}^{-2}$$

and from 250 years of observations of transits of Mercury (Morrison and Ward, 1975)

$$\dot{n} = (-\ 1.29 \pm 0.10) \times 10^{-23}\ \text{rad s}^{-2}. \tag{4}$$

To the change between (3) and (4) corresponds a frequency drift of ETi of

$$\dot{y} = \Delta\dot{n} \,/\, n = 7.5 \times 10^{-19}\ \text{s}^{-1} = 2.4 \times 10^{-9} \text{ per century.}$$

The stability curve of ETi on figure 1 has been drawn with this value and a white noise phase reading of 0.1 s on one-year averages.

3.2.6. The use of ET, its future

The definition of ET is that of a unique ideal time scale: in the absence of any observational errors of the Sun's apparent motion, there should be a unique realization of ET.

However, each ephemeris of a celestial body of the solar system defines implicitly a dynamical time scale which is the approximate representation of the time-like argument of the theory on which it is based. It is, in principle, possible to select the origin and the scale unit of these dynamical time scales so that their readings and rates agree with those of ET *at some instant*, but it is not possible to keep their agreement with ET at any time, for various reasons.

(i)–The dynamical models may differ. In particular ET is defined in Newtonian mechanics, modern ephemeris in relativistic theories.

(ii)–Due to the uncertainty of the second order term of the defining equation of ET and the neglected terms of higher orders, ET is not a strictly uniform time in the Newtonian mechanics.

(iii)–The realization of ephemerides includes a number of error sources: incompleteness of the knowledge of interactions, integration errors, imperfection of the initial conditions.

There was a growing tendency to consider that ET was a generic name for the time scales implicitly defined by the ephemerides., after adjustment at some instant of the origin and unit in order to agree in phase and rate with ET, as defined by (1) and (2), at that instant.

The need for a better definition of the time scales implied in the preparation and publication of ephemerides has been widely debated in a IAU working group created in 1970.

After much discussion in this working group and during many meetings, it was decided to recommend a time-like argument for the independent variable for the theories of motions of solar system bodies. The definition would distinguish between coordinate and proper relativistic time scales, be related to the International Atomic Time TAI at some epoch, and be continuous with Ephemeris Time. After many attempts to word the definition clearly and some last minute modifications at the IAU General Assembly at Grenoble, 1976, the resolution below was adopted.

RECOMMENDATION 5: TIME SCALE FOR DYNAMICAL THEORIES AND EPHEMERIDES (1976)

It is recommended that:

(a) at the instant of 1977 January $01^{d}00^{h}00^{m}00^{s}$ TAI, the value of the new time scale for apparent geocentric ephemerides be 1977 January $1\overset{d}{.}0003725$ exactly;

(b) the unit of this time scale be a day of 86400 SI seconds at mean sea level;

(c) the time scale for equations of motion referred to the barycentre of the solar system be such that there be only periodic variations between these time scales and that for the apparent geocentric ephemerides; and

(d) no time-step be introduced in International Atomic Time.

The resolution was accompanied by Notes which are of an explanatory nature and are not construed to be part of the recommendation; these notes are not reproduced here. In 1979 another IAU resolution gave the names Terrestrial Dynamical Time TDT, and Barycentric Dynamical Time TDB to the time scales defined by Recommendation 5.

These resolutions have been diversely interpreted. Recommendation 5 implies that the unit of TDT be a day of 86400 SI seconds at any instant, and thus that TDT be an ideal form of atomic time, with no relation at all with dynamical theories. However, the adjective 'dynamical' in the designation of the time scales is misleading because it can be understood as 'given by the dynamics', in the same manner as in 'atomic time', 'atomic' means 'given by atomic properties'.

In an attempt to avoid ambiguities, it has been proposed (Guinot and Seidelman, 1988) to re-name TDT and TDB as Terrestrial Time TT and Barycentric Time TB, and to define them more explicitly. In their proposal, TT should be the proper time of a clock at the geocentre, assuming that it does not suffer the gravitational field of the Earth. TB should be the proper time of a clock at the barycentre of the solar system, assuming that it does not suffer the gravitational field of the bodies of the solar system. The origins and units are as in the IAU Recommendation 5 (1976), being understood that the SI second is the atomic second.

One can designate by TT(xxx) some realization of TT; for instance:

$$\mathrm{TT(TAI)} = \mathrm{TAI} + 0.0003725 \text{ d},$$

in the geocentric frame of reference.

For theoretical studies in dynamics, although the aim of the theories is to provide predictions of positions with respect of the time argument for which there are good realizations, i.e. TT or TB, these studies may be developed as a function of time-like arguments which are not equivalent to TT or TB. For the user, these time-like arguments of dynamical studies, their origins and their units do not matter if the conversion into TT or TB can be made available.

3.2.7. Concluding remarks on the Ephemerides Time

Although the concept of dynamical times based on the orbital motions is sound and useful, the realization of a unique dynamical time scale taken as a conventional reference has met many difficulties. The lack of accuracy of reading and the unavoidable delay in availability forbid any practical use outside dynamical astronomy. Using a definition based on the motion of satellites of the Earth could have improved the accuracy of reading and the availability but at the cost of a lack of 'uniformity' due to more complicated models of forces. The dynamical time scale, as realized by Ephemeris Time has also the inconvenience of the dependence on the realization of the space references: the changes of conventional reference frames and astronomical constants have led to excessive complications.

With the development of atomic time scales and modern dynamical theories, Ephemeris Time has lost much of its interest, except, of course, for the pre-atomic time observations of planets and satellites and the study of the Earth's rotation.

3.3. PULSAR TIME

Pulsars were discovered in 1967. They are fast rotating neutron stars emitting beams of electromagnetic radiation which are observed in the 100-1500 Mhz band. Most of the pulsars have periods of the order of one or a few seconds, and have been found to be very stable, with respect to atomic time, except for occasional irregularities (glitches).

In 1982, the pulsar PSR 1937+21 having a period of 1.6 ms was discovered. It is an extremely interesting object, both intrinsically and for the wide range of astronomical and metrological applications which are made possible by the sharpness of its pulses and the stability of its period. Other 'millisecond pulsars' have been subsequently found, with periods of a few milliseconds.

Considered as frequency generators, the pulsars have the defect of being located in space, so that the observed frequency, which is compared to the frequency of laboratory atomic clocks or to terrestrial atomic time scales, is dependent on

- the motion of the laboratory with respect to the barycentre of the solar system,
- the motion of the pulsar with respect to this barycentre,
- the model of propagation of electromagnetic waves,
- the interstellar medium.

Many parameters involved in this dependence are not sufficiently well known a priori, and must be determined by a fit on recorded times of arrival of the pulses. In addition, the period of pulsars is changing, presumably linearly with time, and a drift rate has to be obtained from the observations.

The time of arrival of the PSR 1937+21 pulses can be referred to atomic time scales with an uncertainty smaller than 1 μs. As averaging is possible, this uncertainty is reduced to 0.1 μs in two hours of observations. The increase of the period is 0.5×10^{-19} s/s, i.e., in normalized frequency 2.1×10^{-7} per century (about 10 times larger than the deceleration of the Earth rotation).

On account of this variation of period, which must be calibrated using a better standard of time interval, this pulsar alone cannot give a good measure of time.

The stability of the random component of the pulsar frequency has been compared with the stability of atomic time scales. As the fit of the numerous parameters may absorb some of the long term instabilities of the frequencies of both the pulsars and atomic time scales, the comparison may be misleading. Nevertheless, it appears most probable that some millisecond pulsars have smaller random instabilities than the best atomic time scales, for averaging times of the order of one year or more. Thus:

(i)–The instabilities of atomic time scales impose a limitation on the accuracy of some parameters derived from the pulsar timing over intervals of about one year and more,

(ii)–the pulsars may improve the long term stability of atomic time scales.

Many scientific results can be expected from the timing of several millisecond pulsars. For the metrology of time, the measurement of their relative instabilities and the formation of mean pulsar time scales (Iljin *et al.*, 1986) will be a source of progress.

The frequency stability of pulsar PSR 1937+21 is shown by figure 1.

REFERENCES

Allan, D.W., 1966, 'Statistics of atomic frequency standards', Special Issue on Frequency Stability, *Proc. IEEE*, 54, 199.

Aoki, S., Guinot, B., Kaplan, G.H., Kinoshita, H., McCarthy, D.D. and Seidelmann, P.K., 1982, 'The new definition of Universal Time', *Astron. Astrophys.*, 105, 359.

Bretagnon, P., 1982, 'Théorie du mouvement de l'ensemble des planètes. Solution VSOP 82', *Astron. Astrophys.*, 114, 278.

Dickey, J.O. and Williams, J.G., 1982, 'Geophysical Applications of Lunar Laser Ranging', *Trans. Am. Geophys. Union*, 63, 301.

Guinot, B. and Seidelman, P.K., 1988, 'Time scales - Their history definition and interpretation', *Astron. Astrophys*, 194, 304.

Iljin, V.G., Isaev, L.K., Pushkin, S.B., Palii, G.N., Ilyasov, Yu.P., Kuzmin, A.D., Shabanova, T.V. and Shitov, Yu.P., 1986, 'Pulsar Time Scale - PT', *Metrologia*, 22, 65.

Misner, C.W., Thorne, K.S. and Wheeler, J.A., 1973, '*Gravitation*', W.H. Freeman and Co., San Francisco.

Morrison, L.V. and Ward, C.G., 1975, 'The analysis of the transits of Mercury', *Monthly Notices Roy. Astr. Soc.*, 173, 183.

Rutman, J., 1978, 'Characterisation of phase and frequency instabilities in precision frequency sources: fifteen years of progress', *Proc. IEEE*, 66, N° 9, 1048. This paper gives many references.

Sadler, D.H., 1978, 'Mean Solar Time on the Meridian of Greenwich', *Quat. J. Roy. Astr. Soc.*, 19, 290.

Stephenson, F.R. and Morrison, L.V. 1984, 'Long-term changes in the rotation of the Earth: 700 B.C. to A.D. 1980', *Phil. Trans. Roy. Soc.*, London, A313, 47.

ATOMIC TIME

B. GUINOT
BIPM, Pavillon de Breteuil, 92310 Sèvres, France

1. INTRODUCTION

The first operational caesium-beam frequency standard, built by Essen and Parry in the United Kingdom, began to provide data in mid-1955 and thus opened a new era in the measurement of time.

The atomic frequency standards were first mainly considered as a convenient means to have stable and accurate frequencies, and a standard of time interval, in real time, in the laboratories. Although a few astronomers used their data as soon as they were available to establish uniform time scales for the study of the Earth's rotation, the idea that 'atomic clocks' could give the world time scale raised more reluctance than enthusiasm. This idea finally gained acceptance because the large number of high quality atomic clocks, industrially made and located in many laboratories, could be accurately intercompared. All these clocks could therefore contribute to the establishment of a unique time scale and one could have sufficient reliability of that scale.

The atomic frequency standards and clocks have now numerous and extremely important applications in science and technology. The establishment of the atomic time scales, presented here, is only one of these applications.

We will not elaborate here on the physics of atomic clocks. We intend merely to recall their principles, show how they are used to produce a mean time scale, and provide users with some information they may find useful for timing and interpreting their observations.

More details on the subject of atomic time and atomic clocks can be found in Kartaschoff (1978) and NBS Monograph 140 (1974).

2. CONCEPT, DEFINITION, REALIZATION. THE CRITERION OF ACCURACY

The atomic time is an integrated time (see Section 2.2 of Chapter 15).

J. Kovalevsky et al. (eds.), Reference Frames, 379–415.

The duration standard is provided by the assumed property that the frequency associated to a transition between two selected energy levels of an atom (or a molecule, but only the words 'atom' and 'atomic' will be employed in the following) is the same everywhere and at any time.

Thus the definition of the duration unit, or time unit, comes first. It requires that a transition be selected to this end and that a conventional value be assigned to the corresponding frequency. The ideal associated time scale is defined as the accumulation of ideal time units, under specified conditions to take into account relativistic effects.

Obtaining the time unit requires some instrument which realizes the desired transition, divides its frequency by a known factor, and amplifies it in order to make it usable. We call this instrument an *atomic frequency standard.*

The realization of an atomic time scale requires a continuous count of the time units. It can be obtained either by an *atomic clock*, which is a frequency standard operating continuously and able to count the oscillations or the time units, or by computations based on discontinuous calibrations of secondary clocks such as crystal clocks.

Thus the realizations of a defined atomic time can be as many as there are atomic clocks and frequency standards. In addition, the choice of the origin of each scale is free. There is also an infinite number of possibilities to realize average atomic time scales. For these reasons, the definition must be completed by a conventional choice of a particular realization as the common agreed reference.

The axiom of the invariability of atomic frequencies applies to the atom at rest and unperturbed. But the environment and the device made to observe the atomic frequency add some perturbations which may introduce differences between the ideal atomic frequency and the measured one. An essential requirement is to keep the systematic frequency errors (bias) as small as possible, either by appropriate shielding or by corrections based on the theory of the perturbations and the measurements of the involved parameters.

The *accuracy* of an atomic frequency standard is the aptitude to avoid the biases. One of the tasks of the builders of these standards is to evaluate the possible size of the biases, from theoretical considerations and from the uncertainties of the auxiliary measurements, and to establish an error budget. As a result, the inaccuracy can be specified by its one sigma value in normalized frequency.

In the definition of the time unit and in the conventional adoption of an atomic time scale as time reference, the accuracy of the frequency which can be attained by the realized devices is the main criterion.

In the following sections, we will briefly describe the instrumentation, the methods of intercomparison of atomic clocks, which play an essential role both in ensuring the reliability of the atomic time and its dissemination, and the organization of the atomic time metrology.

3. ATOMIC FREQUENCY STANDARDS AND ATOMIC CLOCKS

The stability of the wavelength of some emissions of light was known well before the end of the 19th century; Maxwell had expressed the view in 1873 that the vibration associated to these emissions might be a better basis for defining the unit of time than the rotation of the Earth.

However, even now the measurement of the optical frequencies remains at the limit of our possibilities. The realization of atomic clocks is a consequence of the discovery of the atomic hyperfine structure and the corresponding wavelength at the centimetre and decimetre level. Although important early work was performed by Rabi and his collaborators in the 1930's it was necessary to wait until the development of radar, to have sufficient knowledge of the microwave techniques allowing the detection of atomic frequencies and their division by known factors, in order to get the usual values in time measurement, such as 1 Hz.

An overview of the development of atomic clocks has been given by Forman (1985). The physics of atomic frequency standards is exposed by Vanier and Audoin (1989).

3.1. AMMONIA CLOCKS

The first atomic clock was realized at the National Bureau of Standards, USA (NBS) in 1948, using the stabilization on an ammonia absorption line. Unfortunately, the necessary high pressure, the collisions and the thermal noise broadened the line and it was only possible to achieve a long-term stability of 2×10^{-8} and also a poor accuracy. Many attempts to improve the ammonia clocks failed to give results which could compete with the atomic beam techique; these efforts were abandoned.

3.2. CAESIUM FREQUENCY STANDARDS AND CLOCKS

The technique of atomic and molecular beams was explored in 1930-1940 by Rabi and his collaborators. It was first used for spectroscopy, but Rabi considered the possibility of reversing the procedure and also recognized the advantages of the hyperfine structure of caesium 133 (frequency 9.2 GHz), as early as 1940. However the work on the caesium beam frequency standard started only in 1948 at NBS, after the proposal by Ramsey to divide in two separate parts the interaction region. The first working caesium beam frequency standard was realized by Essen and Parry at the National Physical Laboratory (NPL), UK, and regular calibrations of frequency of crystal oscillators began mid-1955. Soon afterwards several laboratory caesium frequency standards appeared as well as commercial frequency standards and caesium clocks.

As the definition of the time unit, the second, of the International System of Units (SI) is based on the transition used by the caesium frequency standards, the most accurate of them are qualified as *primary standards*. One uses also the expression *primary clocks* for the primary standards operating continuously and producing a time scale.

The accuracy of the best primary standards seems to reach 2×10^{-14}.

The accuracy of commercial standards (without external calibration) is of the order of 1×10^{-12}.

The stability of the caesium standards is given by figure 1.

The improvement of caesium frequency standards is pursued.

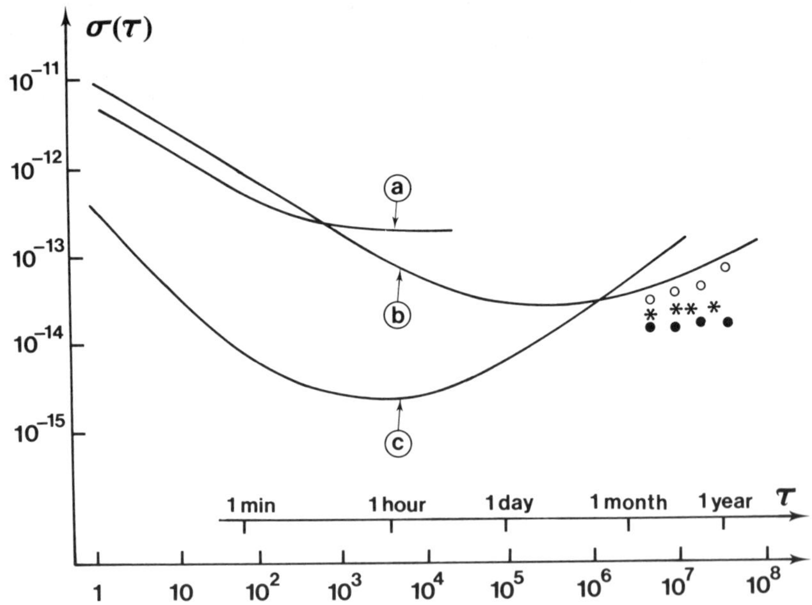

Figure 1. Stability of atomic frequency generators. The curves show typical performances: (a) rubidium clock, (b) industrial caesium standard, (c) hydrogen maser. Some observed values of long-term stability of the best devices are also given: ∘ industrial caesium standard, * hydrogen maser with automatic tuning, • laboratory caesium standard.

3.3. HYDROGEN MASERS

The hydrogen maser was first developed by Ramsey in 1968. A very long interaction time with the microwave frequency at 1.42 GHz is achieved by storing the atoms in a bulb. It is then possible to obtain a much narrower spectral line than with the atomic beams and a better short-term stability (Figure 1). However, the accuracy reaches only about 1×10^{-12} on account of the 'wall shift' (frequency dependence on the coating of the storage bulb) and to the 'cavity pulling' (frequency dependence on the tuning of the resonant cavity). The hydrogen masers are especially useful in astronomy for the very long baseline interferometry (VLBI), where the stability, up to one day, is the most important criterion.

3.4. RUBIDIUM CLOCKS

The use of the Rb 87 hyperfine resonance at 6.8 GHz was initiated by Carver in 1957. The rubidium clocks use the technique of optical pumping. These atomic clocks cannot be accurate but their short-term stability is good (Figure 1). They are available on the market and are much less expensive than hydrogen masers.

3.5. OTHER STANDARDS AND CLOCKS: THE FUTURE

Among several researches and projects, the storage of ions begins to lead to practical realizations of clocks. The optical frequencies (infrared and visible) are possible candidates for accurate standards, but their measurement is still in the domain of sophisticated laboratory experiments or even of tour de force.

The choice of the hyperfine transition of Cs 133 is still the best for the definition of the unit of time and for the atomic time scales. At the moment, no challenger appears.

4. THE ATOMIC SECOND AS TIME UNIT

The early caesium frequency standards had an accuracy of 1×10^{-9} to 1×10^{-10}, while the second of Universal Time, even after removal of the seasonal term, may have variations of several units of 10^{-8} in a decade. It was necessary to express the caesium transition frequency ν_0 with a better unit, the ephemeris second. This was accomplished by Markowitz *et al.* (1958) from 1955 to 1958. The Ephemeris Time was in fact TE0 measured with the Markowitz's Moon camera. They obtained

$$\nu_0 = 9\ 192\ 631\ 770 \text{ Hz} \pm 20 \text{ Hz (in ephemeris seconds)},$$

the uncertainty being almost entirely due to the measurement of the ephemeris second. This value has been confirmed by subsequent observations. It was then adopted as conventional when the 13th Conférence Générale des Poids et Mesures (CGPM), in 1967, adopted the atomic definition of the second in the following terms:

> "The second is the duration of 9 162 631 770 periods of the radiation corresponding to the transition between the two hyperfine levels of the ground state of the caesium-133 atom."

With the additional observations of TE available at the time of writing (1988), the agreement of the SI (atomic) second with the ephemeris second is still satisfactory, no significant bias having been found.

We recall that the 17th CGPM (1983) linked the definition of the metre to the definition of the second as follows:

> "The metre is the length of the path travelled by light in vacuum during a time interval of 1/299 792 458 of a second."

Consequently, the speed of light is strictly

$$299\ 792\ 458 \text{ m s}^{-1}.$$

5. FREQUENCY AND TIME COMPARISONS OF REMOTE TIME STANDARDS

The permanent comparison of frequency standards and clocks is essential for the metrology of time. We briefly describe the methods used in an operational mode. An extensive presentation of the methods, both experimental and operational, more details, and many bibliographic references appear in '*Recommendations and Reports of the CCIR, Vol. VII Standard Frequencies and Time Signals*'* , especially in Reports 363 and 518.

* Comité Consultatif International des Radiocommunications (CCIR) 2, rue de Varembé, 1211 Genève 20, Suisse.

5.1. FREQUENCY COMPARISONS

The atomic definition of the second makes it possible for anyone to realize the second where it is needed. However it is a common practice in metrology to check the standards

- to make sure that they behave properly,
- to evaluate the biases and to develop and test more accurate devices.

The frequency comparisons also provide a means of disseminating the second by calibration of local oscillators such as crystal or rubidium clocks which are cheaper than caesium standards.

An old but still convenient method of frequency comparison is to monitor the phase of radio 'very low frequency' carrier (VLF, 3 to 30 kHz), used for long distance communications and for radio-navigation.

Let us suppose that the nominal frequency of the emission be f_0; its real frequency is $f_{0,E'}$ slightly different. By appropriate frequency multiplication, the local frequency standard S can generate a nomical frequency f_0, with the real value $f_{0,S}$. Relatively simple receivers can measure continously the slowly varying phase difference between the two waves, thus giving $f_{0,E}$-$f_{0,S}$. The simultaneous measure for standard S′ gives $f_{0,E}$-$f_{0,S'}$ and, after exchange of data $f_{0,S}$-$f_{0,S'}$.

A typical example is the use of the OMEGA navigation system, with $f_0 \approx 12$ kHz. The accuracy of the phase measurement is of the order of 2 μs. As there is a diurnal variation of the propagation delays due to the properties of the ionosphere, which depends on the distance and can reach several tens of μs, the mean frequency difference is to be measured over an integral number of days. For instance, over one day, the uncertainty of the mean normalized frequency difference is

$$2\ \mu\text{s}/86400\ \text{s} \approx 2 \times 10^{-11}.$$

The coverage of VLF is worldwide. Over distances of the order of 2500 km or less, the same method applied to low frequency emissions (LF, 30 to 300 kHz) provides a better accuracy.

For a number of stable VLF and LF emissions the true frequency is maintained within $\pm 1 \times 10^{-11}$ or better from its nominal value, and/or the true frequency estimates are published, so that their monitoring is a cheap and convenient method to get the SI second with relative uncertainties of 1×10^{-11} or better. However, this method is not sufficient to intercompare the most accurate atomic frequency standards.

On the other hand, in practice, it is not possible to keep track of the phase differences indefinitely, without slippages or losses of the phase: this method does not provide 'time comparisons' but only 'frequency comparisons'.

5.2. TIME COMPARISONS

We will restrict our discussion here to the most precise and accurate methods of time comparisons, which are used in the establishment and dissemination of a conventional worldwide time scale. The accuracy requires that these methods be considered in the framework of general relativity. However relativistic corrections will not be mentioned here (see Section 7.4).

5.2.1. Clock transportations

Industrial caesium clocks are compact enough and sufficiently insensitive to the accelerations and environmental conditions to be carried while functioning in cars, passager seats of planes, etc. For the time comparisons of a network of fixed clocks A, B, C, ... the travelling clock is successively compared to A, B, C, ... and A again (a round trip is necessary to benefit of the check of the closure), and therefore provides A-B, A-C, ... at the instant of the visit to B, C, etc. The measurement is "absolute", i.e. without biases or systematic errors. The method has been widely used over intercontinental distances, with commercial air-lines, the accuracy being of the order of 10 to 50 ns. Unfortunately, recent air safety regulations practically forbid the continuation of the transportation of operating caesium clocks.

5.2.2. One way transmissions

A signal (radio, optical, etc.) is sent from S and reaches receivers connected to clocks A and B. The recorded times of arrival are ξ_A, ξ_B. If we denote by τ_A and τ_B the delays of propagation, including the instrumental delays of the receivers, we have

$$A - B = \xi_A - \xi_B - (\tau_A - \tau_B),$$

at the instant of reception. The differential delay τ_A-τ_B has to be computed or measured.

The usual applications of this method are summarized hereafter.

(i)-LORAN-C
The LORAN-C is a ground-based system of radio-navigation. A LORAN-C station (master or slave) emits short pulses of a 100 kHz carrier. The ground wave propagation of the pulses, which is very stable, is faster than the sky wave reflected by the ionosphere. Therefore the beginning of the received signal is not perturbed by the sky wave which has a variable propagation delay.

The differential delay τ_A-τ_B of the ground wave has only small variations:

- an annual variation reaching a few 0.1 μs in the worst cases (long overland paths);
- irregular long-term changes due to multipath propagation of also a few 0.1 μs.

The differential delay is calibrated by clock transportation and checked from time to time.

Under normal conditions, *A-B* can be measured permanently with uncertainties of the order of 100 to 200 ns. The coverage extends to about 1000 km from the LORAN-C stations.

As the LORAN-C emissions are monitored by caesium clocks, the strict simultaneity of the receptions is not stringent. The master and slave stations of the same chain being synchronized, *A* and *B* can receive two different stations of the chain with only a minor degradation.

The LORAN-C time receivers are available, convenient, reliable and relatively cheap.

Exchanges of data between *A* and *B*, and a simple processing are needed.

(ii)-Television

The commercial television emissions include in each frame some signals which are good time markers:

- synchronisation pulses;
- some selected pulses in a test line.

These time markers must be received simultaneously, which requires that the time difference of *A* and *B* be a priori known with a few milliseconds uncertainty. As in the previous method, τ_A-τ_B must be calibrated by clock transportation.

In direct view of the same television emitter, the precision of *A-B* can reach 5 ns (the accuracy can reach the same level if the calibration has been made with a matching accuracy).

It is possible to extend the television method over several hundreds of kilometres, using microwave links. But the uncertainties increase and the computations become tedious because the steps in the propagation delay, due to changes in the microwave links, have to be recognized and corrected for.

The receivers are not easily available on the market.

(iii)-Global Positioning System (GPS)

The GPS which is being deployed by the Department of Defense of the United States of America is a satellite positioning system. It will include 18 satellites on 12 sidereal hour orbits distributed so that at least 4 satellites will be visible from any point of the Earth, at any time. The satellites carry atomic clocks.

The positioning is based on range measurements. The satellites emit according to a pseudo random noise code which is correlated with the same code generated by a receiver connected to the local clock. The tracking gives the time difference between the reception time of the satellite signal, recorded on the local clock, and the time of emission recorded on the satellite clock. This difference is the sum of

τ_1, the duration of the propagation from the satellite to the receiver antenna, including refraction,

τ_2, the instrumental and cable delays,

τ_3, the time offset between the satellite and local clocks.

The satellite transmits its ephemeris and a model for the refraction, which are used together with the coordinates of the receiver antenna to compute τ_1. The instrumental delay in τ_2 is measured by the manufacturer and can be checked with a special equipment; the various cable delays can be measured. Thus, one obtains τ_3.

The satellite also transmits an extrapolated value of the difference between the readings of its clock and a common time reference of the GPS called 'GPS time' (to ± a few tens of nanoseconds), so that one gets immediately

local clock - GPS time.

GPS time, disregarding an integral number of seconds, is maintained to about ± 5 μs of the international time scales TAI and UTC defined in sections 7 and 8. For precise time comparisons between two remote clocks A and B, GPS time is eliminated by

$$A - B = (A - \text{GPS time}) - (B - \text{GPS time}).$$

As the rate of the quantities (local clock − GPS time) is small, and can be estimated, it is not necessary, in principle, that the observation be simultaneous, and a way to reduce the random errors and some systematic errors is to average daily over a large number of passes and over many observations for each pass. However, some errors cancel when tracking the same satellite simultaneously with a good geometry (common view method); it is thus possible to reduce the number of records that must be exchanged between the laboratories. The common view schedules for international cooperation are established since 1986 by the Bureau International des Poids et Mesures (previously by the National Bureau of Standards, USA); they are widely used.

Both methods lead to an accuracy of about 10 ns (one sigma) on daily averages, if the calibrations of instrumental delays are sufficiently good. As only the difference of instrumental delays is needed, a calibration method consists in carrying to stations A and B a GPS receiver taken as a transfer standard, and to make observations in colocation. The differential delay can be thus determined with an accuracy of 1 ns.

At distances up to 1000 km, the use of precise differential geodetic coordinates (± 30 cm) seems to provide an accuracy of the time comparisons of 2 ns (Guinot and Lewandowski, 1988).

Commercial GPS time receivers are available. The GPS time comparisons are accurate and convenient. Their use for international cooperation started in 1983 and extended rapidly to the majority of the time services. There is the possibility that the system be voluntarily degraded for the civil users of positioning, but that should not impede accurate time comparisons, although it might complicate them.

5.2.3. Two-way transmissions

Let A and B be two clocks. The principle of the method is to send quasi-simultaneously signals from A to B and B to A along the same path, on the same carrier frequency (or, on frequencies having close values).

Let us call α and β respectively the readings of A and B and designate by indices 1 and 2 respectively the signal from A to B and B to A. Thus:

α_1 is the reading of A when the signal from A to B leaves A,

β_1 is the reading of B when the signal from A to B reaches B, etc.

Neglecting all the instrumental delays which might be involved:

$$A - B = \tfrac{1}{2}(\alpha_1 + \alpha_2) - \tfrac{1}{2}(\beta_1 + \beta_2).$$

This method has often been experimented, using telecommunication satellites, since the first test in 1962, using TELSTAR, between UK and USA. It was used operationally with SYMPHONY, between Canada, Federal Republic of Germany and France, from 1978 to 1982. A new application, employing pseudo-random-noise-modulated signals (MITREX modems) and very low power, requires only small diameter antennas (2 to 5 metres).

The advantage of the two-way time transfer is that it does not require the knowledge of the satellite and station positions and of the refraction. A sub-nanosecond precision can be reached in a operational service. However, the accuracy depends on the calibration of instrumental delays at the emission and reception. It has been possible to check the stability of these delays at the nanosecond level, but the uncertainty of the measurements of their absolute value was much larger, until now.

An experiment of synchronization by laser ranging on a geo-stationnary satellite (LASSO) is expected to give an accuracy of 1 ns.

5.3. THE PRACTICAL SIDE

One can conclude this dicussion on time and frequency comparisons by saying that operationally, with equipment available on the market (GPS time receivers), and without requiring the know-how of experts, the long distance time comparisons can be achieved, in 1987, with a accuracy of 10 ns. For the frequency comparisons, one can take advantage of the stability of the biases, when observing satellites in common view, with the same geometry each day, and the uncertainties are

10^{-14} to 10^{-13} in a day
10^{-15} to 10^{-14} in 10 days.

6. ALGORITHMS FOR ATOMIC TIME SCALES

Some laboratories establish excellent atomic time scales which are the output of a single high-quality caesium clock. However the risk of failure or of unnoticed abnormal functioning is too large when the time scale is the basis of operational systems, such as navigation systems, or used for long-term studies such as occur in astronomy. For establishing a conventional international atomic time scale, an averaging over a large number of clocks has several advantages:

- reliability,
- the benefit of the best clocks, at any time, by appropriate statistical weighting,
- a simple mode of dissemination of the time scale.

6.1. STABILITY ALGORITHMS

Most of the algorithms for atomic time scales are 'stability algorithms' with the aim of optimizing the long-term stability (over weeks, months or years). Each laboratory has its own method, but we can recognize some common features of the algorithms.

Let us call H_i the participating clocks ($i = 1, ..., n$) and $H_i(t_0)$ their readings at some instant t_0. A mean atomic time scale T is obtained by averaging, its reading at t_0 being

$$T(t_0) = \sum_i w_i \, [H_i(t_0) + C_i(t_0)] \, / \sum_i w_i \tag{5}$$

where w_i is the statistical weight assigned to H_i, and $C_i(t_0)$ a correction applied by the computing centre. The role of the C_i is that the corrected reading $H_i(t_0)+C_i(t_0)$ be as close as possible to the reading of an ideal time scale θ of which T is an approximation. Thus the algorithm includes an optimum prediction of the clock behaviour based on its past. However, as this past is referred to T, not to θ which is not known, some feedback is unavoidable. Nevertheless the optimum prediction fulfils two essential roles.

(i)–It avoids phase and frequency steps when the weight system is changed (due to modifications of the clock behaviour), when some clock fails and when new clocks are entered in the algorithm, after a test period.

(ii)–Associated to a good weighting procedure, it optimizes the stability over a selected averaging time.

Usually the components of the C_i are
- a linear function of time, i.e. a phase and frequency correction,
- possible other corrections for deterministic effects such as frequency drift, seasonal variation,
- a prediction of the random component based on a frequency noise model of the clock and past observed frequency samples with respect to T.

In practice, the input is clock reading differences at t_0 such as

$$H_i(t_0) - H_j(t_0) = \xi_{ij} \tag{6}$$

where the numerical value ξ_{ij} is slowly varying, so that there is no need that t_0 be accurately known. If we put

$$x_i = T(t_0) - H_i(t_0), \tag{7}$$

x_i being the *clock correction* to be added to $H_i(t_0)$ in order to get the reading $T(t_0)$ of the mean time scale, (5) and (6) transform into

$$\begin{aligned} \sum_i w_i x_i &= \sum_i w_i C_i(t_0) \\ x_i - x_j &= -\xi_{ij}. \end{aligned} \tag{8}$$

The system of time comparisons x_i-x_j may be redundant, but this is usually avoided. The solution of (8) gives the clock corrections.

This procedure shows the essential role of the clock comparisons:
- they provide the input;

- they make available the mean time scale by issuing corrections to the participating clocks.

Ideally, when dating an event in T with any of the participating clocks, after applying the clock corrections, the same date should be obtained. In practice there are differences of the same order as the uncertainties of the clock comparisons. The ultimate precision of reading of a mean atomic time scale is equal to the accuracy of the best clock comparisons.

6.2. ACCURACY OF THE TIME SCALE INTERVAL

The scale unitary interval of T should not vary with respect to the SI second (under specified conditions for taking into account relativistic effects), and preferably should be close to it.

This condition is not fulfilled by a pure stability algorithm.

The usual method to obtain the accurate time scale T' is to derive it from T. First, the duration of the scale unitary interval of T is evaluated by use of the data of primary frequency standards. This may require a somewhat sophisticated filter (Azoubib *et al.*, 1977). When it is done, T' can be derived from T by a smooth correction which progressively brings the scale unitary interval of T' in conformity with the realizations of the SI second, without affecting noticeably the stability. This method is usually designated as a 'frequency steering'.

Other methods can be considered, such as the introduction of the steering directly in the C_i corrections of Section 6.1.

7. THE INTERNATIONAL ATOMIC TIME TAI

7.1. BRIEF HISTORY

The development of radio time signals at the beginning of the 20th century led to the creation in 1911 of a body in charge of the worldwide unification of time, the Bureau International de l'Heure (BIH).

Until 1955, this unification was resting uniquely on the observations of the rotation of the Earth providing the Universal Time.

After the advent of operational caesium frequency standards, the BIH and some national laboratories began to establish mean atomic time scales.

7.1.1. History of the BIH atomic time scale

From 1955 to 1968, the BIH mean atomic time scale was based on frequency comparisons, as described in 5.1 and was mainly made available by issuing corrections to the times of emission of radio time signals, the accuracy of reading being of about 1 ms. The designation of the time scale was AM in 1955-1957, A3 in 1958-1968.

In 1969 the use of the recently synchronized LORAN-C chains over the Atlantic and the calibration of propagation delays by clock transportation provided accurate time comparisons on a routine basis. Since then, the BIH time scale has become established as an average of clock readings.

From 1969 to 1972, the BIH time scale was an average of 3 to 7 laboratory atomic time scales, each of them being based on several caesium clocks. The scale was called TA or TA(BIH) (or AT, AT(BIH) in English) until 1970, then TAI. The uncertainties of reading amounted to a few 0.1 μs.

In 1973, the processing of data of individual atomic clocks was initiated, with a stability algorithm (as in section 6.1) called ALGOS.

It was subsequently recognized, by comparing TAI with the data of improving primary frequency standards, that the TAI frequency was too high by about 1×10^{-12}, in normalized value.

On 1977 January 1, 0h TAI, the normalized TAI frequency was reduced by strictly 1×10^{-12} and a steering method was adopted (see Section 6.2) in order to avoid future frequency errors. Except for minor improvements, the mode of establishment of TAI has not been changed since then.

In 1983, international time comparisons by GPS (see Section 5.2) appeared and developed rapidly, reducing the uncertainties of reading of TAI.

The various names of the BIH mean atomic time scale reflect changes of the computation procedure or international status in an experimental stage. The scale is nevertheless continuous and will often be referred to as TAI since 1955, in the following.

7.1.2. The administrative aspect

In 1967, the International Astronomical Union recommended that the BIH establish an international atomic time scale. It was, in fact, a recommendation that the already existing BIH atomic time be conventionally adopted as a unique reference for international use.

In 1971, the 14th Conférence Générale des Poids et Mesures (CGPM) requested the Comité International des Poids et Mesures (CIPM) to give a definition of International Atomic Time. This had already been prepared by the CIPM as follows (English translation of the French wording):

"International Atomic Time (TAI) is the time reference

coordinate established by the Bureau International de l'Heure on the basis of the readings of atomic clocks operating in various establishments in accordance with the definition of the second, the unit of time of the International System of Units."

The name 'International Atomic Time' (abbreviated as TAI in all languages) has been adopted as a consequence of those decisions.

In 1985, the time section of BIH was transferred from the Paris Observatory to the Bureau International des Poids et Mesures, in preparation of a taking over of responsibility of TAI by BIPM and CIPM, which was effected in 1988.

7.2. DEFINITION OF TAI

The official 'definition' of TAI is given above. It is complemented as follows.

(i)-The Comité Consultatif pour la Définition de la Seconde (CCDS) gave implementation rules in its 5th session, 1970:

- the duration of the scale unitary interval of TAI is determined by the BIH so that it be in close agreement with the duration of the SI second at a fixed point of the Earth, at sea level;
- the origin of TAI is defined in conformity with the recommendations of the IAU (XIIIth General Assembly, Prague, 1967), i.e. that this scale be approximately in agreement with UT2 at 1958 January 1, 0h UT.

(ii)-In a 'declaration', the CCDS, at its 9th session, 1980, stated that "TAI is a coordinate time scale defined in a geocentric reference frame with the SI second as realized on the rotating geoid as the scale unit".

These complements provide the necessary information for the evaluation of the relativistic terms in the establishment of TAI (see Section 7.4) and for its use in non-terrestrial frames (see Chapter 17).

The so-called CGPM 'definition' of TAI is a conventional agreement to use, as a single worldwide reference, a particular realization of an ideal atomic time scale, which has not been explictly defined and has received no name. This gap between the concept and the realization of atomic time is filled to some extent by the IAU definition of TDT or better by the proposed definition of the Terrestrial Time TT (see Chapter 15, Section 3.2.6), which are ideal forms of atomic time, but with a different unitary scale interval (the day instead of the second) and a different origin than TAI. In addition TT is defined at the geocenter so that it can be considered as a proper time, and TAI is a coordinate time in a geocentric frame.

TABLE I

From the concept to the realization of atomic time

Time unit
o Concept of atomic time interval unit.
o Definition of the (ideal) second.
o Realization of the second by many laboratory instruments.
Time scale
o Concept of the atomic time (integrated time).
o Definition of an ideal atomic time: the Terrestrial Time TT (as a proper time at geocentre, but with the scale unitary interval in agreement with the second on the geoid).
o Realization of a conventional atomic time scale: the International Atomic Time TAI (as a coordinate time in a geocentric non-rotating frame, with the scale unitary interval in agreement with the second in a fixed point on the geoid).

However, in the already complex situation due to the co-existence of several reference time scales, it would not be advisable to introduce new times and time scales for the ideal TAI and the realization of TT, and we consider that the chain from the concept to the realization is as in Table I.

As TAI is not strictly speaking a realization of TT, realizations of TT would be denoted as TT(xxx), xxx being an identification acronym. For instance, for most of the purposes, in a geocentric frame,

$$\mathrm{TT(TAI)} = \mathrm{TAI} + 32.184\ \mathrm{s}$$

can be used. Such a notation is consistent with the definition of TT and distinguishes between the various realizations of TT, when the need of the utmost stability and accuracy (for instance for pulsar studies) requires the use of revised time scales which may have better metrological properties than TAI produced in near-real time.

7.3. ESTABLISHMENT AND PROPERTIES OF TAI

TAI is established according to the method summarized in Section 6. The stability algorithm ALGOS produces the intermediate time scale EAL (Echelle Atomique Libre, designated as T in Section 6.1). It processes a number of atomic clocks increasing from 50 in 1973 to 170 in 1987, distributed all over the world and intercompared mostly using the GPS and, for the remaining ones, by the LORAN-C and television methods. The evaluations of the EAL-clock are made at 10-day intervals. The prediction of the rates and the weighting are based on two-month frequency samples. The computations are made monthly, the results for the month m are made available at the beginning of month $m+2$.

The deviation of the EAL unitary scale interval is given by 10 primary frequency standards (in 1987) having accuracies in the range 10^{-13}-10^{-14}.

The relationship between EAL and TAI is linear, with a relative rate which is changed when necessary by steps corresponding to 2×10^{-14} in normalized frequency, at intervals not shorter than 60 days. These steps occur rarely; reducing their size for a smoother steering is being considered. In processing clock data, seasonal variations of the atomic clocks frequencies have been found. They are not yet fully explained, although the influence of humidity on industrially made clocks appears to be the most probable cause. The effect on TAI does not vanish because most of the participating clocks are located in the Northern hemisphere under similar climatic conditions. The laboratory caesium standards seem to be free from this seasonal variaton. By comparison, the seasonal wave of TAI is of the order of 300 ns peak to peak. National time scales based on industrial caesium clocks show variations of the same order. Correction of this annual wave could be considered when its origin is better understood. For the moment, its possible existence must be kept in mind when the utmost uniformity is required - for the study of pulsar timing, for instance.

The stability of TAI is shown by Figure 1 of Chapter 15. This evaluation assumes of course, that the present level of noise of the instruments does not change. It is somewhat hypothetical for τ over one year, but shows that the degradation of stability by random walk frequency noise is limited, on account of the frequency steering.

On yearly averages the discrepancy of the TAI unitary scale interval from the second at sea level, as given by the primary standards, is maintained well below 1×10^{-13}. For the last few years, one obtains the values of Table II.

The accuracy of reading of TAI is that of the time comparisons: 10 to 50 ns (one sigma) in laboratories having GPS time receivers, 100-200 ns in the other ones. This problem is considered further in Section 10.

7.4. RELATIVISTIC CORRECTIONS

We give some formulae which are useful for clock comparisons and for the establishing of time scales in the terrestrial reference frame of the

TABLE II

Normalized frequency of EAL and TAI according to the data of primary frequency standards. The normalized frequency is 1+y. The uncertainty S is a conservative estimate.

Year	y_{EAL}	y_{TAI}	S	Notes
	10^{-14}	10^{-14}	10^{-14}	
1970	128	128	10	TAI=EAL
1971	119	119	10	
1972	111	111	10	
1973	106	106	7	
1974	100	100	6	
1975	100	100	6	
1976	97	97	6	
1977	89	- 8	5	In 1977.0, frequency adjustment of 100×10^{-14}, then steering of TAI
1978	85	- 7	5	
1979	79	- 6	5	
1980	84	0	5	
1981	82	- 2	5	
1982	79	- 3	5	
1983	78	0	5	
1984	79	- 1	5	
1985	78	- 2	5	
1986	81	1	5	
1987	79	- 1	5	

TAI definition. These formulae appear also in the CCIR Report 439.

We assume that a clock H produces an approximation to its ideal proper time, after the correction due to the motion of the atoms in the clock (including the second order Doppler shift). In the following development, some approximations are made, but the formulae are valid with uncertainties of normalized frequencies lower than 1×10^{-14} up to the altitude of the geostationary satellites. At ground level, the uncertainties are lower than 1×10^{-16}.

We use the Newtonian approximation to the n-body metric. We assume that the gravitational field of the Sun is uniform, i.e. we neglect the tide-generating potential which gives rise to normalized frequency variations of a few units of 10^{-17} on the geiod. The notations are:

τ proper time of H,
t coordinate-time in a non-rotating geocentric frame R,
E centre of mass of the Earth,
$\vec{r}$ $= \vec{EH}$, $r = |\vec{r}|$,

ϕ geocentric latitude of H,
L geocentric longitude of H, positive Eastward,
$\vec{v}$ velocity of H in R, $\vec{v} = d\vec{r}/dt$, $v = |\vec{v}|$,
$\vec{v}_g$ velocity of H with respect to the Earth, $v_g = |\vec{v}_g|$,
U gravitational potential of the location of H (positive sign convention, $\partial U/\partial r < 0$),
$\vec{\omega}$ rotation vector of the Earth ($\omega = 7.292115 \times 10^{-5}\ s^{-1}$),
c speed of light.

From

$$d\tau^2 = (1 - 2\frac{U}{c^2})dt^2 - \frac{1}{c^2}(d\vec{r})^2, \tag{9}$$

one obtains easily, neglecting terms in $1/c^4$:

$$t - \tau = \frac{1}{c^2}\int_{t_0}^{t} Udt + \frac{1}{2c^2}\int_{t_0}^{t} v^2\, dt + constant. \tag{10}$$

As

$$\vec{v} = \vec{v}_g + \vec{\omega} \wedge \vec{r},$$

$$v^2 = v_g^2 + 2\,\vec{v}_g \cdot (\vec{\omega} \wedge \vec{r}) + (\omega r \cos\phi)^2, \tag{11}$$

the modulus of $\vec{\omega} \wedge \vec{r}$ being $\omega r \cos\phi$, this vector being tangential to the parallel of H, directed toward East. Using the clock velocity in longitude $\dot{L} = dL/dt$, the second term of the right-hand member of (11) is $2\dot{L}\,\omega r^2 \cos^2\phi$. The last term of (11) is twice the potential U^c of the centrifugal force at a point coinciding with H, but fixed with respect to the Earth, so that, putting

$$U^* = U + U^c,$$

U^* being constant on the geoid, with the value $(U^*)_0$, (11) becomes

$$t - \tau = \frac{1}{2c^2}\int_{t_0}^{t} U^*dt + \frac{1}{2c^2}\int_{t_0}^{t} v_g^2\, dt$$

$$+ \frac{\omega}{c^2} \int_{t_0}^{t} \dot{L}\, r^2\cos^2\phi dt + constant. \qquad (12)$$

7.4.1. Application to the definition of TAI

For a fixed clock on the geoid, $\vec{v}_g = 0$, $\dot{L} = 0$,

$$t = \tau + \frac{1}{c^2} (U^*)_0 (t - t_0) + constant.$$

The normalized frequency difference between t and τ is

$$(U^*)_0/c^2 = 7 \times 10^{-10}.$$

In order to avoid this inconvenient frequency offset, the definition of TAI, the ideal form of which is designated by t^*, requires that

$$t^* = \tau + constant, \text{ on the geoid,}$$

t^* being obtained from

$$t^* = kt + constant$$

k being a constant close to unity. One finds easily, neglecting again some terms in $1/c^4$

$$k = 1 - (U^*)_0/c^2 \approx 1 - 7 \times 10^{-10}$$

and, for any fixed clock,

$$t^* = \tau + \frac{1}{c^2} \left[U^* - (U^*)_0\right] (t^* - t_0^*) + constant. \qquad (13)$$

Thus, the participation to TAI of atomic clocks and frequency standards fixed with respect to the Earth requires a normalized frequency correction of

$$\Delta f = \left[U^* - (U^*)_0\right]/c^2 \qquad (14)$$

which can be approximated by

$$\Delta f = -\, gh/c^2,$$

where h is the altitude above the geoid and g the acceleration due to gravity (including the rotational acceleration of the Earth). A better value of Δf is

$$\Delta f = \left\{ GM \left(\frac{1}{r} - \frac{1}{a}\right) + \frac{1}{2}\,\omega^2 \left(r^2\cos^2\phi - a^2\right) - \frac{J_2 GM}{2a} \left[1 + \left(\frac{a}{r}\right)^3 \left(3\sin^2\phi - 1\right)\right] \right\} / c^2$$

where

GM	is the geocentric gravitational constant	GM	$= 3.986 \times 10^{-14}\ \mathrm{m^3s^{-2}}$
a,	the equatorial radius of the Earth	a	$= 6\ 378\ 137$ m
J_2,	the quadrupole moment coefficient of the Earth	J_2	$= 1.083 \times 10^{-3}$

The gravitational frequency shift is, close to the geoid, 1.1 10^{-13} for 1000 metres.

Clocks and frequency standards moving with respect to the Earth can also enter in the formation of TAI with

$$t^* = \tau + C + constant \qquad (15)$$

$$C = \frac{1}{c^2} \int_{t_0}^{t} \left[U^*(t) - (U^*)_0\right] dt + \frac{1}{2c^2} \int_{t_0}^{t} v_g^2\, dt + \frac{\omega}{c^2} \int_{t_0}^{t} \dot{L}\, r^2 \cos^2\phi\, dt \qquad (16)$$

t being replaced by t^*, or by the realized time TAI, when evaluating the integrals in C.

In (16), it is sometimes convenient to evaluate the last term by

$$\int_{t_0}^{t} \dot{L}\, r^2 \cos^2\phi\, dt = 2A,$$

A being the area swept by the equatorial projection of EH in an Earth-fixed system, and is reckoned positive when H goes Eastward.

The relativistic definition of TAI requires that the TAI algorithm processes not the readings of the clocks themselves (representing τ), but the corrected readings by addition of the corrections of C of (16).

The clock corrections issued by BIPM include

(i) the departure of the clock from TAI due to physical defects of the clock (instability, inaccuracy) and also to the defects of TAI itself,

(ii) a relativistic part, which, for a fixed clock on Earth, is a linear function of time.

Two clocks are said to be synchronous in the geocentric frame of the TAI definition when their readings are adjusted in order to keep the same approximation of t^* (instead of delivering their proper times τ). The synchronization of fixed ideal clocks with TAI (which in the absence of corrections would produce ideal proper seconds) requires a phase and frequency offset. The synchronization with TAI of moving ideal clocks, such as clocks in satellites, requires, in general, a more complex correction. The synchronization can be checked by clock transportation or by transmission of electromagnetic signals.

7.4.2. Application to the synchronization by clock transport

A and B are two clocks, fixed on the Earth, that should be synchronized by the transport of clock P. Let us assume first that A and B are frequency adjusted so that their outputs are ideal coordinate times in conformity with the TAI definition, but with a phase offset:

$$t_A^* - t_B^* = k_{AB} .$$

The synchronisation consists in determining the constant k_{AB}.

The travelling clock, giving its proper time τ_P, is first placed close to A, and compared to it at date t_1, so that the constant k_{AP} is determined in order to have

$$\tau_P + C_P(t_1) + k_{AP} = t_A^*$$

where $C_P(t_1)$ is computed from (16). Then P is carried close to B and compared to it at date t_2. As, at this instant

$$t_A^* = \tau_P + C_P(t_2) + k_{AP},$$

k_{AB} is obtained, after computation of

$$\Delta C_P = C_P(t_2) - C_P(t_1)$$

by integration from t_1 to t_2 in (16), which requires a knowledge of the path of P.

For instance, in a usual flight by commercial airline, one obtained:

Paris-Montreal	$\Delta C_P = -32$ ns
Montreal-Paris	$\Delta C_P = -42$ ns

In practice, even after conversion into coordinate time, the clock rates differ. The mean rate of P with respect to A is estimated by bringing back P to A, and the reading of A at t_2 is interpolated with the help of P.

7.4.3. Synchronization by electromagnetic signal

The propagation of the signal is characterized by putting $d\tau = 0$ in (9), $\vec{r}$ describing the path. With

$$\vec{v}_e = \vec{\omega} \wedge \vec{r}.$$

and $\vec{r}_g$ representing the path of the signal in an Earth-fixed system:

$$d\vec{r} = \vec{v}_e dt + d\vec{r}_g.$$

then dt is solution of the equation

$$\left(1 - \frac{2U}{c^2} - \frac{(\vec{v}_e)^2}{c^2}\right) dt^2 - \frac{2}{c^2} \vec{v}_e . d\vec{r}_g dt - \frac{(d\vec{r}_g)^2}{c^2} = 0$$

which is, keeping the only solution with a physical meaning, and neglecting terms of the order of $(1/c)^3$ and above,

$$dt = \frac{|d\vec{r}_g|}{c} + \frac{1}{c^2} \vec{v}_e \cdot d\vec{r}_g .$$

The interval of coordinate time elapsed during the propagation from clock A to clock B, t_2-t_1, is given by

$$t_2 - t_1 = \frac{1}{c} \int_{t_1}^{t_2} |d\vec{r}_g| + \frac{1}{c^2} \int_{t_1}^{t_2} \vec{v}_e \cdot d\vec{r}_g.$$

Both terms of t_2-t_1 are path dependent. The first one is not dependent on the direction, while the second is.

In the first term, $d\vec{r}_g$ is the increment of coordinate length, which can be replaced by $d\sigma\ [1-(U^*+(U^*)_0)/c^2]$, $d\sigma$ being the increment of proper length. This term is close to the classical propagation delay, it amounts for instance to about 0.25 s for the synchronisation of two clocks on the ground, using a geostationary satellite. It can be either computed, if the distances are known, or measured by methods such as clock transportation and two-way transmission along the same path.

The second term can be written

$$\frac{\omega}{c^2} \int_{t_1}^{t_2} \dot{L}\ r^2 \cos^2\phi\ dt \quad \text{or} \quad \frac{\omega}{c^2} \cdot 2A$$

with the same sign conventions as in 7.4.1. In a synchronization using a geostationary satellite, it reaches several hundreds of nanoseconds.

7.4.4. Application to the definition of the Terrestrial Time

For a clock at geocentre, assuming that it does not suffer the gravitational potential of the Earth, $t = \tau + constant$. However, the definition of TT (see Chapter 15, Section 3.2.6) requires a frequency offset so that

$$\mathrm{TT} = \tau\left[1 - \frac{1}{c^2}(U^*)_0\right] + constant,$$

which can be written, with sufficient accuracy

$$\mathrm{TT} = \tau - \frac{1}{c^2}(U^*)_0\ (t - t_0) + constant,$$

where $t - t_0$ can be reckoned using a realized time scale such as TAI.

In the geocentric frame of the TAI definition, TT is identical to t^*, the ideal form of TAI, given by (13), but with a phase offset.

In other frames of reference, TT can be considered as a proper time and is not synchronous with clocks which would read t^*. The conversion into TT of these clock readings requires the addition of periodic terms depending on the location of the clock and reaching, for instance, 2.1 μs in a barycentric frame (see Chapter 17).

7.4.5. Application to clocks in satellites of the Earth

When computing t^* for a clock in a satellite, it may be more convenient to refer the velocity to a non-rotating geocentric frame instead of the Earth, as is done in (12). Thus

$$\tau^* = \tau + \frac{1}{c^2} \int_{t_0}^{t} \left(U - (U^*)_0\right) dt + \frac{1}{2c^2} \int_{t_0}^{t} v^2 \, dt.$$

Note that U is the gravitational potential only, which does not include the effect of the Earth rotation.

8. THE COORDINATED UNIVERSAL TIME UTC

Most of the users of UT1, in geodynamics, geodesy, can wait until the publication of tables giving UT1–TAI. A small number of users need precise real time values of UT1–TAI and even predictions (especially for space navigation): their needs can be satisfied by rapid data processing services and fast communication systems. But many users of astronomical navigation and astro-geodesy cannot be reached easily and need an approximation to UT1 in real time.

To serve the interests of these users and of the TAI users, while avoiding to disseminate by time signals two different time scales, a compromise solution was found: the Coordinated Universal Time, which is sufficiently close to UT1 and in strict relationship with TAI. Although the coordination of the time signals, based on the frequency of caesium standards existed beforehand, the strict relationship with TAI was established in January 1965. It is given by Tables III and IV (approximately before 1965), with the maximum departure of UTC with respect to UT1.

In the present system, defined by International Radio Consultative Committee (CCIR), Recommendation 458, and in operation since the 1st of January 1972:

$$\text{TAI} - \text{UTC} = \text{n seconds, n being an integer or zero}$$
$$|\text{UT1} - \text{UTC}| \leq 0.9 \text{ second.}$$

The UTC scale is thus adjusted by insertion or deletion of one second (positive or negative *leap second*), from time to time. The leap second is the last second of a UTC month, but preference is given to the end of December and June. Events can be dated unambiguously in the vicinity of a leap second as shown by the examples of Figure 2. The International

TABLE III

Frequency offsets and steps of UTC, maximum departure from UT1

Date (at 0 h UTC)		Offsets	Steps	maximum |UT1 – UTC|
1961	Jan. 1	-150×10^{-10}		0.05 s
	Aug. 1	"	+0.050 s	"
1962	Jan. 1	-130×10^{-10}		"
1963	Nov. 1	"	-0.100 s	0.10 s
1964	Jan. 1	-150×10^{-10}		"
	Apr. 1	"	-0.100 s	"
	Sep. 1	"	-0.100 s	"
1965	Jan. 1	"	-0.100 s	"
	Mar. 1	"	-0.100 s	"
	Jul. 1	"	-0.100 s	"
	Sep. 1	"	-0.100 s	"
1966	Jan. 1	-300×10^{-10}		"
1968	Feb. 1	"	+0.100 s	"
1972	Jan. 1	0	-0.107 7580 s	0.90 s
	Jul. 1	"	-1 s	"
1973	Jan. 1	"	-1 s	"
1974	Jan. 1	"	-1 s	"
1975	Jan. 1	"	-1 s	"
1976	Jan. 1	"	-1 s	"
1977	Jan. 1	"	-1 s	"
1978	Jan. 1	"	-1 s	"
1979	Jan. 1	"	-1 s	"
1980	Jan. 1	"	-1 s	"
1981	Jul. 1	"	-1 s	"
1982	Jul. 1	"	-1 s	"
1983	Jul. 1	"	-1 s	"
1985	Jul. 1	"	-1 s	"
1988	Jan. 1	"	-1 s	"

Earth Rotation Service (IERS) decides upon and announces the occurrence of leap seconds, such an announcement being made at least eight weeks in advance.

Since 1972, all the radio time signals follow the UTC which is the only time scale directly available to the users and also the basis of public time, after correction of integral numbers of hours required by the time zone system. Many radio time signals include a DUT1

TABLE IV

Relationship between TAI and UTC, until 1988 Dec. 31

Limits of validity (at 0h UTC)	TAI - UTC (in seconds)
1961 Jan. 1 - 1961 Aug. 1	1.422 818 + (MJD − 37 300) × 0.001 296
Aug. 1 - 1962 Jan. 1	1.372 818 + " "
1962 Jan. 1 - 1963 Nov. 1	1.845 858 + (MJD − 37 665) × 0.001 1232
1963 Nov. 1 - 1964 Jan. 1	1.945 858 + " "
1964 Jan. 1 - Apr. 1	3.240 130 + (MJD − 38 761) × 0.001 296
Apr. 1 - Sep. 1	3.340 130 + " "
Sep. 1 - 1965 Jan. 1	3.440 130 + " "
1965 Jan. 1 - Mar. 1	3.540 130 + " "
Mar. 1 - Jul. 1	3.640 130 + " "
Jul. 1 - Sep. 1	3.740 130 + " "
Sep. 1 - 1966 Jan. 1	3.840 130 + " "
1966 Jan. 1 - 1968 Feb. 1	4.313 170 + (MJD − 39 126) × 0.002 592
1968 Feb. 1 - 1972 Jan. 1	4.213 170 + " "
1972 Jan. 1 - Jul. 1	10 s (integral number of seconds)
Jul. 1 - 1973 Jan. 1	11 s
1973 Jan. 1 - 1974 Jan. 1	12 s
1974 Jan. 1 - 1975 Jan. 1	13 s
1975 Jan. 1 - 1976 Jan. 1	14 s
1976 Jan. 1 - 1977 Jan. 1	15 s
1977 Jan. 1 - 1978 Jan. 1	16 s
1978 Jan. 1 - 1979 Jan. 1	17 s
1979 Jan. 1 - 1980 Jan. 1	18 s
1980 Jan. 1 - 1981 Jul. 1	19 s
1981 Jul. 1 - 1982 Jul. 1	20 s
1982 Jul. 1 - 1983 Jul. 1	21 s
1983 Jul. 1 - 1985 Jul. 1	22 s
1985 Jul. 1 - 1988 Jan. 1	23 s
1988 Jan. 1	24 s

MJD = 37 300 corresponds to 1961 Jan. 1
" 37 665 " 1962 Jan. 1
" 38 761 " 1965 Jan. 1
" 39 126 " 1966 Jan. 1

information, which is the value of UT1-UTC rounded to 0.1 s, transmitted by a simple and common audible code (defined by the CCIR). A few time signals give an additional dUT1 information which provides UT1-UTC to ± 0.02 s by

$$UT1 - UTC = DUT1 + dUT1.$$

However DUT1 is little used and tends to disappear.

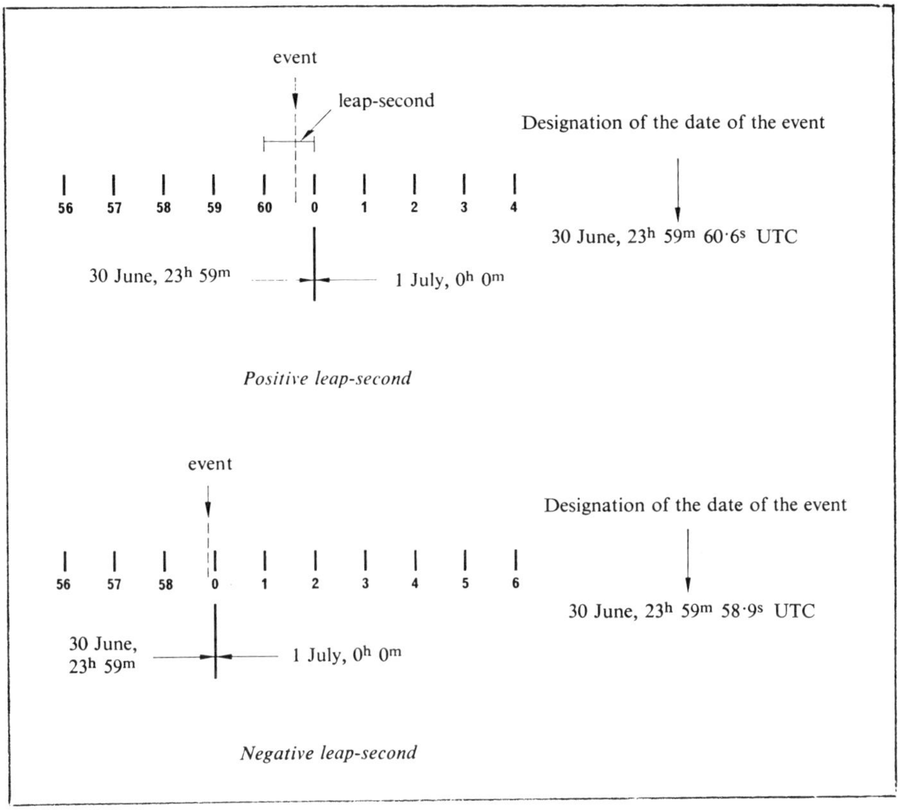

Figure 2. Dating an event in the vicinity of a leap second.

9. THE WORLDWIDE ORGANIZATION OF THE METROLOGY OF ATOMIC TIME

9.1. THE LOCAL APPROXIMATION UTC (k) TO UTC

No physical clock can be settled strictly on UTC or TAI which are known with a delay. To fulfil the immediate needs, the national time services k (k being the acronym of the laboratories) maintain their own approximations to UTC, denoted by UTC(k), by extrapolating the rates of their clocks with respect to UTC.

There are no agreed rules concerning these UTC(k), and the way they are steered on UTC.

In most cases, UTC(k) is the output of a selected clock which is called the master clock of the laboratory; this fact is sometimes emphasized by specifying UTC (k MC), MC for master clock. UTC(k) can also be the output of a physical device averaging the data of several clocks. In some cases, the UTC(k) are computed time scales based on sets of local clocks.

Most of the UTC(k) are frequency steered (without time steps), and their departure from UTC does not exceed 5 μs; this can be achieved, either by adjusting some parameters of the master clock (C-field of a caesium standard, for instance), or by using a "microstepper" external to the clock. Some laboratories, especially those having a primary clock, prefer not to apply frequency correction and they make time step adjustments by several μs when the departure of UTC(k) with respect to UTC is too large.

The CCIR Recommendation 458 states that the time services should relate datings to their own time scale UTC(k). This happens, in practice: clock comparisons, time signal receptions, results of travelling clock visits, etc., are all referred to the UTC(k). The time signal emissions are based on the UTC(k). The BIPM publishes the values of UTC-UTC(k).

It is important to note that the UTC(k), in general, on account of their steering, do not represent the performances of any particular clock. On the other hand, they are not the most stable and accurate time scales. They must be used as intermediate time scales, and the dating of observations must be ultimately referred to TAI or to some atomic time scale defined thereafter, when the reference to 'uniform' time is needed.

9.2. THE LOCAL INDEPENDENT ATOMIC TIME SCALES TA(k)

Several organizations 'k' establish with their own algorithms, a mean atomic time scale using the clocks of a laboratory, or, in some cases, national clocks in several laboratories. In other cases, a national time scale is based on a single high quality primary clock.

All the time scales thus obtained are independent from each other (but not from TAI which uses the data of the same clocks), and are designated by TA(k), an official CCIR notation, in all languages.

Thus the TA(k) are not steered on TAI. They can have a large difference with TAI, up to several seconds. The rate of most of them is not based on accurate realizations of the second. The main purpose of the TA(k) is to provide a stable local reference, with shorter delays than TAI, and with a better accuracy of reading, since they are free of the uncertainties of the long distance time comparisons. The TA(k) are especially useful for maintaining synchronization of systems, and also to steer the UTC(k) on UTC. They can be also used for dynamical studies. They are easily available through the BIPM publications, as shown thereafter in Section 10.

TABLE V

Independent local atomic time scales (1988)

ind Cs : industrially made caesium clock
H maser : hydrogen maser
lab fr st : laboratory caesium frequency standard
prim cl : laboratory caesium standard operating continously

Time scale TA(k)	established by	Source of TA(k)	Accuracy ensured by lab fr st	GPS time link
TA(APL)	Applied Physics Laboratory, Laurel, USA	4 H masers	no	yes
TA(CH)	Office Fédéral de Métrologie, Berne, Suisse	14 ind Cs in 3 lab.	no	yes
TA(CRL)	Communications Research Laboratory, Tokyo, Japan	11 ind Cs 3 H masers 1 lab fr st	no	yes
TA(DDR)	Amt fuer Standardisierung Messwessen und Warenpruefung, Berlin, DDR	2 ind Cs in 2 lab.	no	no
TA(F)	Laboratoire Primaire du Temps et des Fréquences Paris, France	19 ind Cs in 9 lab.	no	yes
TA(NBS)	National Bureau of Standards, Boulder, USA	19 ind Cs 1 H maser 1 lab fr st	yes	yes
TA(NRC)	National Research Council, Ottawa, Canada	1 prim cl	yes	yes
TA(PTB)	Physikalisch-Technische Bundesanstalt Braunschweig, FRG	1 prim cl	yes	yes
TA(SO)	Shanghai Observatory, P. R. of China	4 ind Cs 3 H masers 1 lab fr st	no	no
TA(SU)	Gosstandart, Moscow, USSR	2 ind Cs 8 H masers 2 lab fr st	yes	no
TA(UNSO)= A1(MEAN)	U.S. Naval Observatory, Washington, USA	25 selected ind Cs	no	yes

Table V lists the existing TA(k) in 1987, with some extra information.

9.3. EXAMPLES OF NATIONAL ORGANIZATION

Figure 3 summarizes the clocks and frequency standards equipment which can be found in a time service, and the time scales which can be formed. The local organization can be represented by functional links between the various boxes.

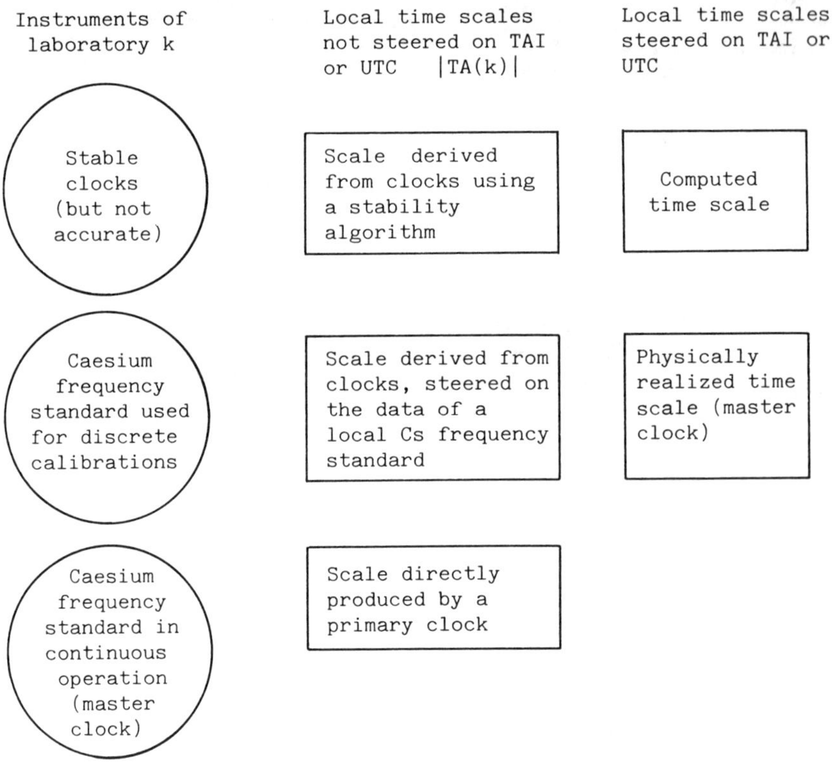

Figure 3. Clocks, frequency standards and time scales which can be found in a time service

Figure 4 drawn with the same pattern of boxes shows the organization of some laboratories, taken as examples.

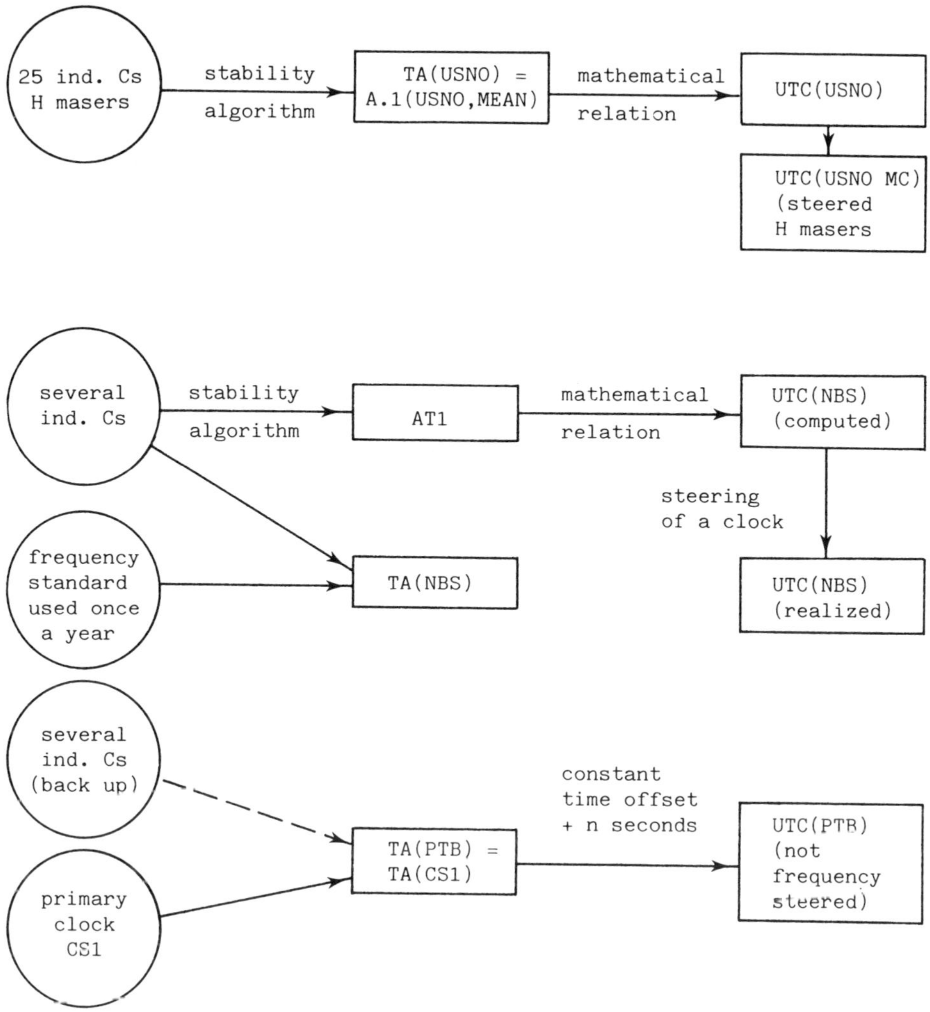

Figure 4. Examples of organization, at US Naval Observatory (USNO), National Bureau of Standards, (NBS), Physikalisch-Technische Bundesanstalt (PTB). Ind. Cs. = industrially made caesium clock.

10. DISSEMINATION OF TAI AND UTC, AND OF THE SI SECOND

Detailed and regularly updated informations on these matters can be found in the *Recommendations and Report of CCIR*, Volume VII, Standards Frequencies and Time Signals. Here are only few notes, with the emphasis on the most accurate methods.

TABLE VI

Excerpt of the BIPM Circular T (or BIH Circular D) showing the list of laboratories where UTC and TAI are accessible.

Date 1988 (0h UTC)	MAR 9	MAR 19	MAR 29
MJD	47229	47239	47249
Laboratory k	UTC-UTC(k)	(Unit = 1 microsecond)	
AOS (Borowiec)	2.16	1.65	0.90
APL (Laurel)	0.03	0.04	0.03
ASMW (Berlin)	0.08	0.07	0.03
AUS (Canberra)	-12.79	-12.93	-13.07
BEV (Wien)	-4.93	-5.44	-6.09
CAO (Cagliari)	1.08	1.32	1.54
CH (Berne)	1.49	1.47	1.44
CRL (Tokyo)	-1.69	-1.74	-1.78
CSAO (Shaanxi)	0.63	0.60	0.57
FTZ (Darmstadt)	16.12	16.28	16.46
IEN (Torino)	-1.35	-1.29	-1.07
IFAG (Wettzell)	-2.04	-1.34	-0.88
ILOM (Mizusawa)	-35.73	-35.75	-35.77
INPL (Jerusalem)	56.74	57.83	58.98
JATC (Xian)	1.11	0.96	0.82
KSRI (Daejeon)	-5.47	-5.64	-5.80
NBS (Boulder)	-0.62	-0.67	-0.75
NIM (Beijing)	9.93	9.80	9.67
NPL (Teddington)	4.15	4.17	4.20
NPLI (New-Delhi)	-9.90	-10.38	-10.91
NRC (Ottawa)	-9.49	-9.43	-9.41
NRLM (Tsukuba)	-23.05	-23.21	-23.38
OMSF (San Fernando)	3.59	3.71	3.82
OP (Paris)	-0.52	-0.57	-0.59
ORB (Bruxelles)	-48.92	-9.16	-9.23
PKNM (Warsaw)	-1.15	-0.77	-
PTB (Braunschweig)	4.27	4.29	4.30
SO (Shanghai)	1.99	2.05	2.10
STA (Stockholm)	0.48	0.48	0.44
SU (Moscow)	20.76	20.56	20.38
TAO (Tokyo)	-1.90	-1.93	-1.97
TL (Taiwan)	268.95	270.41	271.86
TP (Praha)	0.24	0.56	0.86
TUG (Graz)	-2.04	-1.77	-1.51
USNO (Washington) (MC)	-4.25	-4.17	-4.11
VSL (Delft)	3.91	3.88	3.84
YUZM (Beograd)	-0.18	-0.71	-0.97
ZIPE (Potsdam)	0.43	0.46	0.46

TABLE VI continued

Date 1988 (0h UTC)	MAR 9	MAR 19	MAR 29
MJD	47229	47239	47249
Laboratory k	TAI-TA(k)	(Unit = 1	microsecond)
AOS (Borowiec)	-88.40	-90.31	-92.46
APL (Laurel)	0.03	0.04	0.03
CH (Berne)	-47.95	-48.16	-48.37
CRL (Tokyo)	-3.44	-3.47	-3.52
CSAO (Shaanxi)	39.61	39.58	39.55
DDR (Berlin)	-26.07	-26.30	-26.56
F (Paris)	52.13	52.55	52.94
JATC (Xian)	-0.27	-0.06	-0.22
NBS (Boulder)	-45108.02	-45108.39	-45108.78
NIM (Beijing)	-7.61	-7.67	-7.79
NRC (Ottawa)	21.58	21.64	21.65
PTB (Braunschweig)	-359.13	-359.11	-359.10
SO (Shanghai)	-45.69	-45.66	-45.65
SU (Moscow)	2827270.76	2827270.56	2827270.38
USNO (Washington)	-34555.12	-34555.67	-34556.24

10.1. HIGH ACCURACY DISSEMINATION

The primary dissemination is the publication by BIPM of the corrections UTC-UTC(k) and TAI-TA(k) at 10-day intervals, in monthly Circular T. Until the end of 1987, these data were published in the same form by the Bureau International de l'Heure (BIH) Circular D. Table VI is a excerpt of one of these circulars. Past data appear also in annual reports issued by BIH, then by BIPM. The data are also in files with direct access by telephone (consult BIPM).

A user who wishes to date an event in the UTC and TAI scales with the highest accuracy must first establish a time link with one of the laboratories listed in Circular T. To this end, the methods described in Section 5.2 are convenient. Let 'j' be this laboratory. He has then access to TAI and UTC through the listed values of UTC-UTC(j). The uncertainty of the UTC/TAI date is the quadratic sum of the uncertainty of the user-'j' link and of the UTC-UTC(j). The latter is in the range 10 to 50 ns (one sigma) when 'j' uses a GPS time receiver, in the range 100-200 ns for a LORAN-C receiver. With the data of Circular T, the user can also refer the event to any of the UTC(k) and TA(k) of other laboratories.

Alternatively, the user who has a GPS time receiver obtains, for his clock H, the difference

$$H - \text{GPS time}$$

He has then

$$\text{UTC(USNO)} - \text{GPS time}$$

by the *USNO Daily time differences and relative phase values, series 4,* and

$$\text{UTC} - \text{UTC(USNO)}$$

by BIPM Circular T. To simplify the procedure, the BIPM plans to publish directly

$$\text{UTC} - \text{GPS time}$$

10.2. OTHER MEANS OF DISSEMINATION

(i)–The GPS time is maintained to ± 1 μs from UTC(USNO) (plus an integral number of seconds), and UTC–UTC(USNO) is normally smaller than 5 μs. Thus GPS furnishes UTC and TAI in real time to about ± 5 μs. A closer agreement of GPS time with UTC is planned.

(ii)–In the range 1 μs-1 ms, when the GPS is not used, one has to apply the same method as described in Section 10.1: a link with a national laboratory. In this range of accuracy, there is a need for cheaper and more convenient methods. Experiments of general dissemination of UTC by satellites, using TRANSIT and GOES, have shown the capability of providing an accuracy of 50 μs, but no system operates in an official time and frequency mode at present.

(iii)–At the millisecond level, many radio time signal emissions from ground stations provide UTC by transmitting second pulses with minute and hour markers. Some emissions include a code for automatic dating of observations and public dissemination of time. The CCIR requires that the time signals should not deviate from UTC by more than one millisecond; in practice the deviation at the emission is much less, but at the reception one has to apply a correction for propagation delay, which has an uncertainty of 1 ms on account of the irregularities of the propagation by reflexion on the ionosphere. Lists of time signals can be found in the CCIR documents, in the Annual Report of the Section of Time of BIPM (previously BIH Annual Report) updated yearly, in nautical documents, etc.

10.3. DISSEMINATION OF THE SI SECOND

The second can be derived from the UTC receptions. But for many technical applications, it is more convenient and less expensive to receive one of the many broadcast standard frequencies. According to CCIR rules, the deviation from the UTC and TAI frequency should not exceed 10^{-10}; in fact, the deviation is smaller than 10^{-11} for most of them, and smaller than 5×10^{-13} for a few. Lists of standard frequencies, with the characteristics of their emission are found in the CCIR and BIPM documents.

REFERENCES

Azoubib, J., Granveaud, M. and Guinot, B., 1977, 'Estimation of the scale unit duration of time scales', *Metrologia*, 13, 87.

Forman, P., 1985, 'Atomichron: the atomic clock from concept to commercial product', *Proc. IEEE*, *73*, n° 7, 1181.

Guinot, B. and Lewandowski, W., 1988, 'Accurate time comparisons and positioning by GPS over distances up to 1000 km', *Bulletin géodésique* (in preparation).

Kartaschoff, P., 1978, '*Frequency and Time*', Academic Press, New York.

Markowitz, W., Hall, R.G., Essen, L. and Parry, J., 1958, 'Frequency of cesium in terms of Ephemeris Time', *Phys. Rev. Letters*, 1, 105.

National Bureau of Standards (USA), 1974, '*Time and Frequency, Theory and Fundamentals*', NBS Monograph 140.

Vanier, J., Audoin, C., 1989, '*The quantum physics of atomic frequency standards*', Adam Hilger ed., Bristol.

TIME SYSTEMS IN GENERAL RELATIVITY

TOSHIO FUKUSHIMA
Hydrographic Department, Tokyo, Japan

As is shown clearly in the preceding chapters, the measurement of time is the most precise and reliable observation we can make at present. In utilizing such precise observational data to establish and maintain a highly uniform time system, we confront ourselves with some basic questions such as 'What is a time system?', 'What is its uniformity?' and 'How can we relate two different time systems with each other?'

It is clear that the introduction of general relativity is essential in order to answer the above questions correctly because the error of the latest clock is much smaller than the magnitude of the main correction term due to general relativistic theories. In this chapter, we will discuss the concepts, definitions and related formulae of time systems from the viewpoint of general relativistic theories.

As for the symbols, notations, terminology and other basic concepts concerning general relativistic theories, we follow mainly the book of Misner, *et al.*, (1970). The readers can consult Chapter 5 of this book.

1. BASIC DISCUSSION ON TIME SYSTEMS

First, we consider some basic questions such as what a time system means in Newtonian mechanics and in relativistic theories, respectively.

1.1. TIME IN NEWTONIAN MECHANICS

In Newtonian mechanics as well as many other classical philosophies, time is absolute. Namely time is thought to be unique and uniform all over the universe at any instant. The time interval of any pair of events is constant in whatever frame of reference is used for its measurement. Time is then always quoted with the definite article i.e. 'the time'.

Since we cannot travel from the future to the past, we can only

J. Kovalevsky et al. (eds.), Reference Frames, 417–444.

once record the happening of a specific event. Therefore it is impossible to compare two time intervals directly with each other. Thus the repeatability of measurements, which is the basis of the metrology, is not applicable to the time itself.

In this sense the time has been regarded as a holy concept which man cannot measure easily. This implies that the thing which we measure by a man-made tool called 'clock' may not be the true time itself. Such an ambiguous situation on the measurement of time has produced the expression 'time-like argument' for a scientist to express his vague doubt whether the result he obtained is the true time or not. The word 'argument' shows that the time often means not the object of observation but a tag to label the observed event.

Thus in Newtonian mechanics, we have only one idealization of time reference called 'the time' and multiple realizations of it called 'time-like arguments'. People use the word 'time scale' in place of 'time-like argument' when they emphasize the aspect of time as a measuring reference.

1.2. TIME-LIKE ARGUMENTS IN RELATIVISTIC THEORIES

The appearance of relativistic theories has brought a drastic change to the interpretation of time and space. The relativistic principles make the Newtonian difference between time and space ambiguous. Relativistic theories permit one to produce any kind of mixture of the Newtonian time and space. Therefore we need new concepts of time as well as of space to specify such mixtures one by one.

We especially need a concept representing idealized times in relativistic theories, which is simply represented by the expression 'the time' in the Newtonian mechanics.

From now on we will use the word 'time-like argument' in the sense of an ideal frame of reference in the time dimension while the expression 'time scale' introduced in the preceding chapter is reserved for a realization of it.

It should be stressed that there can be many time-like arguments in the relativistic theories. Although a simple explanation of the expression 'time-like argument' is already given, it is very difficult to answer in a few sentences the question of what 'time-like argument' actually means. Then we prefer to define a time-like argument step by step in the next subsection.

1.3. RELATIVISTIC DEFINITION OF TIME-LIKE ARGUMENT

In relativistic theories, the law of causality becomes more limited than that in Newtonian mechanics. This change requires new concepts on the relation between time and space.

Let us fix a point in the spacetime and name it *O*. Then the whole spacetime is split into the following six subsets of spacetime from the viewpoint of causality relation with the point (see figure 1):

(i)–The set constructed by the point *O* itself called the 'present' of *O*.
(ii)–The set of points to which a certain kind of particle with nonzero rest mass can attain from the point *O*. This set is called the 'time-like future' of *O* and is shown by T^+.
(iii)–The set of points to which a certain kind of particle with zero rest mass such as a photon can attain from the point *O*. This set is called the 'null future' of *O* and is shown by N^+.
(iv)–The set of points from which a certain kind of particle with nonzero rest mass can attain to the point *O*. This set is called the 'time-like past' of *O* and is shown by T^-.
(v)–The set of points from which a certain kind of particle with zero rest mass can attain the point *O*. This set is called the 'null past' of *O* and is shown by N^-.
(vi)–The set of points from which no particle can attain the point *O* and to which no particle can attain from the point *O*. This set is called the 'space-like region' with respect to *O* and is shown by *S*.

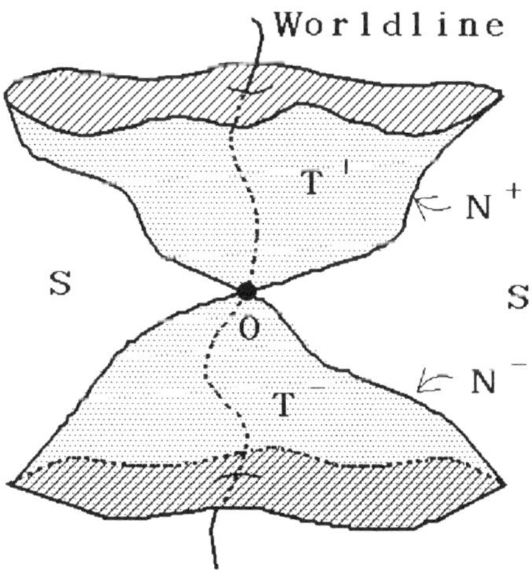

Figure 1. Six Subregions of Spacetime

Using these subregions, we can state the relativistic definition of a time-like argument as follows:

A time-like argument is a quantity defined on a subset of the whole

> spacetime so that its value at any point P in the subset is greater (less) than that at any other point O in the same subset as long as the point P is located in the time-like future (past) of O.

For example, a proper time is a time-like argument partly because it is defined on the worldline of a particle with nonzero rest mass, which is a subset of spacetime in fact, and partly because its value is clearly ordered in an ascending way along the worldline from the past to the future.

It is well known that the relativistic theories have pulled down the time from its throne as an absolute measure into one of four coordinates in the four dimensional spacetime. This change of view caused the question of which coordinate is suitable to be called the time coordinate from the four coordinates defined by the given coordinate system. The relativistic theories have answered this question as follows:

> A coordinate is suitable to be called the time coordinate if and only if the corresponding diagonal component of the given metric tensor is negative.

We say that such a coordinate is 'time-like'. It is easy to show that this definition of 'time-like coordinate' is compatible with the above definition of 'time-like argument'. In other words, thus defined a 'time-like coordinate' is also a time-like argument.

1.4. UNIFORMITY OF TIME-LIKE ARGUMENTS

It is also difficult to define clearly the uniformity of time-like arguments. We must remark that there are two kinds of uniformity in time-like arguments. One is the uniformity with respect to time and the other is the uniformity with respect to space. Illustratively speaking, the former is whether a sufficiently small time interval, say a second, of a clock today is the same as a second of the same clock tomorrow. While an example of the latter is whether a second of a clock at Paris is the same as a second of a clock at Tokyo at the same instant in a sense.

In Newtonian mechanics, time is thought to be unique. Therefore the uniformity of time with respect to the time itself is trivial. On the other hand the uniformity of time with respect to space was a hypothesis on which physical theories are constructed. Then realizing an ideal clock, which seems to tick uniformly, is attained from a group of clocks distributed on some region of space by a suitable statistical procedure such as taking a weighted mean of their readings.

In relativistic theories, the uniformity of time-like arguments cannot be defined globally. The gravitational field or the relative velocity between a pair of separately located clocks causes a difference in their readings even after variations due to local environments are removed. Thus we cannot take any statistical average of readings of clocks distributed in a finite spacetime unless the relativistic effects are corrected properly.

2. VARIOUS TIME SYSTEMS

Next we will pigeonhole a group of well-known time systems from the general relativistic point of view.

2.1. ATOMIC TIME

As has already been described in the preceding chapter, an atomic time is a time scale. It is a realization of a certain time-like argument which will be discussed in Section 2.2.

It is well known that the concept of atomic time is based on quantum physics, the physics thought to be the most reliable in a small size spacetime such as a laboratory. Although the region which the atomic clock occupies is finite, we will ignore its dimensions from now on. Thus, throughout this chapter, we shall assume that an atomic clock occupies a point.

From the above assumption, we can say that the definition of atomic time is local. More precisely, the atomic time is defined only by physical quantities defined at the point it is located: quantities such as its velocity, its acceleration, the temperature, the pressure, the electromagnetic field and other environmental factors. It should be noted that an ideal observation of atomic time is never perturbed by or interfered with remote events.

If we can remove all the variations due to local phenomena from the obtained readings of the atomic clock, we will think them to be uniform. Of course this assertion is an empirical one and should be criticized. However, we must remark that, even if this uniformity is realized for each atomic clock, it never ensures that different atomic clocks located at different places show the same reading.

2.2. PROPER TIME

The concept of proper time is obtained as a relativistic idealization of atomic time. In other words, a proper time is a time-like argument to which the corresponding atomic time scale shall be made as close as possible.

In fact, proper time is defined as follows:

> Let us consider a clock located in a shelter which shuts out any physical disturbance such as the geomagnetic variation, the temperature fluctuation, and others. Assume that the local variations are removed from the readings of the clock sufficiently so that the corrected results are thought to be uniform. Call such an idealized clock a standard clock. Then the proper time of that clock is defined by the integrated time constructed only by ticks of the clock.

Here we must note that the hypothesis that an atomic time is a realization of a proper time is valid if quantum physics is compatible with the general relativistic theories.

Generally the numerical value of proper time depends on the history of the standard clock to which the proper time is referred. In terms of relativistic theories, the proper time constructed by a standard clock is a function of its worldline. This reveals that there are as many proper times as there are different standard clocks. The spirit of the principle of relativity tells that there is no privileged proper time and we must deal with all proper times equally.

The proper time thus defined is a physical observable in relativistic theories. In other words, its measured value is not dependent on the coordinate system chosen for measurement. However, we cannot compare two different proper times directly. This is because the worldlines, along which these proper times are defined respectively, have no common region in general and such common area is inevitable for comparison of different proper times.

2.3. COORDINATE TIME

The concept of a coordinate time comes from regarding time as one of the coordinates which describe events in the four dimensional spacetime. Strictly speaking, a coordinate time is a time-like component of four coordinates in a coordinate system as described in Section 1.3. Therefore there exist as many coordinate times as there are coordinate systems.

A coordinate system is nothing but a system of assigning a name tag to all the points in the given spacetime. That is it is just an artificial tool for men to grasp the world around them. Since a coordinate time is a part of a coordinate system, it is also a man-made tool and has no physical substance.

Usually, in general relativistic theories, a coordinate time is defined as the proper time of a standard clock which rests at the space origin of chosen coordinate system. For example, the coordinate time of a geocentric coordinate system is defined as the proper time of a clock which rests at the geocentre while the gravitational effect of the Earth itself is ignored in computing the proper time (see a detailed discussion in Section 3.2).

Since there is no privileged proper time in the general relativistic theories, we need a certain tool to which we refer in comparing different proper times. Coordinate times are the very tools necessary for connecting many proper times. In fact, when we need to find the connection between two proper times, we first obtain their relation with a common coordinate time separately. Then, we compare these two relations directly. Therefore it is important to give the relation between an arbitrary pair of proper and coordinate times, which is the main object of this chapter.

On the other hand, it is often difficult to find a common coordinate time to which a given pair of proper times are related explicitly. Then we use multiple coordinate times one by one as a sequence of linkage between such a pair of proper times. Thus it is also important to clarify the relation between two arbitrary coordinate times. We remark that such a relation is given by a fully four dimensional coordinate transformation formula since we cannot separate time-time relations from the fully four dimensional formula in the relativistic theories (a detailed discussion is given in Section 4).

2.4. DYNAMICAL TIME

The concept of dynamical time is based on a vague confidence that the time as an independent variable of dynamical equations, which is the definition of a dynamical time, must flow continuously and constantly.

Very often a certain coordinate system is chosen and fixed to describe the equation of motion on which a dynamical time is based. In this sense, we can say that a dynamical time is a realization of coordinate time. Strictly speaking, therefore, a dynamical time is not a time-like argument but a time scale.

Practically speaking, a dynamical time is determined by comparing observations and theoretical predictions of dynamical phenomena in a certain coordinate system. Thus a dynamical time depends on three factors: the dynamical theory giving a method of reduction, the phenomenon to be observed and on the chosen coordinate system in which the observed value is expressed.

2.5. COORDINATED TIME

There is another realization of coordinate time called coordinated time. The name 'coordinated time' is very similar to 'coordinate time'. However, we must discriminate these two. The coordinated time is a time scale constructed by an appropriate coordination of many time scales.

A typical example of coordinated time is the TAI maintained by the BIPM. In reality, TAI is obtained from the reading of many atomic clocks widely spread over the Earth through a certain statistical procedure after subtracting the systematic differences among them (see details in Chapter 16). We should note that any coordinated time as well as the TAI is neither an atomic time nor a proper time but a certain average of them. The value of a coordinated time is dependent on the chosen physical and statistical models and on the choice of a coordinate system as well as in the case of a dynamical time.

Theoretically the idealization of TAI called TT is expected to be equal to the idealization of many TDTs, dynamical times of the Earth. These two idealizations represent the same coordinate time and are different only in the method of determination and in the basic assumptions.

2.6. BROADCAST TIME

A broadcast time is a time scale which differs in principle from an atomic time, a dynamical time or a coordinated time. In fact, a broadcast time of a receiver is defined as an integrated time scale constructed by received time signals which are broadcasted using ticks of a certain time scale.

The time broadcasted is based on an atomic clock in some cases and on a coordinated (i.e. virtual) clock in others. For example, the GPS time broadcasted from a certain NAVSTAR satellite corresponds to the former while various UTC time signals are examples of the latter. The pulsar time (see Chapter 15, Section 3.3) is a sort of broadcast time from a clock called a pulsar.

We must remark that both a coordinate time and a realization of broadcast time can be defined over a region of spacetime spread in the space-like directions while a proper time is defined only along a purely time-like domain called a worldline. In this sense, a realization of broadcast time can be a certain kind of coordinate time. In fact, the time coordinate of the optical coordinate system of Synge corresponds to a realization of this broadcast time (Synge, 1960) although such a coordinate system as well as such a coordinate time has not been used widely.

In Newtonian mechanics and in the special theory of relativity, a broadcast time is in principle equivalent with a broadcasted time scale. That is there should be only a constant difference between these two time scales as long as the relative position of the receiver with respect to the station transmitting time signals does not change. However, these two time scales must be discriminated in the general relativistic theories since the signal propagation depends on the curvature of spacetime and the curvature itself varies with time and place. In other words, the hyperplane of constant broadcasted time signals varies with time due to the varying gravitational field.

3. RELATION BETWEEN COORDINATE AND PROPER TIMES

3.1. EQUATION OF PROPER TIME

Let us fix a finite region of spacetime and choose a certain coordinate system which covers the region. Let us consider a pair of points in the given region of spacetime and assume that they are very close to each other so that we can ignore the second and higher order effects in evaluating any difference of a quantity between these two points as a function of dx^μ, the infinitesimal difference in coordinates between them.

Let us now consider a symmetric 4×4 tensor function of coordinates $g_{\mu\nu}(x^\lambda)$ and name it the metric. Consider a quantity $g_{\mu\nu}(x^\lambda)dx^\mu dx^\nu$. This quantity will be positive, zero or negative, depending on the variation of the metric tensor with respect to x^λ and the combination of metric and dx^μ. As has already been mentioned in Section 1.3, we say that the relation between these two adjacent points is space-like, null and time-like according to whether the above quantity is positive, zero or negative, respectively. From now on we will restrict ourselves to the case of time-like relation.

Then the infinitesimal interval of proper time $d\tau$ is defined as

$$(cd\tau)^2 = - g_{\mu\nu}\, dx^\mu dx^\nu > 0 \qquad (1)$$

where c is a constant for the conversion of units called the speed of light in vacuum.

Since the right hand side of the above equation is positive, we can take the square root of the equation as

$$cd\tau = (- g_{\mu\nu}\, dx^\mu dx^\nu)^{1/2} \qquad (2)$$

Now we introduce the concept of coordinate velocity defined by

$$v^j = dx^j/dt = cdx^j/dx^o \tag{3}$$

where $t=x^o/c$ is the time coordinate of the chosen coordinate system.

Then we can rewrite equation (2) as

$$d\tau/dt = (-g_{oo} - 2g_{oj}v^j/c - g_{jk}v^jv^k/c^2)^{1/2} \tag{4}$$

where Einstein's summation convention is used.

This first order ordinary differential equation with respect to τ is called the equation of proper time. This equation determines a proper time τ as a function of the given coordinate time and the position of the clock in the chosen coordinate system.

In the special theory of relativity, we can set the metric to be Minkowskian as

$$g_{00} = -1, \quad g_{jj} = +1, \quad \text{others} = 0 \tag{5}$$

by a suitable choice of coordinate system. We remark that such coordinate system is called to be inertial. Also we assume the coordinate velocity of the clock in the chosen inertial coordinate system is constant. Then we can integrate the equation of proper time as

$$\tau = \tau(\boldsymbol{v}) = t/\gamma + \tau_o \tag{6}$$

where τ_o is a constant and γ is the famous Lorentz factor defined by

$$\gamma = (1 - v^2/c^2)^{-1/2} > 1 \tag{7}$$

Equation (6) shows that the proper time of a clock moving constantly relative to the chosen coordinate system runs slower than the proper time of a clock which rests in the inertial coordinate system. This is the correct expression of the well-known Lorentzian time dilatation effect. We should note that a coordinate time t can be replaced by the proper time of any resting clock $\tau(0)$.

3.2. NEWTONIAN APPROXIMATION

Now we proceed to the next stage in solving the equation of proper time. We note that the spectial theory of relativity gives us no information about the gravitational forces. In other words, the Minkowskian metric describes a universe without any gravitational field.

The fact that the Newtonian law of universal attraction gives a good model of the real universe implies that the next step to extend the Minkowskian metric is to include the Newtonian gravitational potential into the metric tensor as follows

$$g_{oo} = -1 + 2U/c^2, \quad g_{jj} = +1, \quad \text{others} = 0 \tag{8}$$

where U is the Newtonian (or scalar) force function. Here we use the word 'force function' in the sense of a negative gravitational potential. The above metric is called the Newtonian metric.

It is noted that not only the derivatives of U but also the value of U itself are meaningful in the general relativistic theories. That is much different from the Newtonian common sense. It is clear that to change the constant part of U is to adopt another metric tensor. This is to choose another coordinate time.

With use of the Newtonian metric, we obtain the Newtonian approximation of equation of proper time as follows

$$d\tau/dt = [1 - (2U+v^2)/c^2 + \ldots]^{1/2} \tag{9}$$

where we do not write explicitly terms whose order of magnitude are equal to or greater than the square of that of the Newtonian part, i.e. $o(v^4/c^4)$ (see Section 3.7).

If we ignore higher order terms, we can expand the square root in the above equation as

$$d\tau/dt = 1 - (U + v^2/2)/c^2 \tag{10}$$

This is the Newtonian equation of proper time.

From this approximate equation, we know that the ticks of a standard clock slow down compared with the coordinate time when the clock moves around or when it is influenced by a gravitational field. As a coordinate acceleration is equivalent to a kind of uniform gravitational field, it also changes the rate of a proper time.

From the form of equation (10), it is clear that as for resting clocks, the proper times are the same in the Newtonian approximation if and only if they undergo the same Newtonian gravitational field. Thus the clocks resting on an equipotential surface such as the geoid can be synchronized easily. This fact brought the qualifier 'at mean sea level' in the definitions of units of the TDT and the TDB in the IAU recommendation, which is given in Chapter 16. We must note that to include a rotational velocity in the Newtonian potential means to adopt a rotating coordinate system as the basic coordinate system. Therefore it corresponds in principle to choose a new coordinate time (see also Sections 3.5 and 3.6).

3.3. SOLUTION FOR THE CASE OF KEPLERIAN MOTION

In order to understand how a proper time varies as a function of coordinate time in a practical physical model, we illustrate here its functional form explicitly under an ideal situation.

Let us assume that there is only one gravitating body and call it the central body. Let us assume also that the central body is spherically symmetric and adopt a coordinate system where the central body rests at its space origin. Then the metric tensor is given by equation (8) in the Newtonian approximation and the force function U is written as

$$U(t, \boldsymbol{x}) = GM/r \tag{11}$$

where G is the universal constant of gravitation, M is the mass of the central body and r is the distance between the central body. A point is specified by its coordinates $(t, \boldsymbol{x})$ in the coordinate system adopted.

Further, we assume that a clock freely moves in the gravitational field produced by the central body. In this case a free motion is a Keplerian motion. Then the following equation of energy integral holds in the Newtonian approximation

$$- U(t, \boldsymbol{x}) + \boldsymbol{v}(t, \boldsymbol{x})^2/2 = E = \text{constant} \tag{12}$$

where E is the total energy of the clock divided by its mass.

Using the above equation we can rewrite the Newtonian equation of proper time as

$$d\tau/dt = 1 - [E + 2U(t, \boldsymbol{x})]/c^2 \tag{13}$$

Now, if we assume that the clock moves on a Keplerian ellipse around the central body, the expressions of E and U are given as

$$E = - GM/(2a) \tag{14}$$

$$U = GM/r = (GM/a)/(1 - e \cos u) \tag{15}$$

where a is the semi-major axis, e is the eccentricity and u is the eccentric anomaly of the orbit, respectively. Note that a and e are constants with time. Numerically the eccentric anomaly u is easily obtained by solving the following Kepler's equation

$$u - e \sin u = l \tag{16}$$

Here l is the mean anomaly of the Keplerian motion given by

$$l = n\,t + l_o \tag{17}$$

where l_o is a constant and n is a constant of time called the mean motion of the Keplerian orbit. In the following, we will set l_o zero. Namely we will count t from a certain pericenter passage of the Keplerian orbit.

By means of equation (16) and the differential formula with t and u

$$n\,a\,dt = r\,du \tag{18}$$

equation (13) is easily integrated as

$$\tau - \dot{\tau}_o = [1 - (3GM/2c^2a)]t + (2GM/c^2a)(e/n)\sin u \tag{19}$$

The quantity $2GM/c^2$ is called the gravitation radius of the central body, which is 2.954 km for the Sun and is 8.868 mm for the Earth.

From the above solution we obtain the following informations:

(i)-The relation between a proper time and a coordinate time consists of two parts; the change in rate and the periodic difference.

(ii)-A proper time of a clock freely moving in a finite region delays when compared to a coordinate time. The amount of delay rate is dependent on the typical strength of the gravitational field on the clock's orbit and on the total energy of the clock. It is noteworthy that the general relativistic theories predict a delay three times larger than does the special theory of relativity.

(iii)-The periodic difference has the same period as does the orbit and its magnitude is proportional to the eccentricity of orbit.

When the eccentricity is small, we can expand the right hand side of equation (19) as

$$(\tau - \tau_o) = t + (GM/c^2a)[- (3/2)t$$

$$+ (e/n) \{(2 - e^2/4) \sin l + e \sin 2l + (3/4)e^2 \sin 3l + ...\}] \tag{20}$$

To see the size of this difference, we list in table 1 the magnitudes of each term for some typical cases; the solar system around our Galaxy, some planets around the Sun, the Moon, a GPS/NAVSTAR satellite, and LAGEOS around the Earth. We added the case of a clock resting on the geoid for comparison.

It should be noted that the difference in rate is inversely proportional to the semi-major axis a while the main periodic difference is directly proportional to the quantity e/na or $ea^{1/2}$. Thus, if the eccentricity is kept the same, the difference in rate decreases while that of periodic term increases slowly as the orbit of the clock expands.

Although the solution obtained here is an approximate one and much care must be taken in applying this to a practical problem, this formula is so simple and effective that it has been widely used. For example the periodic part of above formula is used to correct the transmitted time signal in the GPS/NAVSTAR system (Van Dierendonck *et al.*, 1980).

TABLE 1. Difference between coordinate and proper times for some Keplerian orbits

Location of clock	gravitating body	difference in rate (10^{-9})	periodic difference: amplitude	periodic difference: period
Sun	Galaxy	1000000	—	—
Mercury	Sun	38.3	12.68 ms	0.24 y
Venus	ibid.	20.5	0.57	0.62
Earth (Moon)	ibid.	14.8	1.66	1.00
Mars	ibid.	9.71	11.42	1.88
Jupiter	ibid.	2.85	10.92	11.9
Saturn	ibid.	1.55	17.15	29.5
Moon	Earth	0.0173	47 ns	27.3 d
GPS/NAVSTAR	ibid.	0.251	6.8-23	0.499
LAGEOS	ibid.	0.545	7.0	0.157
Geoid*		0.697	—	—

* In this case a clock is assumed to be fixed on the geoid and not moving on a Keplerian orbit.

3.4. IGNORANCE OF SELF-GRAVITATIONAL EFFECTS

In the preceding subsection, we discussed the proper time of a (massless) clock moving freely in a given gravitational field. The next step may be to extend the relation (19) to a more complicated case such as the case where perturbations due to other gravitating bodies are taken into account. This will be discussed in Section 4.

Before doing this, however, we must solve a fundamental problem on clocks comoving with massive bodies as follows:

Let us consider a clock moving with a massive body such as the Earth and assume that the Earth is moving around the Sun, for simplicity, in a Keplerian orbit. If the gravitational field of the Earth itself is ignored and a clock is assumed to be located at the geocentre, it is easy to express its proper time as a function of the chosen coordinate time as we have already done. But how should we take into account the gravitational effect of the Earth itself?

Of course we should drop no gravitational effects in obtaining the relation between the true proper time and the chosen coordinate time. However, we can omit any effect in obtaining the relation between a coordinate time and another since any coordinate time is the time coordinate of a certain coordinate system which we can define in the way we want. For example, it makes the situation simple to ignore the direct gravitational effect of the Earth in defining a coordinate time representing the Earth as an approximated proper time of the Earth.

If we adopt this 'way of ignorance', it is clear that

(i)-a coordinate time of the whole solar system is never equal to the proper time of the clock resting at the barycentre of the solar system, and
(ii)-a coordinate time representing the Earth is never the same as the true proper time of a standard clock at the geocentre in the true spacetime.

To see how this 'way of ignorance' works, let us use it in constructing a coordinate time representing the Earth as follows:

a) Split the full Newtonian force function U into a sum of that due to the Earth U^{+} and to the other bodies U^{*} as

$$U = U^{+} + U^{*} \tag{21}$$

If the solar system is assumed to be a system of mass points, the above components are written as

$$U^{+}(\boldsymbol{x}) = GM_E/|\boldsymbol{x}-\boldsymbol{x}_E| \tag{22}$$

$$U^{*}(\boldsymbol{x}) = \sum_{J \neq E} GM_J/|\boldsymbol{x}-\boldsymbol{x}_J| \tag{23}$$

where $\boldsymbol{x}$, $\boldsymbol{x}_E$ and $\boldsymbol{x}_J$ are the positions of an arbitrary point, the Earth and a body J in the solar system barycentric coordinate system, respectively, and M_E and M_J are the masses of the Earth and a body J.

b) Consider a fictitious metric tensor g^{*} composed only by U^{*} as

$$g^*_{oo} = -1 + 2U^*/c^2, \quad g^*_{jj} = +1, \quad \text{others} = 0 \tag{24}$$

c) Solve the following Newtonian equation of proper time for a clock resting at the geocentre in the fictitious spacetime characterized by the above metric tensor:

$$d\tau^*/dt = 1 - (U^*_E + \boldsymbol{v}_E^2/2)/c^2 \tag{25}$$

where

$$U^*_E = U^*(\boldsymbol{x}_E) \tag{26}$$

is the Newtonian force function which the Earth feels and $\boldsymbol{v}_E$ is the coordinate velocity of the Earth at the coordinate time t.

d) Define a coordinate time representing the Earth t' by a proper time τ^* as

$$t' = \tau^* \tag{27}$$

From the above mathematical derivation, it is clear that this 'way of ignorance' of the self-gravitational effect is equivalent to choosing a new coordinate system at whose space origin the considered clock rests.

3.5. EQUATION OF PROPER TIME IN A RELATIVE COORDINATE SYSTEM

Even in the relativistic theories, the expression of equations and quantities by coordinates will be changed by the choice of coordinate system. Thus, as for the equation of proper time, we need its transformation law among various coordinate systems to find the relation among different coordinate times. As a first example we will consider the case of a shift of the space origin.

In the Newtonian approximation, the shift of space coordinate origin leads to a concept of relative coordinate systems. It is well known that the Newtonian gravitational potential in a relative coordinate system becomes the sum of the direct self-gravitational part and the tidal potential due to the external bodies.

For example, consider the case that the space coordinate origin is changed from the barycentre of the solar system to the geocentre. The coordinate system is transformed from the solar system barycentric coordinate system to the geocentric one. Then the Newtonian force function in the geocentric coordinate system is given as

$$V(\boldsymbol{y}) = U^{+}(\boldsymbol{x}_E+\boldsymbol{y}) + [U^{*}(\boldsymbol{x}_E+\boldsymbol{y}) - U_E^{*}] \tag{28}$$

where $\boldsymbol{y}$ is the position of an arbitrary point in the geocentric coordinate system, and U^{+}, U^{*} and U_E^{*} are those defined in section 3.4.

Then the equation of proper time in the new coordinate system has the same form as that in the old coordinate system as long as the metric tensor is assumed to be Newtonian. For example the Newtonian equation of proper time in a geocentric coordinate system is written as

$$d\tau/dt' = 1 - [V(\boldsymbol{y}) + \boldsymbol{u}^2/2]/c^2 \tag{29}$$

where t' is the coordinate time of the geocentric coordinate system and $\boldsymbol{y}$ and $\boldsymbol{u}$ are the position and coordinate velocity of a clock in the same coordinate system.

3.6. EQUATION OF PROPER TIME IN A ROTATING COORDINATE SYSTEM

Next, let us consider the case that a new coordinate system is obtained by a space rotation from the old one. Among many coordinate systems transformed by the space rotation only, the rigidly rotating one is the most important. A geocentric coordinate system rotating constantly with the period of a mean sidereal day is of a special importance since it is a very good approximation of the coordinate system moving with the Earth, which is the frame of reference for many of our measurements.

Let us consider such a rotating coordinate system. Then the relations between the rotating and non-rotating geocentric coordinate systems are given as

$$\boldsymbol{z} = \underset{\sim}{R}\, \boldsymbol{y} \tag{30}$$

$$\boldsymbol{w} = \underset{\sim}{R}\, (\boldsymbol{u} - \boldsymbol{u}_{rot}) \tag{31}$$

$$V'(z) = V(\boldsymbol{y}) + \boldsymbol{u}_{rot}{}^2/2 \tag{32}$$

where $\underset{\sim}{R}$ is the rotation matrix of frame and

$$\boldsymbol{u}_{rot} = \boldsymbol{w} \wedge \boldsymbol{z} \tag{33}$$

is the rotational velocity caused by the rotation of the frame and evaluated at $\boldsymbol{z}$, where $\boldsymbol{w}$ is the angular velocity of the frame rotation, and the symbol $\wedge$ denotes the vector product.

Here $\boldsymbol{z}$ and $\boldsymbol{w}$ are the position and coordinate velocity of a point in the rotating coordinate system, respectively, while $\boldsymbol{y}$ and $\boldsymbol{u}$ are the position and coordinate velocity of the same point in the non-rotating one, respectively. Also V' denotes the Newtonian force function in the rotating coordinate system while V does that in the non-rotating one. Sometimes V' is called the effective gravitational force function in a rotating coordinate system.

It should be noted that the time coordinate is not changed by a rigid space rotation of a constant angular velocity. Then the Newtonian equation of proper time in this rigidly rotating coordinate system is given as

$$d\tau/dt' = 1 - [V'(\boldsymbol{z}) + \boldsymbol{w}^2/2]/c^2 \tag{34}$$

If a clock rests in the rotating coordinate system, this equation becomes

$$d\tau/dt' = 1 - V'(\boldsymbol{z})/c^2 \tag{35}$$

At a given instant t', let us consider a space-like plane on which V' has the same value as that at mean sea level and call it the geoid at the instant. Then it is clear that all clocks on the geoid can be set to have the same proper time as long as the Newtonian approximation holds. This is because the right hand side of equation (35) becomes the same for all such clocks at the same t'.

Further, by an appropriate change of unit of time, we can synchronize these proper times to be the same as t', the coordinate time of geocentric coordinate system itself. Thus we can realize a geocentric coordinate time t' by taking an average of atomic times distributed on the geoid. This is exactly the principle on which the maintenance procedure of TAI is based.

3.7. HIGHER ORDER CORRECTIONS

Postponing giving the Newtonian relation between a proper time and a coordinate time in a general case, we will develop further the approximation of the metric. To do this simply, we must choose a specific theory from many general relativistic theories. However, we cannot specify what is the correct one from the present knowledge. Also it is too difficult to include all theories proposed in discussing the relation between a proper time and a coordinate time. Thus we will deal with only a group of relativistic theories, which are called 'metric theories'.

The common point to all the members of this group is that a theory is constructed by means of a metric tensor in the four dimensional Riemannian spacetime. Usually the difference among the members is found in the functional form of the metric tensor. Some constraints on the form of metric tensor lead to a unified treatment of metric theories called the Parametric Post-Newtonian (PPN) formalism. As for the PPN formalism, the readers can consult with the textbook of Will (1981). Then the next stage is to solve the equation of proper time based on a post-Newtonian metric such as the PPN metric.

A sufficiently general expression of post-Newtonian metric is given by

$$g_{oo} = -1 + 2U/c^2 + 2W/c^4, \quad g_{oj} = (\mathbf{U})_j/c^3, \text{ and}$$

$$g_{jk} = \delta_{jk} + 2(\underset{\sim}{U})_{jk}/c^2 \tag{36}$$

where W, $\mathbf{U}$ and $\underset{\sim}{U}$ are the non-linear part of scalar force function, the vector force function and the tensor force function, respectively (see Chapter 5).

Substituting these post-Newtonian expressions of metric tensor into the rigorous formula (4), we obtain the following post-Newtonian equations of proper time:

$$d\tau/dt = 1 - (U + \boldsymbol{v}^2/2)/c^2 - (U^2/2 + U\boldsymbol{v}^2/2$$

$$+ \boldsymbol{v}^4/8 + W + \boldsymbol{U} \bullet \boldsymbol{v} + \boldsymbol{v} \bullet \underset{\sim}{U}\, \boldsymbol{v})/c^4 \tag{37}$$

where the dots mean the inner product of vectors.

As a first approximation, we solve the above equation under the assumption that a clock moving on a Keplerian orbit around the non-rotating central body as in Section 3.3. We also assume that the metric is that of the Einstein's general theory of relativity in the isotropic coordinate system. In this case

$$W = -U^2, \quad \boldsymbol{U} = 0, \quad \text{and } \underset{\sim}{U} = U\,\underset{\sim}{1} \tag{38}$$

where $\underset{\sim}{1}$ is the 3×3 unit tensor.

Then the post-Newtonian part of (37) becomes

$$(d\tau/dt)_2 = -[3U^2 + 4E\,U + E^2/2]/c^4 \tag{39}$$

Using the differential formula

$$n\ a^2\ (1 - e^2)^{1/2}\ dt\ =\ r^2\ df \qquad (40)$$

we can integrate equation (39) as

$$(\tau - t)_2 = -\ (GM/c^2a^2)\ [\{3f\ (1 - e^2)^{-1/2} - 2u\}/n\ +\ t/8] \qquad (41)$$

where f is the true anomaly of the Keplerian motion whose value is numerically given by the formula

$$f\ =\ 2\ \tan^{-1}[\{(1 - e)/(1+e)\}^{1/2}\ \tan\ (u/2)] \qquad (42)$$

When the eccentricity of orbit e is small, we can expand the right hand side of equation (41) as

$$(\tau - t)_2 = -\ (GM/c^2a)^2\ [\{(9/8)\ +\ (3/2)e^2\}\ t$$

$$+\ (e/n)\ \{(4\ +\ (5/2)e^2)\ \sin\ l\ +\ (11/4)\ l\ \sin\ 2l$$

$$+\ (7/4)\ l^2\ \sin\ 3l\ +\ ...\}] \qquad (43)$$

At present, it seems meaningless to solve equation (37) more rigorously than we did here and discuss its effect. This is because even the latest planetary and lunar ephemerides cannot provide the correct information for the argument of the Newtonian term (i.e. position and velocity of the Earth, the Sun, the Moon and other planets) up to more than the order of v^2/c^2 which we need in order to make the solution of the above equation to be fully correct up to the order of v^4/c^4.

4. RELATION BETWEEN TDT AND TDB

In the following, we shall give an example of practical solution of the Newtonian equation of proper time. The example is the relation between a coordinate time representing the Earth and a coordinate time representing the whole solar system.

Although both of these time-like arguments are coordinate times, we borrow the word 'dynamical time' to name them: the former as the Terrestrial Dynamical Time (TDT) and the latter as the solar system Barycentric Dynamical Time (TDB). We do not discuss here whether these names are appropriate or not. However many people have misused the term 'dynamical time' as if it is a coordinate time. Thus we hope that our names are not so far from what the people imagine about the title of this subsection, "the relation between TDT and TDB".

4.1. FORM OF TDT-TDB RELATION

From now on, we will use TDT in the sense of the time coordinate of a sort of geocentric coordinate system called the Terrestrial Coordinate System (TCS) while TDB in the sense of the time coordinate of one of solar system barycentric coordinate systems called the Barycentric Coordinate System (BCS). As for the details of the TCS and the BCS the readers can consult with the paper of Fukushima *et al.* (1986a).

We define the TDT as the proper time of a clock at the geocentre in a fictitious spacetime, where the metric is given by the Einstein-Infeld-Hoffmann (EIH) one caused by the Sun, planets and the Moon as mass points and no direct contribution from the Earth is taken into account. We also define TDB as the coordinate time when the solar system spacetime is described by the above EIH metric.

We note that this definition suits well the present style of creation of the planetary and lunar ephemerides such as DE200/LE200 of JPL (Standish and Williams), VSOP82/ELP2000 of the Bureau des Longitudes (Bretagnon; Chapront-Touze and Chapront) and the new series of Japanese Ephemeris of JHD (JHD, 1984) since they are all constructed with the assumption that each celestial body moves on a geodesic in its own fictitious spacetime such as the one described above.

As has been stated in Section 1.5, the relation between these two coordinate times is a part of the fully four dimensional coordinate transformation between the TCS and BCS. In fact, the TDB of an event is written in the following form:

$$\mathrm{TDB} = \mathrm{TDB}\ (\mathrm{TDT}, \boldsymbol{y}) \tag{44}$$

where $\boldsymbol{y}$ is the space coordinate vector in the TCS of the same event. Note that $\boldsymbol{y}$ becomes 0 at the geocentre, the space coordinate origin of the TCS.

Usually we consider the case that a clock is located near the Earth. Then the magnitude of $\boldsymbol{y}$ is so small that we can expand the above expression as

$$\mathrm{TDB} = \mathrm{TDB}_E(\mathrm{TDT}) + [\partial \mathrm{TDB}/\partial \boldsymbol{y}(\mathrm{TDT})]_E\ \boldsymbol{y} + \ldots \tag{45}$$

where TDB_E (TDT) is the TDB of the geocentre at the given TDT and $[\partial \mathrm{TDB}/\partial \boldsymbol{y}]_E$ is the partial derivative of TDB with respect to $\boldsymbol{y}$ evaluated at the geocentre.

The functional form of TDB_E (TDT) will be given in Section 4.3. while the term dependent on the location of clock will be discussed in Section 4.4. The TDT discussed here is almost the same as the time-like argument t' defined in Section 3.4. The difference between these two coordinate times is due to the change of unit of time which will be explained in the next subsection.

4.2. ADJUSTMENT OF UNIT OF TIME

Until now we have dealt with only the relation among various time-like arguments as a fully metrological quantity, i.e. the product of a numerical value and a unit. The relation among their numerical values may differ from the corresponding relation as a fully metrological quantity since the former is affected by the difference in chosen units of time among time-like arguments.

From the relativistic point of view, we can choose any function of coordinates as the unit of time for each time-like argument. However, it is natural to choose such units of time that the ratio of any pair of these time units is a constant over the whole spacetime where they are well defined. In fact the IAU recommended in 1976 that the difference in readings (i.e. numerical values) of time scales between the TDT and the TDB must be only periodic (see Chapter 15).

The spirit of the above recommendation is that it is more convenient to choose the units of all time-like arguments so that there will be no secular differences among the numerical values. If we follow this recommendation, we must multiply the whole formula of TDB-TDT relation after we solved the equation of proper time such as (23) by a conversion factor close to unity.

In the Newtonian approximation, the above adjustment procedure is equivalent to solving the following modified equation instead of the true equation of proper time:

$$d\tau/dt = 1 - \langle U + \boldsymbol{v}^2/2 \rangle_p / c^2 \tag{46}$$

where $\langle \; \rangle_p$ is an operator which removes a constant part and other secular terms. For example, applying this operator to equation (16) becomes

$$\langle \tau - t \rangle_p = (2GM/c^2a)\,(e/n)\,\sin u \tag{47}$$

Here we stress that this adjustment of the unit of time inevitably requires the adjustment of the unit of length since the units of time and length, second and meter for example, are connected tightly by the defining constant of c, the speed of light. For example, in order to obtain the correct relation on unit of length, we need the value of the average of $d\tau/dt$. For further discussion on the problem, see the paper of Fukushima *et al.* (1986b).

4.3. RELATION BETWEEN TDB AND TDT AT THE GEOCENTRE

According to the discussion on the adjustment of unit of time in Section 4.2, we must take a long time average of the equation of proper time to obtain the correct TDB-TDT relation at the geocentre. To do this, we need an analytical way of solving the equation since numerical ones are not appropriate for such a purpose.

As an example, we introduce here the analytical solution of TDB_E-TDT by Hirayama *et al.* (1988). Their result is more accurate and is expressed with more digits than the work of Moyer (1981) as the pioneer in this field.

Their solution is obtained by the formula processing of the latest semi-analytical ephemeris of planets VSOP82 (Bretagnon) and that of Moon ELP2000/82 (Chapront-Touze and Chapront). Comparison with the numerical planetary/lunar ephemerides DE200/LE200 (Standish and Williams) in the value of dTDT/dTDB shows that the error of their solution amounts to about a few nanoseconds. More than 130 terms are needed to compute the series to a precision of ten nanoseconds.

The first 12 terms of their solution are as follows:

$$
\begin{aligned}
TDB_E - TDT &= 1656.675 \sin (E-102^{\circ}.9377) \\
&+ 22.418 \sin (E-J-179^{\circ}.916) + 13.840 \sin (2E+154^{\circ}.124) \\
&+ 4.770 \sin (J-8^{\circ}.888) + 4.677 \sin (E-S-179^{\circ}.995) \\
&+ 2.257 \sin (S-92^{\circ}.481) + 1.686 \sin (4E-8M+3J+106^{\circ}.882) \\
&+ 1.555 \sin D + 1.277 \sin (2V-2E-179^{\circ}.893) \\
&+ 1.193 \sin (E-2J+177^{\circ}.357) + 1.115 \sin (V-E+0^{\circ}.009) \\
&+ 10.216\, T \sin (E+142^{\circ}.985)
\end{aligned}
\tag{48}
$$

where the unit of coefficients is microsecond, the symbols V, E, M, J and S are the mean longitude of Venus, the Earth-Moon barycentre, Mars, Jupiter and Saturn, respectively, the symbol D is the mean elongation in longitude of the Moon from the Sun, and T is the TDB in Julian centuries from J2000.0. "See Table 2 for more detailed results."

Of course one can replace the TDB by the corresponding TDT in evaluating these angle arguments and T since the error caused by this replacement is of higher order and negligible. We mention that the recent work of Fairhead *et al.* (1987) is almost the same as that of Hirayama *et al.*.

"TABLE 2. Analytical solution of $TDB_E - TDT$. The series is a summation of terms: amplitude * $\sin(nT+\alpha)$. The amplitude is measured in microseconds, n in degrees/Julian century and α in degrees."

No.	Amplitude	Period	Phase	Arguments											n	α
	µs	years	°	Me	V	E	Ma	J	S	U	N	D	F	l	°/36525d	°
1	1656.675	1.00	-102.9377	0	0	1	0	0	0	0	0	0	0	0	35999.3729	357.5287
2	22.418	1.09	-179.916	0	0	1	0	-1	0	0	0	0	0	0	32964.467	246.199
3	13.840	0.50	154.124	0	0	2	0	0	0	0	0	0	0	0	71998.746	355.057
4	4.770	11.86	-8.888	0	0	0	0	1	0	0	0	0	0	0	3034.906	25.463
5	4.677	1.04	-179.995	0	0	1	0	0	-1	0	0	0	0	0	34777.259	230.394
6	2.257	29.46	-92.481	0	0	0	0	0	1	0	0	0	0	0	1222.114	317.596
7	1.686	1783.39	106.882	0	0	4	-8	3	0	0	0	0	0	0	-20.186	288.336
8	1.555	0.08	0.000	0	0	0	0	0	0	0	0	1	0	0	445267.114	297.852
9	1.277	0.80	-179.893	0	2	-2	0	0	0	0	0	0	0	0	45036.886	343.134
10	1.193	1.20	177.357	0	0	1	0	-2	0	0	0	0	0	0	29929.562	209.120
11	1.115	1.60	0.009	0	1	-1	0	0	0	0	0	0	0	0	22518.443	81.522
12	0.794	0.55	0.829	0	0	2	0	-2	0	0	0	0	0	0	65928.934	133.059
13	0.600	3.98	90.894	0	2	-3	0	0	0	0	0	0	0	0	9037.513	153.454
14	0.495	1.01	179.989	0	0	1	0	0	0	-1	0	0	0	0	35570.906	326.400
15	0.486	1.07	179.733	0	0	2	-2	0	0	0	0	0	0	0	33718.147	29.799
16	0.468	1.01	-179.998	0	0	1	0	0	0	0	-1	0	0	0	35780.887	336.120
17	0.447	238.92	57.379	0	8	-13	0	0	0	0	0	0	0	0	150.678	207.154
18	0.435	15.78	139.592	0	0	1	-2	0	0	0	0	0	0	0	-2281.226	249.192
19	0.431	84.02	-174.646	0	0	0	0	0	0	1	0	0	0	0	428.467	139.409
20	0.376	1.14	91.104	0	3	-4	0	0	0	0	0	0	0	0	31555.956	235.178
21	0.243	8.10	165.628	0	3	-5	0	0	0	0	0	0	0	0	-4443.417	209.235
22	0.238	93462.09	0.000	0	0	8	-16	4	5	0	0	0	0	0	0.385	184.592
23	0.231	1.07	-86.792	0	0	1	0	0	-2	0	0	0	0	0	33555.145	273.520
24	0.204	0.52	81.739	0	0	2	0	-1	0	0	0	0	0	0	68963.840	248.320
25	0.173	0.33	51.183	0	0	3	0	0	0	0	0	0	0	0	107998.119	352.582
26	0.159	0.57	10.414	0	0	2	0	-3	0	0	0	0	0	0	62894.029	108.292
27	0.144	7.89	122.139	0	0	2	-4	0	0	0	0	0	0	0	-4562.452	341.339
28	0.138	0.53	-179.452	0	3	-3	0	0	0	0	0	0	0	0	67555.328	65.088
29	0.120	164.77	-43.575	0	0	0	0	0	0	0	1	0	0	0	218.486	260.774
30	0.119	1.15	150.028	0	0	3	-4	0	0	0	0	0	0	0	31436.921	109.694
31	0.117	5.93	-19.410	0	0	0	0	2	0	0	0	0	0	0	6069.811	49.293
32	0.102	1.13	180.000	0	0	0	0	0	0	0	0	1	0	-1	-31931.980	342.653
33	0.098	2.47	150.680	0	0	2	-3	0	0	0	0	0	0	0	14577.848	5.313
34	0.080	1.34	171.569	0	0	1	0	-3	0	0	0	0	0	0	26894.656	168.981
35	0.075	2.14	-179.774	0	0	1	-1	0	0	0	0	0	0	0	16859.074	285.259
36	0.073	30.47	-110.457	0	0	0	0	2	-4	0	0	0	0	0	1181.356	117.936
37	0.064	1.09	-171.313	0	0	1	0	1	-5	0	0	0	0	0	32923.710	73.118
38	0.064	1.09	-8.846	0	0	1	0	-3	5	0	0	0	0	0	33005.225	238.953
39	0.064	302.43	-32.843	0	3	-7	4	0	0	0	0	0	0	0	119.034	151.564
40	0.059	14.73	176.132	0	0	0	0	0	2	0	0	0	0	0	2444.228	276.287
41	0.054	0.36	-2.910	0	0	3	0	-3	0	0	0	0	0	0	98893.402	195.435
42	0.048	40.43	-23.544	0	0	8	-15	0	0	0	0	0	0	0	890.493	128.688
43	0.048	2.93	121.450	0	0	3	-5	0	0	0	0	0	0	0	12296.622	85.683
44	0.043	9.93	-0.736	0	0	0	0	2	-2	0	0	0	0	0	3625.584	327.812
45	0.043	0.04	0.000	0	0	0	0	0	0	0	0	1	0	1	922466.208	73.050
46	0.042	1.00	-114.374	0	0	1	0	2	-5	0	0	0	0	0	35958.615	164.408
47	0.041	0.51	77.416	0	0	2	0	0	-1	0	0	0	0	0	70776.632	228.271
48	0.040	0.40	179.890	0	4	-4	0	0	0	0	0	0	0	0	90073.771	145.943
49	0.038	883.27	26.390	0	0	0	0	2	-5	0	0	0	0	0	-40.758	204.706
50	0.037	1.99	165.492	0	4	-6	0	0	0	0	0	0	0	0	18075.026	290.613
51	0.037	1.23	121.222	0	0	4	-6	0	0	0	0	0	0	0	29155.695	190.488
52	0.037	7.84	-108.106	0	5	-8	0	0	0	0	0	0	0	0	4594.096	358.061
53	0.035	12.02	85.859	0	0	0	0	3	-5	0	0	0	0	0	2994.148	298.526
54	0.034	0.67	-88.459	0	4	-5	0	0	0	0	0	0	0	0	54074.398	137.128
55	0.033	1.04	8.783	0	0	1	0	2	-6	0	0	0	0	0	34736.501	237.488
56	0.032	0.71	2.420	0	0	3	-3	0	0	0	0	0	0	0	50577.221	317.520
57	0.032	1.03	170.862	0	0	1	0	-2	4	0	0	0	0	0	34818.017	42.935
58	0.030	0.89	-11.869	0	5	-7	0	0	0	0	0	0	0	0	40593.468	194.765
59	0.030	0.52	0.633	0	0	2	0	0	-2	0	0	0	0	0	69554.518	101.411
60	0.028	1.00	-150.192	0	0	3	-8	3	0	0	0	0	0	0	-36019.559	290.796
61	0.027	1.00	7.270	0	0	5	-8	3	0	0	0	0	0	0	35979.187	289.190
62	0.026	0.09	77.062	0	0	1	0	0	0	0	0	-1	0	0	-409267.741	239.677
63	0.025	3.59	96.108	0	0	4	-7	0	0	0	0	0	0	0	10015.396	169.941
64	0.025	5.26	88.361	0	0	3	-6	0	0	0	0	0	0	0	-6843.677	57.161
65	0.023	1.02	-3.842	0	0	1	0	0	0	-2	0	0	0	0	35142.439	188.514

No.	Amplitude μs	Period years	Phase °	Arguments Me	V	E	Ma	J	S	U	N	D	F	l	n °/36525d	α °
66	0.023	0.60	20.185	0	0	2	0	-4	0	0	0	0	0	0	59859.123	83.712
67	0.022	0.44	77.564	0	2	-1	0	0	0	0	0	0	0	0	81036.259	341.057
68	0.021	0.75	-27.343	0	0	4	-5	0	0	0	0	0	0	0	48295.995	37.356
69	0.020	14.98	-63.746	0	0	0	0	2	-3	0	0	0	0	0	2403.470	214.725
70	0.018	0.92	47.555	0	0	1	0	1	0	0	0	0	0	0	39034.279	182.373
71	0.017	85.72	180.000	0	0	0	0	0	0	2	-2	0	0	0	419.962	199.413
72	0.016	0.62	-105.732	0	1	0	0	0	0	0	0	0	0	0	58517.816	76.248
73	0.016	28.51	101.516	0	0	0	0	2	-6	0	0	0	0	0	-1262.871	229.754
74	0.016	2.67	11.053	0	1	-2	0	0	0	0	0	0	0	0	-13480.930	352.100
75	0.015	25.56	-90.657	0	3	-5	0	1	0	0	0	0	0	0	-1408.512	347.302
76	0.015	0.32	179.912	0	5	-5	0	0	0	0	0	0	0	0	112592.214	227.479
77	0.015	4.65	63.383	0	0	5	-9	0	0	0	0	0	0	0	7734.170	246.816
78	0.014	0.38	8.221	0	0	3	0	-4	0	0	0	0	0	0	95858.496	172.214
79	0.014	0.35	-112.127	0	0	3	0	-2	0	0	0	0	0	0	101928.307	120.569
80	0.014	11.70	-162.680	0	0	0	0	1	-5	0	0	0	0	0	-3075.663	341.284
81	0.013	60.95	162.670	0	0	0	0	1	-2	0	0	0	0	0	590.678	96.867
82	0.012	1.34	93.677	0	0	5	-8	0	0	0	0	0	0	0	26874.470	272.543
83	0.012	1.11	5.755	0	0	1	0	0	-3	0	0	0	0	0	32333.031	315.989
84	0.012	0.78	-56.610	0	0	5	-7	0	0	0	0	0	0	0	46014.769	117.689
85	0.012	1.34	-170.312	0	2	-4	0	0	0	0	0	0	0	0	-26961.860	151.782
86	0.012	1750.00	90.000	0	0	4	-8	1	5	0	0	0	0	0	20.571	93.138
87	0.011	1.23	-179.750	0	0	1	0	-3	2	0	0	0	0	0	29338.883	277.817
88	0.011	11.35	4.055	0	0	7	-13	0	0	0	0	0	0	0	3171.719	46.688
89	0.011	6.60	34.680	0	0	6	-11	0	0	0	0	0	0	0	5452.944	327.713
90	0.010	1.07	-72.528	0	0	1	0	-2	3	0	0	0	0	0	33595.903	109.468
91	0.010	22.13	-0.711	0	3	-5	0	2	0	0	0	0	0	0	1626.394	111.599
92	0.010	1.00	170.646	0	8	-12	0	0	0	0	0	0	0	0	36150.051	60.887
93	0.010	42.01	10.978	0	0	0	0	0	0	2	0	0	0	0	856.934	279.088
94	0.009	0.04	0.000	0	0	0	0	0	0	0	0	3	0	-1	858602.249	38.356
95	0.009	1597.90	180.000	0	0	0	0	2	-6	3	0	0	0	0	22.530	170.403
96	0.009	1.88	-145.657	0	0	0	1	0	0	0	0	0	0	0	19140.299	209.776
97	0.009	46.52	131.903	0	0	3	-6	2	0	0	0	0	0	0	-773.866	189.406
98	0.008	0.53	10.332	0	0	4	-4	0	0	0	0	0	0	0	67436.294	70.465
99	0.008	0.47	-89.473	0	5	-6	0	0	0	0	0	0	0	0	76592.841	217.627
100	0.008	171.44	0.000	0	0	0	0	0	0	1	-1	0	0	0	209.981	9.706
101	0.008	19.86	155.557	0	0	0	0	1	-1	0	0	0	0	0	1812.792	139.831
102	0.008	0.52	79.636	0	0	3	-2	0	0	0	0	0	0	0	69717.520	30.169
103	0.008	1.20	70.152	0	0	1	0	-4	5	0	0	0	0	0	29970.319	283.600
104	0.008	165.23	113.436	0	6	-10	0	3	0	0	0	0	0	0	217.883	303.705
105	0.007	1.00	160.261	0	8	-14	0	0	0	0	0	0	0	0	-35848.695	209.569
106	0.007	1.01	-161.719	0	0	1	0	0	0	0	-2	0	0	0	35562.400	50.050
107	0.007	25.89	-51.246	0	0	9	-17	0	0	0	0	0	0	0	-1390.733	210.586
108	0.007	0.94	-7.158	0	0	0	2	0	0	0	0	0	0	0	38280.599	343.709
109	0.007	0.82	-87.050	0	0	6	-9	0	0	0	0	0	0	0	43733.543	196.849
110	0.006	1.20	-168.558	0	0	1	0	0	-5	0	0	0	0	0	29888.804	41.521
111	0.006	0.27	179.941	0	6	-6	0	0	0	0	0	0	0	0	135110.657	309.021
112	0.006	1.51	163.535	0	0	1	0	-4	0	0	0	0	0	0	23859.750	126.596
113	0.006	1.11	-174.113	0	0	1	0	-2	2	0	0	0	0	0	32373.789	317.805
114	0.006	0.53	92.477	0	0	2	0	0	-3	0	0	0	0	0	68332.404	143.177
115	0.006	3.95	66.123	0	0	4	-8	0	0	0	0	0	0	0	-9124.903	144.523
116	0.006	6.58	23.009	1	0	-4	0	0	0	0	0	0	0	0	5475.183	233.394
117	0.006	0.27	-2.873	0	0	4	0	-4	0	0	0	0	0	0	131857.869	261.587
118	0.006	1.05	12.284	0	0	1	0	-1	1	0	0	0	0	0	34186.581	128.476
119	0.005	1137.66	180.000	0	5	-6	-4	0	0	0	0	0	0	0	31.644	145.367
120	0.005	1.46	66.060	0	0	6	-10	0	0	0	0	0	0	0	24593.244	354.526
121	0.005	0.55	-21.992	0	0	5	-6	0	0	0	0	0	0	0	65155.068	147.741
122	0.005	0.98	-172.329	0	0	9	-15	0	0	0	0	0	0	0	36889.866	80.370
123	10.216T	1.00	142.985	0	0	1	0	0	0	0	0	0	0	0	35999.373	243.451
124	0.171T	0.50	40.047	0	0	2	0	0	0	0	0	0	0	0	71998.746	240.980
125	0.027T	29.46	144.584	0	0	0	0	0	1	0	0	0	0	0	1222.114	194.661
126	0.027T	11.86	-58.290	0	0	0	0	1	0	0	0	0	0	0	3034.906	336.061
127	0.026T	1783.39	-172.072	0	0	4	-8	3	0	0	0	0	0	0	-20.186	9.382
128	0.007T	1.20	-126.852	0	0	1	0	-2	0	0	0	0	0	0	29929.562	264.911
129	0.006T	238.92	-90.000	0	8	-13	0	0	0	0	0	0	0	0	150.678	59.775
130	0.005T	3.98	-166.535	0	2	-3	0	0	0	0	0	0	0	0	9037.513	256.025
131	0.043T	1.00	50.655	0	0	1	0	0	0	0	0	0	0	0	35999.373	151.121

T in 36525 days from J2000.0

As is seen in Section 3.5, the post-Newtonian effect is so small in the case TDB_E - TDT than 0.5ns. Therefore we need no additional terms to the above series to compute TDB_E - TDT at a precision of a nanosecond.

4.4. LOCATION DEPENDENT TERM IN TDB-TDT

Next we discuss the location-dependent term in (45) the relation between the TDT and the corresponding TDB. To give the explicit form of this term, we need a post-Newtonian coordinate transformation formula connecting the TCS and the BCS. From the formula of Fukushima *et al.* (1986a), we obtain the following expression of partial derivatives:

$$[\partial \mathrm{TDB}/\partial \boldsymbol{y}]_E = \boldsymbol{v}_E/c^2 + [(2U_E^* + \boldsymbol{v}_E^2/2)\,\boldsymbol{v}_E + \underset{\sim}{U}_E^*\,\boldsymbol{v}_E + \mathbf{U}_E^*]\;/c^4 \qquad (49)$$

where the force functions with the superfix * mean that the contribution of the Earth itself is removed from them while those with the suffix E denote their values at the geocentre. The first term of the above partial derivative causes a well-known Doppler shift of frequency.

If we choose the Einstein-Infeld-Hoffmann metric, the above force functions are explicitly written as

$$U_E^* = \sum_{J \neq E} GM_J/r_{EJ}, \quad \mathbf{U}_E^* = -4 \sum_{J \neq E} GM_J \boldsymbol{v}_J/r_{EJ},$$

$$\text{and } \underset{\sim}{U}_E^* = U_E^* \,\underset{\sim}{1} \qquad (50)$$

where r_{EJ} is the distance between the Earth and a body J at the same TDB.

Then the expression (45) is rewritten explicitly for this case as

$$\mathrm{TDB} = \mathrm{TDT} + (\mathrm{TDB}_E - \mathrm{TDT}) + \boldsymbol{y} \bullet \boldsymbol{v}_E/c^2$$

$$+ [(\boldsymbol{v}_E^2)(\boldsymbol{y} \bullet \boldsymbol{v}_E)/2 + \boldsymbol{y} \bullet \{ \sum_{J \neq E} GM_J(3\boldsymbol{v}_E - 4\boldsymbol{v}_J)/r_{EJ}\}]/c^4 \qquad (51)$$

where TDB_E - TDT is already given in Section 4.3. This is the most general transformation formula between the TDT and the corresponding TDB. In the above formula, the typical order of magnitudes for a clock on the Earth are 1.7ms, 2.1μs, and 0.2ps for the second, third and fourth terms, respectively. Thus the last term is negligible on the Earth as well as the post-Newtonian correction to the amount TDB_E - TDT, which is already discussed in Section 3.7.

4.5. EQUATION OF PROPER TIME IN THE TERRESTRIAL COORDINATE SYSTEM

Finally we will give the equation of proper time with respect to the TDT which is related to the TDB in the preceding subsections.

The unit of TDT is defined to be a second on the geoid. This means that the effective gravitational force function in the Terrestrial Coordinate System is set to be zero on the geoid. Then the Newtonian equation of proper time is written in this non-rotating coordinate system as

$$d\tau/d\mathrm{TDT} = 1 - [V(\mathbf{y}) - V'(\mathrm{geoid}) + \mathbf{u}^2/2]/c^2 \qquad (52)$$

and is written in the rigidly rotating geocentric coordinate system as

$$d\tau/d\mathrm{TDT} = 1 - [V'(\mathbf{z}) - V'(\mathrm{geoid}) + \mathbf{w}^2/2]/c^2 \qquad (53)$$

where V, $\mathbf{y}$ and others are just the same as those in sections 3.5. and 3.6. Here V' (geoid) is the value of V' on the geoid and is given in the Geodetic Reference System 1980 as 6.263686×10^7 m^3/s, which leads to the change of rate of 6.97×10^{-10}.

For example, the frequency or all atomic clocks equipped with GPS/NAVSTAR satellites is tuned by the amount of 4.48×10^{-10} which is an upper limit of the averaged value of the correction in the above equation (Spilker, 1980). This amount comes from the difference in rate between on the satellite and on the geoid (see table 1).

REFERENCES

Bretagnon, P., Planetary Ephemeris VSOP82 [magnetic tape].

Chapront-Touzé, M., and Chapront, J., Lunar Ephemeris ELP2000/82 [magnetic tape].

Fairhead, L., Bretagnon, P., and Lestrade, J.-F., 1987, '*Proceedings of the IAU Symposium No.* 128 *held in Coolfont, U.S.A., October* 20-24, 1986', to be published.

Fukushima, T., Fujimoto, M.-K., Kinoshita, H., and Aoki, S., 1986a, '*Proceedings of the IAU Symposium No.* 114 *held in Leningrad, U.S.S.R., May* 28-31, 1985', D. Reidel Publ. Co., Dordrecht, p. 35.

Fukushima, T., Fujimoto, M.-K., Kinoshita, H. and Aoki, S., 1986b, *Celestial Mechanics*, **36**, 215.

Hirayama, Th., Kinoshita, H., Fujimoto, M.-K., and Fukushima, T., 1988, '*Proceedings of the IAG Symposia* (*IUGG XIX General Assembly, Vancouver, Canada, August* 10-22, 1987) *Tome I, Relativistic Effects in Geodesy*', p. 91.

JHD, 1984, *Basis of the New Japanese Ephemeris*, Supplement to the Japanese Ephemeris for the year 1985.

Misner, C.W., Thorne, K.S., and Wheeler, J.A., 1970, *Gravitation*, W.H. Freeman and Co., San Francisco.

Moyer, T.D., 1981, *Celestial Mechanics*, **23**, 33.

Spilker, J.J., 1980, *Global Positioning System - Papers published in NAVIGATION*, Inst. of Navigation, Washington, D.C., U.S.A., Vol. I.

Standish, E.M., and Williams, J.G., Planetary and Lunar Ephemerides DE200/ LE200 [magnetic tape].

Synge, J.L., 1960, *Relativity: the general theory*, North-Holland Publ. Co., Amsterdam.

Van Dierendonck, A.J., Russell, S.S., Kopitzke, E.R. and Birnbaum, M., 1980, *Global Positioning System - Papers published in NAVIGATION*, Inst. of Navigation, Washington, D. C., U.S.A., Vol. I, p. 55.

Will, C.M., 1981, *Theory and experiment in gravitational physics*, Cambridge Univ. Press, Cambridge.

PART 6

STANDARDS

STANDARDS FOR TERRESTRIAL AND CELESTIAL REFERENCE SYSTEMS

G. A. WILKINS
Royal Greenwich Observatory,
Herstmonceux Castle, Hailsham, East Sussex BN27 1RP
United Kingdom

1. THE USE OF STANDARDS

The purpose of this chapter is to review the characteristics of the standards that are required for the specification and use of terrestrial and celestial reference systems. It is clear from the preceding chapters that a wide variety of techniques can be used to determine the positions and velocities of points on the surface of the Earth and of objects in space. The results obtained by different techniques (and by different implementations of one technique, see Chapter 12, Section 2) will differ from each other, but the differences can be reduced by the use of appropriate systems of standards in the derivation of the required results from the observational data. Ideally, there would be a comprehensive, self-consistent system of internationally adopted standards that would be generally applicable, but practical considerations prevent the realisation of such a system. It is therefore usually necessary to use a combination of sub-systems, each of which is appropriate for a limited field of application.

Standards must be specified in such a way that they can be used directly for practical purposes, such as those listed in the next section. In some cases it may be necessary to specify secondary standards in order that the primary standard may be made more readily accessible, even though this may introduce some loss in precision. It is important that the standards that are used are clearly identified so that differences between results from different techniques that are due to differences in standards may be recognised, and so that the results may be corrected later if better standards are introduced.

A standard may be specified so that it corresponds as closely as possible to a conceptual definition, but it remains valid even if it is subsequently

J. Kovalevsky et al. (eds.), Reference Frames, 447–460.

found that it departs from the concept by a measurable amount. The specification of a standard need only be changed when it is clear that a new standard would have significant advantages that would compensate for the disadvantages involved in the change. In general it is better to continue to use the same system of standards for long periods, even if marginal improvements in the results could be obtained by modifying the system. The system should be changed, however, if it is clear that the results are being seriously degraded by the continued use of poor standards. Changes in a system of standards should, if possible, be made in such a way as to preserve the self-consistency of the system and to maintain its compatibility with other related systems. The name of the system should also be clearly changed.

Before considering particular systems of standards it is useful to notice that standards may be used for a variety of purposes and that there are many different types of standards. A system of standards may consist of several types of standards and may be intended for use in a wide range of applications or for a specific purpose.

2. THE PURPOSES OF STANDARDS

The principal purposes for which standards are intended to be used are as follows:

(i) to allow measurements made at different times and places to be compared or combined without bias;

(ii) to make available a set of high-quality (accurate, consistent) reference data for use in practical applications;

(iii) to provide a satisfactory basis for the comparison of the results from different techniques (of observation and of analysis); and

(iv) to provide an appropriate basis for the specification of geometrical positions and of the times of events.

Standard terrestrial or celestial reference systems are required mainly for the latter purposes, but they may also be used for the other purposes.

3. THE TYPES OF STANDARDS

The principal types of standards that are in common use are as follows:

(i) numerical constants that define units of measurement or that specify the relationships between different units for the same quantity;

(ii) numerical quantities that specify the best values of particular physical properties or phenomena;

(iii) mathematical models that represent the properties or behaviour of particular materials, objects or phenomena;

(iv) measurement and reduction procedures that are to be followed in order to obtain the best result; and

(v) physical objects that provide primary or secondary standards for use in the measurement of particular quantities.

Standard reference systems that are to be used to specify the time and spatial coordinates of terrestrial and celestial events must include all these types of standard if they are to be appropriate for all the purposes for which they are to be used. Even this list is not complete since the following points should also be noticed.

In the publication and exchange of data it is very useful to use standard formats that specify not only the layout of the numbers but also the units in which the quantities are to be expressed. Such formats may also include codes that indicate the quality of the measurements or the use of particular forms of reduction or correction.

The specification of a set of standards should itself use the general standards for the notation and symbols for the physical quantities concerned and for the units in which they are expressed. In particular, the units and symbols of the general international system of SI units (BIPM, 1985; Markowitz, 1976) should be used, except where there are recognised alternative units that are more appropriate for use in astronomy and geophysics. All symbols for physical and mathematical quantities should be unambiguously identified and care should be taken to ensure that the terminology used will not be confusing or misleading, especially when the standards will be of interest to scientists of a wide variety of disciplines. An attempt is made in later sections of this chapter to set out appropriate meanings for terms that are often used in this context in inconsistent ways.

4. SYSTEMS OF CONSTANTS

One of the important elements in the specification of a standard or a conventional reference system is the system of constants that is used in the development of the mathematical models of the physical system concerned and in the derivation of the positions of stations or objects from the observational data. It must be recognised that the word 'constants' is used rather loosely for the values of several types of quantity, and it is not always made clear that a quantity may vary with time and that the given value refers to a particular epoch. The following distinctions may be made:

(i) *a defining constant* has a value that is fixed arbitrarily in such a way as to define a unit of measurement; e.g. the value of the velocity of light defines the unit of length in terms of the unit of time- interval;

(ii) *a primary constant* has a value that is adopted after consideration of the various determinations of the quantity concerned; and

(iii) *a derived constant* has a value that is calculated from theoretical relationships between the values of the defining, primary and derived constants.

A quantity will normally only be treated as a defining constant, rather than as a primary constant, if thereby it is possible to define the unit more precisely than by other means or if thereby it is possible to avoid the necessity for changing a large quantity of derived data whenever a new determination of the constant is made. The adoption of a fixed value for the Gaussian gravitational constant is an example of the latter type; it serves to define the astronomical unit of length (au) for use with the astronomical units of mass and time to simplify the computation of orbits in the solar system. The adopted value of a defining constant is usually the best available estimate at the time of its adoption so that new and old units are then virtually indistinguishable.

The choice as to which quantities will be considered to be primary and which derived (or secondary) is to some extent arbitrary, but it is usual to take as primary constants those quantities that are most directly determined with high precision. The values of the primary constants are to be treated as exact to the stated precision in order to ensure the self-consistency of the system. The values of the primary constants may be chosen so that the calculated and measured values of the derived constants are in best agreement with each other.

An alternative approach to the generation of a system of constants is to treat all (except the defining constants) in the same way and to determine the values that give a best fit to the measured values (and their variances) within the constraints imposed by the requirement that the adopted values satisfy the theoretical relationships between them. This approach has been used in the CODATA 1986 adjustment of the fundamental physical constants (Cohen and Taylor, 1986). In both approaches there may also be auxiliary constants whose uncertainties are so small that they may be ignored.

5. STANDARDS FOR REFERENCE SYSTEMS

The establishment of a terrestrial or celestial reference system requires the use of a set of standards made up of these various types. In particular it will be necessary to adopt:

(i) conceptual definitions of the time-scales and spatial coordinate frames to which measurements of time and position are to be referred;

(ii) mathematical models for the transformation between the time and space coordinates in different frames;

(iii) formal procedures by which measurements are to be made and analysed in order to determine positions or orientations;

(iv) mathematical models for use in the reduction and analysis of measurements, together with appropriate numerical values of the parameters;

(v) lists of objects/sites that will be used as primary and secondary reference points in each reference frame; and

(vi) coordinates at some epoch, together with their rates of change, for the primary and secondary reference points.

It is important that any standard reference system be fully documented in such a way as to eliminate, as far as possible, any ambiguity or uncertainty about the specification of the standard. The documentation should cover all aspects of the system, although it is sufficient to give references to other standards provided that the documents concerned are readily accessible and that any inconsistences in, for example, terminology or notation are noted.

6. TERMINOLOGY FOR REFERENCE SYSTEMS

There are at present no universally adopted meanings for the following terms which are used in the context of reference systems for use in astronomy and geodesy:

Term 1: coordinate frame
Term 2: coordinate system
Term 3: reference frame
Term 4: reference system.

In this chapter they are used with the following meanings:

(1) **Coordinate frame:** a set of rectangular coordinate axes (or other geometrical construction) with respect to which the position of a point may be specified. The geometrical relationship between two coordinate frames may be expressed by the combination of a translation vector, which specifies the position of the origin of one frame with respect to the origin of the other, and a rotation matrix, which specifies the orientation of one frame with respect to the other.

(2) **Coordinate system:** a method of specifying the position of a point with respect to a particular coordinate frame. The position may be specified by linear and/or angular coordinates, while a frame may be specified conceptually in a wide variety of ways. For example, in the system of geodetic coordinates: longitude, latitude and height are measured with respect to a spheroid that approximates to the surface of the Earth in a specified manner.

(3) **Reference frame:** a catalogue of the adopted coordinates of reference points that serves to define, or realize, a particular coordinate frame. The coordinates of other points may be determined by making differential measures of their positions with respect to the reference points. The term reference frame is often used as if it were synonomous with the coordinate frame that it defines. (See the Introductory Chapter.)

(4) **Reference system:** the totality of procedures, models and constants that are required to establish and maintain one or more reference frames, i.e. to determine from observations the coordinates of the reference points that define the frames and to improve the adopted values as new observational data are accumulated. (See the Introductory Chapter.)

A reference system may be dependent on observations that are made by a particular technique, or it may be 'conventional' and be intended for general use. For example, the lunar-laser-ranging (LLR) reference system is used to determine a terrestrial reference frame that is based on the coordinates of the telescopes that make laser-ranging observations of retroreflectors on the surface of the Moon; it is also used to establish a lunar reference frame that is based on the coordinates of the retroreflectors; and it includes a celestial reference frame that is, in effect, defined by the adopted ephemerides, or models, of the motion and rotation of the Moon (Chapter 3, Section 5). On the other hand, the reference system of the new International Earth Rotation Service (IERS) will be a conventional system in which the coordinates of the points of the terrestrial reference frame will be derived by combining data from several techniques (Chapter 7, Section 3; Wilkins and Mueller, 1986). Initially this will be done in such a way that the coordinate frame corresponds as closely as possible to the frame of the BIH system for the previous year while the adopted variations of the coordinates of the reference points are consistent with an appropriate model of the motions of the tectonic plates on which the points are situated.

The celestial reference frame of the IERS reference system is to be realized initially by the adoption of a catalogue of the coordinates of quasars that are observed by the VLBI networks that participate in the monitoring of the rotation of the Earth. The quasars are to be chosen so that their directions can be regarded as fixed, although the observational data will be carefully examined to verify that there are no relative movements between them because of changes in the distribution of intensity across the radio sources concerned (Chapter 2, Section 6). The coordinates in the catalogue are to be such that the frame corresponds as closely as possible to the FK5 reference frame for the standard epoch of J2000.0, and hence to the reference frame used in current fundamental ephemerides of the motions of the planets of the Solar System.

The IERS terrestrial and celestial reference frames are intended to be 'standard' reference frames, and the observational data will be analysed to determine from further observations the coordinates and motions of other points with respect to these frames and also the geometrical relationship between the frames as a function of time. Any differences between the technique-dependent frames and the standard frames will also be determined. When sufficient new data have been obtained and analysed it is to be expected that it will be necessary to change the adopted coordinates and motions of the reference points, and eventually to modify the procedures, models and constants of the system. This process is known as the maintenance of the system.

7. TERMINOLOGY FOR EARTH-ROTATION PARAMETERS

The monitoring of the relative orientation of the standard terrestrial and celestial frames provides information on:

(i) the motion with respect to the celestial frame of the axis of rotation of the Earth;

(ii) the motion of the axis of rotation with respect to the terrestrial frame; and

(iii) the variations of the rate of rotation around the axis of rotation.

These three components of the relative orientation have physical causes that are largely independent of each other; moreover, the first can be modelled with high accuracy using the theories of precession and nutation of the Earth, while at present the other two components cannot be predicted accurately over significant intervals. It is therefore convenient to introduce the concept of an intermediate coordinate frame for which the z-axis may be regarded for most purposes as the axis of rotation of the Earth, although it is not quite the same as the instantaneous axis of rotation. The chosen axis defines the direction of the 'celestial ephemeris pole', otherwise known as the 'true pole of date' (see Chapter 11).

The motion of the true pole of date may be represented as a relatively short-period motion, or nutation, of small amplitude around the 'mean pole of date', which precesses around the pole of the ecliptic. The z-axis of the standard celestial frame corresponds to the direction of the mean pole at the standard epoch.

The second component is usually represented by the coordinates of the celestial ephemeris pole with respect to the terrestrial frame, and is referred to as 'polar motion'. It must be recognised that the axis of the terrestrial frame is defined in a conventional way; it corresponds to the pole implied by the adopted coordinates of the observatories that first monitored the polar motion, and is known as the 'conventional international origin' (Chapter 7, Section 2). This choice implies that the axis of the terrestrial frame is not the same as the axis of principal moment of inertia of the Earth (or axis of figure).

Analysis of the observed polar motion shows that it has two principal components, which have periods of one year and about 14 months. The term 'Chandler motion' is most appropriately used for the second component (Chapter 10, Section 3), rather than for the total motion. There is also a small secular drift.

The rotation of the Earth on its axis was once used to realize a standard timescale, which became known as 'universal time', UT, and which was expressed in terms of the concept of the Greenwich hour angle of a 'fictitious mean sun'. Now, however, UT is best regarded as an intermediary measure of the angle through which the Earth has rotated, and it is determined as a function of coordinated universal time, UTC (Chapter 15, Section 3.1). The difference between UT and UTC is a measure of the departure of the rotation from uniformity, but it must be recognised that UT is not a direct measure of the orientation of the terrestrial frame with respect to the standard celestial frame. This difference in orientation is also expressed in two parts: the angle represented by Greenwich mean sidereal time, GMST, gives the orientation of the terrestrial frame with respect to the intermediate celestial reference frame of date; while the orientation of this frame relative to the standard frame is given by the adopted theories of precession and nutation. UT and GMST are themselves related by an adopted formula; GMST is sometimes called by the more appropriate name 'Greenwich sidereal angle'.

The variations in the rate of rotation are usually represented, in an inverse manner, by the variations in the length of the universal day (the interval of time in which the angle represented by UT increases by one cycle) with respect to an SI day. Again it must be recognised that the universal day is not equal to the period of rotation of the Earth with respect to the standard celestial frame; even the sidereal day differs from this period by a small amount because of the effects of precession and nutation.

The quantities that are determined from observations and tabulated as a function of time to represent the changing orientation of the Earth are usually the two coordinates of the celestial ephemeris pole with respect to the conventional terrestrial pole and the difference UT − UTC; observed corrections to the calculated values of the precession and nutation may also be given. These observational quantities are often called the 'earth-rotation parameters' or the 'earth-orientation parameters'. The former term is also used in a wider context to refer to the much larger set of quantities that are required to represent both the orientation of the Earth in space and the model of the physical system.

8. DEVELOPMENT OF STANDARDS FOR ASTRONOMY AND GEODESY

As an aid to the understanding of the present position in respect of the systems of standards required in the establishment of terrestrial and celestial reference systems it is useful to review briefly the way in which these systems have evolved. At first the geodetic and astronomical constants were adopted independently in small groups, but by 1960 the inaccuracies and inconsistences of the constants then in use were very apparent, as indicated in the short account of the system of astronomical constants given in the

Explanatory Supplement (NAO, 1961).

The IAU (1964) System of Astronomical Constants (Fricke et al, 1966) and the Geodetic Reference System 1967 (IAG, 1971) were the first attempts to develop international standards that were appropriate to the higher precision of measurements by the new techniques of satellite geodesy and radio astronomy. Even so these systems omitted many quantities that are now regarded as being essential to the specification of a comprehensive system. These systems were extended and modified in the IAU (1976) System of Astronomical Constants (Duncombe et al, 1977) and its subsequent developments and in the Geodetic Reference System 1980 (Moritz, 1984).

The experiences of the MERIT Short Campaign (Wilkins, 1982) confirmed the need for the use of a comprehensive system of standards in order to simplify the comparison and evaluation of the many techniques of observation and analysis that would be used during the MERIT Main Campaign. The policy that was stated in the introduction to the lists of the MERIT Standards (Melbourne et al, 1983) is of more general applicability and is as follows:

"This document is intended to define the standard set of constants, models, and reference systems to be used by the observing stations and by the operational and analysis centers during the MERIT Campaign. To meet the objective of intercomparing current and new methods for the determination of the Earth Rotation Parameters (ERP), it is essential that all techniques use compatible, if not identical, standards in the reduction of observations and in the subsequent comparison of the results. In consonance with the IUGG 1983 Resolution on MERIT Standards, and with the policy set by the MERIT Steering Committee, all participants in the MERIT campaign are urged to comply with the Standards in their entirety wherever practicable. If an individual operational or analysis center cannot fully comply with these guidelines, the center must carefully identify the exceptions. In these cases, the center is obliged to provide an assessment of the effects of the departures from the standards so that intercomparison of results by the MERIT Working Group will be possible. In the case of models, the use of models equivalent to those specified herein is acceptable, provided that the center establishes equivalency."

The list of numerical values of the constants was accompanied by 15 appendices giving details of specialised numerical models and other data to be used in the reductions. An update was issued in 1985, but unfortunately it did not include any instructions on how best to indicate that the new material on tides and tectonic motions had been used in reductions.

Some of the values in the MERIT Standards differ from those in the IAU and IAG systems since for the particular purposes of Project MERIT it was considered that it would be necessary to use the best available data. These values have now been used in other contexts with the result that there is sometimes ambiguity about which values have been used to obtain published results. The results obtained since 1983 show that the MERIT Standards are no longer adequate for the reduction of the most precise measurements, and so a new set of standards is being developed for use by the organisations that participate in the new International Earth Rotation Service (IERS).

It is already necessary to take into account relativistic effects, such as the deflection of light in a gravitational field, in the reduction of the most precise astrometric measurements; further improvements are to be expected when the Hubble Space Telescope and the Hipparcos satellite come into use (Murray, 1983). Relativistic effects in radio-wave propagation are also important in the reduction of VLBI data. The IAU system of astronomical constants is under review to ensure that the definitions and values of the constants are compatible with relativistic theory; agreement on the form of the theory to be used is also necessary.

9. A SUMMARY OF THE STANDARDS FOR THE REFERENCE SYSTEMS

The preceding sections of this chapter show that the establishment and maintenance of standard terrestrial and celestial frames of high precision requires the adoption of a consistent system of parameters, models and procedures for a very wide variety of bodies and phenomena. The notes that follow are intended to summarise the principal constituents of such a system, and to indicate some of the current problems. It is hoped that these problems will be resolved in the new system of IERS standards and that it will be free from ambiguities and inconsistencies, as well as being as accurate as possible in the present circumstances.

9.1. UNITS AND DEFINING CONSTANTS

SI units, in which the velocity of light in m/s defines the metre, are used. The astronomical unit of time is a day of 86 400 s. The Gaussian gravitational constant then serves to define the astronomical unit of length (au). There is some controversy at present about the way in which these astronomical units should be used with relativistic theories (Murray, 1983, p.28). Other constants give the au in metres and light-seconds.

9.2. GEODETIC CONSTANTS FOR THE EARTH

For some purposes it is useful to approximate to the shape of the surface of the Earth by an ellipsoid of revolution, whose ellipticity is computed from an adopted value of the C_{20} coefficient in the spherical harmonic expansion for the gravity field of the Earth (Chapter 4, Section 3). At present different expansions of the gravity field are used for different satellites, but it is desirable that only one model be used in future.

9.3. PRECESSION, NUTATION AND ABERRATION

It is necessary to adopt a precise theory for precession and nutation, i.e. for the rotation of the celestial reference frame of date with respect to the standard frame for the adopted epoch (Chapter 11). A value for the obliquity of the ecliptic at the epoch is also required; there is some dispute about how the ecliptic is best defined. This theory requires the adoption of theories of the motions of the Sun, Moon and planets, although their accuracy is not critical. The accurate representation of the nutation does, however, require the adoption of a detailed model of the Earth, including the shape and viscosity of the core (Chapter 8, Section 5).

The aberration that arises from the annual and diurnal motions of the observing instrument with respect to the direction of arrival of the radiation from the source must be computed by standard formulae that included relativistic effects. The deflection and retardation of light and radio waves in the gravitational field of the Sun must also be computed from standard formulae (Chapter 1, Section 2).

9.4. UT / GMST RELATION

The nature of the relationship between UT and GMST is to some extent arbitrary, but it is usually accepted that continuity with previous practice is important. In particular, the rate of rotation of the Earth should appear to be constant with respect to the scale of universal time.

9.5. TIDAL PARAMETERS

It is important that the periodic motions of the stations with respect to the terrestrial reference frame be allowed for as accurately as possible so that the earth-rotation parameters will not be affected by them. It is necessary to take into account the changing distortions of the Earth's crust due to the changes in the load of the oceans, as well as the direct effects of the changing gravitational forces due to the Sun and Moon on the elastic Earth. Similarly in computing the orbits of satellites it is necessary to take into account both the direct and indirect changes in the gravitational field; for low satellites the effects of the ocean tides are determined from the observed perturbations.

9.6. ATMOSPHERIC PARAMETERS

The presence of the atmosphere affects the different types of observations in different ways. Refraction affects both direction and time-delay; the humidity of the atmosphere is particularly important for radio waves and is to be measured directly by the use of water vapour radiometers when possible. In all cases standard techniques for the measurement of the atmospheric parameters and for the computation of refraction must be specified. The effects of the ionosphere on radio waves must also be taken into account.

9.7. PERTURBATIONS ON SATELLITE ORBITS

There are several non-gravitational forces that perturb the orbits of satellites. These include the effects of air-drag and radiation pressure, and standard techniques for modelling them are required. Some of these effects may have to be represented by empirical expressions whose coefficients are to be determined from observations. Standard corrections must also be specified for such matters as the distance of the laser retroreflectors from the centre of mass of the satellite.

9.8. PARAMETERS FOR THE MOON

Lunar laser ranging requires the adoption of many other parameters and models for the Moon to specify such quantities as: its moments of inertia and gravity field; the ephemerides of its motion and rotation; the coordinates of the retroreflectors and the effects of tidal forces on them. The orbit of the Moon is computed simultaneously with the orbits of the planets, and so values for the masses of the planets must also be included in the system (Chapter 3, Section 4.3). The equations of motion must take into account relativistic effects.

9.9. TIMESCALES

Coordinated universal time UTC is used as the reference timescale during the making of observations and in the reporting of the variations of UT, but it is subject to occasional discontinuities of exactly 1 second. The standard time-scale from which UTC is derived is international atomic time TAI. The timescales now used in dynamical astronomy for the computation of the ephemerides of the Moon and planets are also derived from TAI. The adopted expressions for their differences from TAI must form part of the systems of standards that specify the associated technique-dependent reference frames.

9.10. COORDINATES OF THE STATIONS

The terrestrial reference frame is defined by the adopted catalogue of the coordinates of the stations and of their rates of change. Such a frame can be derived for each technique, but each involves one or more arbitrary parameters. These frames can be combined together by making appropriate translations and rotations while taking into account the constraints imposed by the results from colocated observations. The term colocation is used when instruments for two different techniques are located at the same or nearby sites, so that the differences in their coordinates may be measured accurately by terrestrial surveying techniques.

There is no point or area on the Earth's surface that may be regarded as fixed, and so there must be an element of arbitrariness about the adoption of the standard terrestrial reference frame. It has been agreed that the frame should be such that there should be 'no net rotation or translation between the reference frame and the surface of the Earth'. Even this criterion is not without ambiguity as to how it should be implemented. A further constraint is that it is desirable that the new frame should not produce any significant discontinuity in the scale of UT, and this implies that the zero of longitude must not change significantly. (The prime meridian no longer passes exactly through the Airy transit circle at Greenwich since for many years it has, in practice, been defined indirectly by the procedures used to determine UT.) It is, however, not clear that the coordinates of the reference stations should be chosen so as to ensure that the pole of the new standard terrestrial system is as close as possible to the conventional international origin now used. It might be better to choose them so that the new pole would correspond more closely to the principal axis of inertia of the Earth.

9.11. COORDINATES OF THE QUASARS

The standard celestial reference frame is defined by the adopted coordinates of the quasars that are observed by the various VLBI networks. Again this catalogue is to be formed by combining together the catalogues for the individual networks while taking into account sources that are common to two or more networks. The coordinates in this catalogue should be chosen so that the frame corresponds as closely as possible to that of the current standard, which is that of the FK5 star catalogue for the standard epoch of J2000.0, by, for example, taking into account the optical positions of the quasars. This will then ensure that the frame corresponds closely to the frame used in the development of the ephemerides of the Moon and planets. The stellar and dynamical reference frames are nominally identical, but the match between them can only be determined from analysis of new observational data. Systematic differences between the technique-dependent reference frames may be seen as systematic differences between the sets of earth- orientation parameters that are obtained by the different techniques.

9.12. THE NEW IERS STANDARDS

The generation of the new set of IERS standards that will be the basis of the new conventional terrestrial and celestial reference frames will itself represent an important contribution to geodesy and astronomy, and to the many fields of application of the regular monitoring of the rotation of the Earth. It is important that these standards should be used consistently in the reduction and analysis of observations for the determination of earth-rotation parameters and station coordinates, so that the results from different techniques, or different sets of observations, can be properly compared and combined. It is hoped that these standards will be of general application in the areas that they cover.

REFERENCES

BIPM, 1985, Le Système International d'Unités, 5th edition, in French and English, Bureau International des Poids et Mesures, Sevres, France.

Cohen, E.R., and Taylor, B.N., 1986, The 1986 adjustment of the fundamental physical constants, CODATA Bulletin no.63.

Duncombe, R.L., Fricke, W., Seidelmann, P.K., and Wilkins, G.A., 1977, Trans. Int. Astron. Union 15B, 56-67.

Fricke, W., Brouwer, D., Kovalevsky, J., Mikhailov, A.A. and Wilkins, G.A., 1966, Trans. Int. Astron. Union, 12B, 593-598.

IAG, 1971. Geodetic Reference System 1967, Publ. Spec. no. 3 du Bull. Geod., Paris.

Markowitz, W.M., 1976, Guide on the use of SI, Bull. Geod. 50, 3-7.

Melbourne, W., (Ch.), 1983, Project MERIT standards. U.S. Naval Observ. Circ. no. 167. See also Update No.1 (Dec. 85).

Moritz, H., 1984, Geodetic Reference System 1980, Bull. Geod. 58, 388-398.

Murray, C.A., 1983, Vectorial astrometry, Adam Hilger Ltd., Bristol, UK.

NAO, 1961. Explanatory Supplement to the Astronomical Ephemeris..., Her Majesty's Stationery Office, London.

Wilkins, G.A., 1982, Project MERIT: Report on the Short Campaign and Grasse Workshop, Royal Greenwich Observatory, Herstmonceux, U.K.

Wilkins, G.A., and Mueller, I.I., 1986, Joint Summary Report of the IAU/IUGG Working Groups on the rotation of the Earth and the terrestrial reference system, in J.-P. Swings, (ed), Highlights of Astronomy 7, 771-788, Reidel, Dordrecht, Holland. Also in Bull. Geod. 60, 85-100.

APPENDIX

ASTRONOMICAL AND GEODETIC FUNDAMENTAL CONSTANTS

Table 1 : The successive IAU Systems of Astronomical Constants and their values adopted in MERIT Standards

The year of the IAU System of Astronomical Constants. Names of Constants	1896 and 1911 (Paris international conferences)	IAU 1964	IAU 1976	MERIT Standards-1983
		DEFINING	CONSTANTS	
Number of ephemeris seconds in 1 tropical year (1900)		$s = 31\ 556\ 925.974\ 7$ (1900)		
Gaussian gravitational constant, defining the A.U.		$k = 0.017\ 202\ 098\ 95$	$k = 0.017\ 202\ 098\ 95$	IAU 1976
		PRIMARY	CONSTANTS	
Measure of A.U. (Light time for unit distance in 1976)		$A = 149\ 600\ .x\ 10^6$m	$\tau_A = 499.004\ 782$ s	$\tau_A = 499.004\ 783\ 70$ s
Velocity of light		$c = 299792.5\ \ 10^3 ms^{-1}$	$c = 299\ 792\ 458\ \ ms^{-1}$	IAU 1976
Equatorial radius of the Earth	$a_e = 6\ 378\ 388$ (Hayford)	$a_e = 6\ 378\ 160$ m	$a_e = 6\ 378\ 140$ m	$a_e = 6\ 378\ 137$ m
Earth dynamical form-factor		$J_2 = 0.001\ 082\ 7$	$J_2 = 0.001\ 082\ 63$	IAU 1976
Earth dynamical flattening (Constant of gravitation in 1976)	$H = 0.0032\ 825$		$G = 6.672 \times 10^{-11} m^3 kg^{-1} s^{-2}$	IAU 1976

J. Kovalevsky et al. (eds.), Reference Frames, 461–466.

	1896-1911	IAU 1964	IAU 1976	MERIT 1983
Geocentric gravitational constant		GE=398 603 .x $10^9 m^3 s^{-2}$	GE=3.986 005 x $10^{14} m^3 s^{-2}$	GE=3.986 004 48 x 10^{14} $m^3 s^{-2}$
Ratio of the masses of the Moon and Earth		μ = 1/81.30	μ = 0.012 300 02	μ = 0.012 300 034
Sidereal mean motion of Moon		n = 2.661 699 489 x x 10^{-6} rad s^{-1} (1900)		
Newcomb luni-solar precession in longitude	P_i = 5037".08 (1900)			
General precession in longitude per tropical century	P = 5025".64 (1900)	P = 5025".64 (1900)	P = 5029".0966 (2000)	IAU 1976
Obliquity of the ecliptic	ε= 23°27'08".26(1900)	ε = 23°27'08".26 (1900)	ε = 23°26'21".448 (2000)	IAU 1976 (for optical techniques)
Constant of nutation	N = 9".21	N = 9".210 (1900)	N = 9".2025 (2000)	Wahr model (1980)
		AUXILIARY CONSTANTS AND FACTORS		
k/86400, for use when the unit of time is 1 second		k' = 1.990 983 675 x 10^{-7}		
Number of sec. of arc in one rad.		206 264.806		
Factor for aberration constant		F_1 = 1.000 142		

	1896-1911	IAU 1964	IAU 1976	MERIT 1983
Factor for mean distance of the Moon		$F_2 = 0.999\ 093\ 142$		
Factor for parallactic inequality		$F_3 = 49\ 853''.2$		
		DERIVED	CONSTANTS	
Solar parallax	$\pi_0 = 8''.80$	$\arcsin(a_e/A)=\pi_0=8''.794\ 05$ (8".794)	$\arcsin(a_e/A)=\pi_0=8''.794\ 148$	8".794 144
Light-time for unit distance (unit distance in 1976)		$A/c = \tau_A = 499^s.012 = 1^s/0.002\ 003\ 96$	$c\tau_A=A=1.495\ 978\ 70 \times 10^{11}m$	$c\tau_A=A=1.495\ 978\ 7066 \times 10^{11}m$
Constant of aberration	$k = 20''.47$	$F_1 k'\tau_A = k = 20''.4958$	$k = 20''.495\ 52$	Direct computation
Flattening of the Earth	1 : 297.00 (1911)	$f = 0.003\ 352\ 9 = 1/298.25$	$f = 0.003\ 352\ 8 = 1/298.257$	IAU 1976
Heliocentric grativational constant		$A^3k'^2=GS=1.32718 \times 10^{20} m^3s^{-2}$	$A^3k^2/D^2=GS=1.327\ 124\ 38 \times 10^{20} m^3s^{-2}$	$GS=1.327\ 124\ 40 \times 10^{20} m^3s^{-2}$
Ratio of masses of Sun and Earth		(GS)/(GE) = S/E = 332 958	(GS)/(GE) = S/E = 332 946.0	(GS)/(GE)=S/E= =332 946.038
Ratio of masses of Sun and Earth + Moon		$S/E(1+\mu) = 328\ 912$	$S/E(1+\mu) = 328\ 900.5$	328 900.55
Mass of the Sun			$(GS)/G=S = 1.9891 \times 10^{30}kg$	IAU 1976

	1896-1911	IAU 1964	IAU 1976	MERIT 1983
Perturbed mean distance of the Moon, in meters		$F_2(GE(1+\mu)/n^2)^{1/3} =$ $= a_M = 384\ 400 \times 10^3$m		
Constant of sine parallax for Moon		$a_e/a_M = \sin \pi_M = 3422''.451$		
Constant of lunar inequality	$L = 6''.465$	$\mu a_M/A(1+\mu) = L = 6''.439\ 87$		
Constant of parallactic inequality		$F_3 a_M(1-\mu)/A(1+\mu)=P=124''.986$		
RECIPROCAL MASSES OF PLANETS (mass of Sun = 1)				
Mercury		6 000 000	6 023 600	IAU 1976
Venus		408 000	408 523.5	IAU 1976
Earth + Moon		329 390	328 900.5	328 900.550
Mars		3 093 500	3 098 710	IAU 1976
Jupiter		1 047.355	1 047.355	1 047.350
Saturn		3 501.6	3 498.5	3 498.000
Uranus		22 869	22 869	22 960
Neptune			19 314	IAU 1976
Pluto			3 000 000	130 000 000

REFERENCES FOR TABLE 1

Procès verbaux de la conférence internationale des étoiles fondamentales de 1896, Bureau des Longitudes, Paris, 1896.
Transaction of the IAU Vol. XIII B, pp. 593-598.
Transaction of the IAU Vol. XVI B, pp. 52-60.
U.S. Naval Circular 168, 1983.

Table 2 : Defining constants of the geodetic reference system

Year of the system	1967	1980
Names of constants	Numerical values	Numerical values
Semi-major axis(m)	a = 6 378 160	a = 6 378 137
Geocentric gravitational constants (m^3s^{-2})	GE = 398 603 x 10^9	GE = 3 986 005 x 10^8
Dynamical form factor	J_2 = 10 827 x 10^{-7}	J_2 = 108 263 x 10^{-8}
Earth angular velocity($rads^{-1}$)	= 7.292 115 1467 x 10^{-5}	= 7.292 115 x 10^{-5}
Orientation of the reference ellipsoid	The minor axis is parallel to the direction of the Conventional International Origin (CIO) The primary meridian is parallel to the zero meridian of the BIH adopted longitudes	

REFERENCES FOR TABLE 2

Geodetic system 1967, Publication spéciale n°3 du Bulletin Géodésique, IAG, 1971, Paris.
Moritz, H., 1979, Report of special study group n°5.39 of IAG. Fundamental geodetic constants presented at XVIIth General Assembly of IUGG, Canberra (reprinted at the Geodesist's Handbook 1988, Bull. Géodésique, vol. 62, n°3, 1988, IAG, Paris).

Table 3 : Derived constants of the geodetic reference systems

Name of constant	1967 Numerical values	1980 Numerical values
Derived Geometric constants :		
Semi-minor axis	b = 6 356 774.5161 m	b = 6,356 752.3141 m
Linear eccentricity	E = 521 864.6732 m	E = 521 854.0097 m
Polar radius of curvature	c = 6,399 617.4290 m	c = 6 399 593.6259 m
e = first eccentricity	e^2 = 0.006 694 605 328 56	e^2 = 0.006 694 380 022 90
e'= second eccentricity	e'^2 = 0.006 739 725 128 32	e'^2 = 0.006 739 496 775 48
Flattening	f = 0.003 352 923 712 99	f = 0.003 352 810 68118
Reciprocal flattening	f^{-1}= 298.247 167 427	f^{-1}= 298.257 222 101
$m' = (a^2-b^2)(a^2+a^2)$	m' =0.003 358 544 730 00	-
$n' = (a-b)(a+b)$	n' = 0.001 679 277 100 50	-
Meridian quadrant	Q =10 002 001.2313 m	Q = 10 001 956.7293 m
Mean radius $R_1=(2a+b)/3$	R_1= 6 371 031.5054 m	R_1= 6 371 008.7714 m
Radius of sphere of same surface	R_2= 6 371 029.9148 m	R_2= 6 371 007.1810 m
Radius of sphere of same volume	R_3= 6 371 023.5234 m	R_3= 6 371 000.7900 m
Derived Physical constants :		
Normal potential at ellipsoid	U_0= 6 263 703.0523 kgal.m	U_0=6 263 686.0850 x x 10 $m^2.s^{-2}$
	J_4 = -0.000 002 371 264 40	J_4 =-0.000 002 370 912 22
Spherical-harmonic coefficients	J_6= 0.000 000 006 08516 J_8= -0.000 000 000 01428	J_6 = 0.000 000 006 083 47 J_8 =-0.000 000 000 014 27
$m = \omega^2 a^2 b/GE$	m = 0.003 449 801 434 30	m = 0.003 449 786 003 08
Normal gravity at equator	γ_e = 978.031 845 58 gal	γ_e=9.780 326 7715 $m.s^{-2}$
Normal gravity at pole	γ_p = 983.217 727 92 gal	γ_p=9.832 186 3685 $m.s^{-2}$
$f^* = (\gamma_p-\gamma_e)/\gamma_e$	f^*=0.005 302 365 523 30	f^*= 0.005 302 440 112
$k = (b\gamma_p-a\gamma_e)/a\gamma_e$	k =0.001 931 663 383 21	k = 0.001 931 851 353

INDEX